Sustainable Development Goals Series

The Sustainable Development Goals Series is Springer Nature's inaugural cross-imprint book series that addresses and supports the United Nations' seventeen Sustainable Development Goals. The series fosters comprehensive research focused on these global targets and endeavours to address some of society's greatest grand challenges. The SDGs are inherently multidisciplinary, and they bring people working across different fields together and working towards a common goal. In this spirit, the Sustainable Development Goals series is the first at Springer Nature to publish books under both the Springer and Palgrave Macmillan imprints, bringing the strengths of our imprints together.

The Sustainable Development Goals Series is organized into eighteen subseries: one subseries based around each of the seventeen respective Sustainable Development Goals, and an eighteenth subseries, "Connecting the Goals", which serves as a home for volumes addressing multiple goals or studying the SDGs as a whole. Each subseries is guided by an expert Subseries Advisor with years or decades of experience studying and addressing core components of their respective Goal.

The SDG Series has a remit as broad as the SDGs themselves, and contributions are welcome from scientists, academics, policymakers, and researchers working in fields related to any of the seventeen goals. If you are interested in contributing a monograph or curated volume to the series, please contact the Publishers: Zachary Romano [Springer; zachary.romano@springer.com] and Rachael Ballard [Palgrave Macmillan; rachael.ballard@palgrave.com].

Arjen E. J. Wals • Birgitte Bjønness
Astrid Sinnes • Ingrid Eikeland
Editors

Whole School Approaches to Sustainability

Education Renewal in Times of Distress

Springer

Editors
Arjen E. J. Wals
Wageningen University & Research
Wageningen, The Netherlands

Birgitte Bjønness
Norwegian University of Life Sciences
Ås, Norway

Astrid Sinnes
Norwegian University of Life Sciences
Ås, Norway

Ingrid Eikeland
Norwegian University of Life Sciences
Ås, Norway

ISSN 2523-3084 ISSN 2523-3092 (electronic)
Sustainable Development Goals Series
ISBN 978-3-031-56174-0 ISBN 978-3-031-56172-6 (eBook)
https://doi.org/10.1007/978-3-031-56172-6

This Springer imprint is published by the registered company Springer Nature Switzerland AG
The registered company address is: Gewerbestrasse 11, 6330 Cham, Switzerland

Paper in this product is recyclable.

Acknowledgements

Our first token of appreciation goes to our managing editor Stine Marie Fyske Haraldsen who organized and checked all the files, communicated with authors, and liaised with the people of Springer who supported the publishing process. Clearly without someone who is organized, watches time schedules and responds to questions, we would not have a book now. Then, of course, we thank the authors for their willingness to contribute to the book. We celebrate associated desires for institutional change, and collegial openness to share and add to ongoing, professional conversations concerning schools within the urgency of the ecological crisis. We also appreciate their contributions to the 'blind' peer review process that was used. We as editors, with some of us also being connected to peer reviewed journals, realize how difficult it is to get people to review work, especially in a meaningful and attentive way. At the same time, we know how important such a review of peers is as with every round of feedback the quality of the work improves.

We also wish to acknowledge the Dutch Government-Supported Programme for Learning for Sustainable Development (LvDO) for providing the funding that makes this book available 'open access' so that anyone with an Internet connection can download the entire book or specific chapters for free. We wish to acknowledge Roel van Raaij who has been an advocate of Education for Sustainable Development, both nationally and internationally, from within the Dutch government for decades, and Ellen Leusink who facilitates and supports ESD through the LvDO program and paved the way for the open access.

We also want to acknowledge the Department of educational science at NMBU led by Hans Erik Lefdal together with Akershus county municipality (the school district South of Oslo) for the support to establish and run the University—School partnership for the last 6 years or so, continuing to this day. The partnership with schools, especially the collaboration with school coordinators and principals, have supported the development of the Whole School Approach through research and praxis. Without this foundation and support, we would not have had the knowledge, time, and confidence to work on this book.

Ås, Norway, October 2023

Contents

Contributors

Erika Åberg is the Head of the Nature school in the municipality of Umeå, Sweden. She has extensive experience of in-service training for teachers and principals in environmental and sustainability issues, as well as school developmental work.

Michèle Andrews is the co-Founder and Executive Director at DoorNumberOne.org. She has a Bachelor of Commerce from Queen's University and a Masters of Education from the University of Toronto. She is passionate about creating a path for schools and communities to engage in aspirational, hope-filled, regenerative climate action.

Shakeel M. C. Atchia is a lecturer in the Science Education department of the Mauritius Institute of Education. His main fields of action are science education, STEM education and quality assurance.

Barbara Bader is a professor of science, environment, and sustainable development education. She studies the relationship between scientific knowledge and the engagement of young people regarding climate change. Inspired by a socialized conception of science, critical pedagogy, and cultural studies, she works on climate change education and practical training in elementary and secondary education.

Anne-Marie Ballegeer is an assistant professor at the department of Mathematics and Science Education of Salamanca University. Her current research interests are natural sciences education and education for sustainable development and Climate Change. She is a member of the "Education Mathematics Science and Climate Change" research group.

Genendorj Battsetseg is Programme Manager of the Information and Training Centre for Nature and Environment. Her research area is Eco-school's development.

Chrysanthi-Kadji Beltran is an Associate Professor, at Frederick University, Education Department. Her research evolves around the development of ESD competences through appropriate teacher education and connecting ESD to quality education and transformative learning. She also explores the integration of the SDGs in higher education and teacher education for ESD.

Teresa Berglund is senior lecturer in biology and environmental and sustainability education at the Department of Environmental and Life Sciences, Karlstad University, Sweden. Her research area is education for sustainable development, with a focus on teaching, learning, and implementation in local school organizations.

Farhana Borg is Associate Professor of Educational Work at Dalarna University (DU), Sweden. She is a Coordinator for DU's *Research Collegium for the Advancement of Sustainability*. Her research interests include children's learning and teachers' pedagogical practices related to sustainability. She was Head of the Early Childhood Education Collegium between 2018 and 2020.

Alison Elliott is the Trinity College School Sustainability Programs Director, and has been leading environmental work there since 2002. She is a sought-out mentor and has led her school to national-level awards and certifications. Her vision is a regenerated world where we all do "more good" instead of "less bad."

Ellen Field is an Assistant Professor in the Faculty of Education at Lakehead University. She conducts research on policy and practice of climate change education, is Associate Editor at the Canadian Journal of Environmental Education, and co-chair of the Canadian hub for the Monitoring and Evaluating Climate Communication and Education project.

Niklas Gericke is professor of science education and director and research leader of the research center Science, Mathematics, and Engineering Education Research (SMEER) at Karlstad University, Sweden. His main research areas are biology education and sustainability education including education for sustainable development from conceptual, teaching, and implementation perspectives.

Carmen Gloder is an Early Childhood Education Teacher, ICT Coordinator, and Head of Studies at Ceip Enrique Tierno Galván in San Sebastián de los Reyes, Madrid. Member of the coordination team of Teachers for Future Spain, and Coordinator of the school program "Recreos Residuos Cero" zero waste break time.

Mita Nangia Goswami is a consultant for the Swiss Agency for Development and Cooperation's ESD Project Phase 2. Her research area is Education for Sustainable Development.

Stella Hadjiachilleos is an educator and researcher at the Unit of EESD of the Cyprus Ministry of Education, Sport and Youth. She has produced numerous education tools and materials for ESD in nonformal settings, offered in-service teacher training and taught ESD courses in education departments of universities in Cyprus.

John Hannah is an Innovation Leader and Learning Landscapes Teacher, Hillfield Strathallan College. He is an avid naturalist and an award-winning teacher and advocate for learning outdoors as a way to connect students to the natural world while making them happier, healthier, and smarter.

Kristen Hargis is a postdoctoral scholar with the Monitoring and Evaluating Climate Communication and Education Project. Her expertise is in the analysis of education policy for the inclusion of climate change content, as well as the use of practice theory to analyze climate change education practices.

Patrick Howard is the Dean of the School of Education and Health and a Professor of Education at Cape Breton University. His interests are in topics related to environmental and sustainability education and the intersections of environment and well-being in people and communities.

Aksel Hugo is an associate professor at the Norwegian University of Life Sciences. His main expertise is in integrative life science teacher education, agroecology education and ecological entrepreneurship education. He was also an upper secondary school science teacher and a principal at a vocational school of agriculture.

Elisabeth Iversen is an associate professor at the Norwegian University of Life Sciences. Her research and development interests revolve around environmental and sustainable education, in-service teacher education and vocational education. She is currently also teaching science and environmental/agricultural education at Tana Upper Secondary School in Finnmark, Norway.

Gerd Johansen is an associate professor at the Faculty of Science and Technology, Norwegian University of Life Sciences (NMBU), Ås, Norway. She teaches science and technology education in teacher education. Her research interests include sustainability and knowledge practices in science and technology education.

Eleonor Kerr Director of Operations and Human Resources, at Hillfield Strathallan College, is part of the Hillfield Leadership Team, and one of the drivers in making sustainability and regeneration part of the school's strategic and operational plans. She has worked at the school since 2004.

Eva Knekta is an associate professor at the Department for Science and Mathematics Education at Umeå University, Sweden. Her main expertise is in educational measurement, motivation and science education for sustainability.

Enzo Rainiero Ferrari Lagos is a chemist who has created an educational competence through his initial research (PhD) to mobilize society to tackle climate change. His research interests focus on Natural Science Education, pro-environmental behaviour, and Climate Change education. He is currently a postdoctoral researcher at Bath University.

Miriam Leirós is a primary school teacher and coordinator of the Teachers for Future network.

Stephanie Leite (she/her) is a PhD candidate and Vanier Scholar at McGill University. Her background in art, conflict transformation, and sustainability influences her research on transformative climate change education. She is affiliated with the Leadership and Learning for Sustainability Lab, studying transitions toward more collaborative, transdisciplinary research and practice models.

Jos J. N. Lichtenberg his professional career is built on innovation in the building industry. He retired as a professor in 2016 but is still active in the development of new building concepts, including schools.

Annika Manni is an assistant professor at the Department of Applied Educational Sciences at Umeå University, Sweden. Her expertise is in educational work focusing environmental and sustainability education in practice.

Rosalie Gwen Mathie is a PhD candidate and Lecturer at the Norwegian University of Life Sciences working with sustainability-oriented professional development. Her main expertise and interests are concerning the role of reflexive and participatory processes in whole institution staff professional development, and co-developing methods to support sustainability-oriented education and institutional transitions.

Øyvind Mathisen is a public PhD student at the Norwegian University of Life Sciences (NMBU). His PhD thesis revolves around deliberative communication in classrooms when students handle renewable energy technology. Additionally, he is a teacher at an upper secondary school. He teaches the subjects mathematics, physics, natural science, geography and history.

Anna Mogren holds a PhD in the field of Biology didactics and is a lecturer in the field of Educational work at the Faculty of Arts and Social Sciences, Karlstad University. Her main research interest is school improvement and education for sustainable development (ESD), and she is a member of the research groups SMEER and SOL- FORSK. She has a long experience of practical school improvement in ESD and collaborates with national school authorities.

Masi Mohammadi is a full professor of Smart Architectural Technologies at the Eindhoven University of Technology and the leading professor in the Academy of the Built Environment at the HAN University of Applied Sciences. Her work focuses on 'empathic environments'.

Émilie Morin Ph D., is a professor in psychopedagogy of practical education in preschool and elementary school at the Université du Québec à Rimouski (UQAR). Her research focuses on climate education for empowerment. Based on the youth voices, she described the conditions to develop this sense of empowerment in school.

Álvaro Lozano Murciego holds a PhD in Computer Engineering from the University of Salamanca. He is currently an assistant professor at the Department of Computer Science, University of Salamanca. His research interests include machine learning, recommender systems, IoT sensors, data mining, route optimization, and edge computing.

Birgitta Nordén is an Assistant Professor at the Department of Science, Mathematics and Society, Malmö University. Her main expertise is in systemic design towards sustainability including creativity and innovation, organizational change, future-oriented methodologies, linked to environmentally and socially sustainable society building, critical eco-reflexive Bildung, and transformative teaching in higher education.

Daniel Olsson is senior lecturer in biology education at Karlstad University, Sweden. His research includes how education can be improved to empower people's action competence for sustainability, and he has recently worked in close collaboration with schools in a Swedish municipality to develop a whole school approach toward ESD.

Orsoo Oyuntungalag is a consultant for planning and monitoring of the Information and Training Centre for Nature and Environment. Her research area is Teachers Professional Development.

Jelle Boeve-de Pauw is an assistant professor in Biology Education at the Freudenthal Institute (Utrecht University). He teaches in the master's program on Science Education and Communication. His research focuses on measurement and evaluation in environmental education and ESD. He is affiliated to Edubron at the University of Antwerp.

Christina Phillips is a Lecturer and Practicum Coordinator at the Ontario Institution for Studies in Education, University of Toronto. Her interests are science education, teacher training, and sustainability education.

Brita Phuthi has worked for over a decade as an advisor to the Norwegian Folk High Schools. Her main expertise is in global citizenship education, sustainable development, and transformative learning. Phuthi is currently a PhD candidate at Oslo Metropolitan University. Her chapter is not part of the PhD project.

Marco Rieckmann is Professor of Higher Education Development at the University of Vechta (Germany). His main areas of research are education for sustainable development, global learning, and sustainable development of higher education institutions. He served as the main author of the UNESCO (2017) publication "Education for Sustainable Development Goals. Learning Objectives."

Katarina Rončević has more than 10 years of experience in education for sustainable development and inclusive education and has been working on whole school approaches. Since 2019, she works for Greenpeace Germany in the education team and is a PhD student at the University of Vechta (Germany). Katarina is a member of the international ESD Expert-Net and cofounder of The Turquoise Change e.V.

Camilo Ruiz is a physicist working on Climate Change and Education and topics related to science and education. He belongs to the Research group EMC^3 at the Salamanca University, Spain.

Anwar Rumjaun is an Associate Professor in the Science Education Department at the Mauritius Institute of Education. His research areas are in science and STEM education and education for sustainability.

Ingrid Pramling Samuelsson is Professor in Early Childhood Education at Gothenburg University, Sweden. She holds a UNESCO Chair in Early Childhood Education and Sustainable Development and was World President of the World Organization for Early Childhood Education between 2008 and 2014 and still acts as one of the coordinators in the organization concerning sustainability.

Anette Forssten Seiser is associate professor in educational work at Karlstad University, Sweden. Her research focus on school leadership and school improvement, conducted at local schools as well as in government projects, often addresses the role and function of school leaders in different teaching research initiatives.

Hiroko Shibakawa is an Assistant Professor at ESD Promotion Centre, Graduate School of Education, Okayama University. She is engaged in teacher education and research on ESD from the perspective of lifelong learning and social education, designing teacher training programmes and learning materials in collaboration with the board of education, schools and teachers.

Erdenebayar Shinetsetseg is Executive Director of the Information and Training Centre for Nature and Environment. Her research area is Education for Sustainable Development.

Drew Stephens has been Head of School at Southridge School since 2007. He began his career in outdoor education and athletics after completing his Bachelor of Physical Education, Bachelor of Education, and Master of Science degrees from McMaster and Queen's Universities in Ontario, Canada.

Lundaajantsan Temuujin is the Chairman of the Information and Training Centre for Nature and Environment. His research area is Climate Change Education.

Geneviève Therriault Ph.D., is a professor in educational theory at the Université du Québec à Rimouski. She is interested in the treatment of environmental issues at school. As holder of the Research Chair in Environmental Education and Sustainable Development UQAR-Desjardins, she aims to develop the empowerment of young people and teachers.

Anna W. Vanderveen works as a senior policy advisor on sustainability education at SME and GDO, the Dutch Association for Environmental Education. Her main expertise is in bridging gaps between theory, policy and practice on diverse scale levels.

Peter Van Petegem is a full professor of education sciences in the Faculty of Social Sciences at the University of Antwerp. He teaches education and school policies in the master's program on Training and Education Sciences. His research interests include the evaluation of education innovations. He is a member of Edubron.

Roel van Raaij is a Dutch senior policy officer, environmentalist, and teacher—working for the Ministry of Agriculture, Nature and Food Quality, leading the Netherlands National ESD program. Roel has been involved in UNECE since the beginning of the ESD strategy writing and as chair of the ad hoc expert group for indicators.

Dries Verhelst is a post-doc researcher in the Faculty of Social Sciences and research coordinator for TIMSS 2023 at the University of Antwerp. His research includes school policy and school organizational management educational effectiveness and school effectiveness. He is a member of the Edubron research group.

Aravella Zachariou is the Head of the Unit of EESD and Vis. Ass. Prof. on ESD at Frederick University. She is the Chair of the UNECE ESD Steering Committee and her research focuses on the ESD implementation in formal and nonformal education as well as in teachers' professional development.

Editors

Arjen E. J. Wals is a professor of transformative learning for socio-ecological sustainability at Wageningen University where he also holds the UNESCO Chair of Social Learning and Sustainable Development. Furthermore, he is a guest professor at the Norwegian University for the Life Sciences (NMBU). He holds an Honorary Doctorate from Gothenburg University, Sweden.

Birgitte Bjønness is an associate professor in science education at the Norwegian University of Life Sciences. Her research interests include sustainability education, health and wellbeing in schools, inquiry-based learning, and action research in education.

Astrid Sinnes is a professor in science education and vice-rector for sustainability at the Norwegian University of Life Sciences. Her expertise is in sustainability transitions in education.

Ingrid Eikeland holds a PhD in science center education based on the dissemination of controversies. In the last 4 years, she has managed the university–school partnership at the Norwegian University of Life Science.

1 It Takes a Whole School: An Introduction

Arjen E. J. Wals, Ingrid Eikeland, Birgitte Bjønness, and Astrid Sinnes

Most reports, articles and chapters, written by our students or peers, and, indeed, by ourselves, start by pointing at the dismal state people and the entire planet Earth are in. Three of the four editors of this book are more or less of the same 'post-midlife' generation. We were high school students in the seventies and eighties of the last century, around 40 years ago in the Global North, in Norway and the Netherlands to be more specific. Back then there were also researchers and policymakers, writing alarming texts but they were often marginalized and not taken too seriously. As environmental and science educators in our early careers, we were part of a growing group of concerned scientists who felt it was important that our schools engaged young people in environmental issues, connected them to nature and the outdoors, and even made them more 'action competent' (Jensen & Schnack, 1997). Most schools, Ministries and Departments of Education, were not so inclined to create space for this, pointing at an overcrowded curriculum and many other competing topics that other interest groups felt were important enough to be taken up by general education (e.g. citizenship, health, human rights, peacebuilding, biodiversity, water, food security). Sure, there were some visionary school leaders and teachers who saw both the importance and the benefits of making education more enriching, engaging and empowering by connecting more to existential issues (see, for instance, Greig et al., 1987, 1989). Indeed, there were schools, around the world, that were at the forefront of environmental education already in the seventies and eighties, even before 'sustainable development' as a concept existed.

It's now 2023, more than 50 years after the infamous Stockholm conference on the Human Environment (United Nations, 1972) and the Limits-to-Growth report of the Club of Rome (Meadows et al., 1972). Since those early years we have participated in numerous professional networks of environmental education, STEM education and sustainability education, and dialogued with researchers and practitioners from all over the world, although mainly from countries from the so-called 'Global North'. Together we have written, literally, hundreds of papers, and attended a similar number of meetings, conferences, symposia and workshops. So where are we now? What has changed? Has education found a way to enrich, engage and empower? As always, the answer is not that simple.

On the one hand, one could say 'no'! Schools, more than ever, have become an extension of the globalizing economy with a focus on efficiency,

A. E. J. Wals (✉)
Wageningen University & Research, Wageningen, The Netherlands
e-mail: arjen.wals@wur.nl

I. Eikeland · B. Bjønness · A. Sinnes
Norwegian University of Life Sciences, Ås, Norway
e-mail: ingrid.eikeland@nmbu.no; birgitte.bjonness@nmbu.no; astrid.sinnes@nmbu.no

A. E. J. Wals et al. (eds.), *Whole School Approaches to Sustainability*, Sustainable Development Goals Series, https://doi.org/10.1007/978-3-031-56172-6_1

accountability, skills for employability, cognition and disciplinary understanding. Schools are nowadays, mostly unintentionally and in subliminal ways, mainly preparing young people for a life of work and consumerism (Nussbaum, 2012; Biesta, 2023). This is the bleak and sobering analysis that does not do justice to the many teachers and school leaders who, often against the grain, try to pay attention to the whole child and the unfolding of children's own identity and potential not just in the context of work. But still, many are unable to due to constraints imposed upon them.

On the other hand, one could say: 'yes'! Schools, more than during the times of our early careers, are looking to re-orient themselves to the time humanity finds itself, referred to sometimes as the Anthropocene (Crutzen, 2006) or the Capitolocene (Pedersen et al., 2023). They want to be relevant and responsive to what young people need but also to what the Earth needs. Rather than the question of 'what does the economy demand from education,' the question is slowly but surely becoming: 'what does the Earth demand from education?' Engaging in the latter question opens other perspectives and pathways for education.

Mind you, that first question was not always what education was about. Depending on how far back one goes, historically education was essentially about being guided into the world with the qualities and abilities needed to unfold one's potential and to live a dignified and meaningful life. That original intent needs to be revived at this time of systemic global dysfunction and ecological collapse. What is the mission of education in, the Anthropocene or Capitolocene? According to the British economist Kate Raworth, the existential challenge facing us as a species is learning to live within ecological limits (climate, biodiversity, ozone layer, water, soil, air, etc.) in a humane way with a strong social foundation (e.g. good and accessible education, justice, fair distribution, good governance, attention to democracy, human rights). She uses the metaphor of the doughnut to describe this safe space for humanity (Raworth, 2017).

Education can make an excellent contribution to learning to live within the doughnut. Not by dogmatically, indoctrinating and teaching with a raised finger towards children and conditioning them to show certain behaviour, but by actively involving them in the search for a sustainable world for humans and other species, and by equipping them with the qualities and competencies they need. Think of a caring and critical attitude, empathy, solidarity with people far away, future generations, other species and so-called 'sustainability competencies' (Brundiers et al., 2021), think of dealing with complexity, ambiguity, uncertainty, learning to think in relationships and systems, imagining alternative more hopeful futures, thinking creatively about and working on local solutions that are mindful of the wider world.

The question of what do young people need in times of high anxiety and rather bleak prospects for the future? This is a question that concerns the three 'senior' editors but the fourth, youngest editor, in particular. After all, she just gave birth to a beautiful daughter who still has a full life ahead of her that might take her into the year 2100 and beyond.

The current prospect for living in the twenty-second century is not good when we continue our current pathway of unsustainable development, according to the latest IPCC report. This notion and the worries it triggers invites educational policymakers, schoolteachers and leaders to look for ways to pay attention to emotional and mental well-being in difficult times. Providing hope and meaning in times of doom, gloom and despair is critical (Frumkin, 2022; Ojala, 2023). Some refer to this as nurturing inner sustainability (Woiwode et al., 2021) which calls for certain qualities like mindfulness, an ethic of care and empathy and reflexivity. In relation to what the Earth needs, education is exploring ways to help young people understand planetary boundaries and find ways to stay within them. We might call this, somewhat artificially, outer sustainability which calls for certain competencies like systems thinking, dealing with ambiguity and uncertainty, but also taking a moral position considering ethics (Brundiers et al., 2021). More and more schools, networks of schools and education-related research and professional development are looking for ways to pay attention to both inner and outer sustainability. They seem to agree, as we do, with the statement that just like business-as-usual is no longer an option, neither is education-as-usual.

Accepting this premise also implies that we cannot view emergent topics like sustainability, climate change and justice, global citizenship and inclusion as topics that can literally be taught, like mathematics, photosynthesis or Newton's laws, topics that, when we fully understand them and 'didacticize' and make them 'teachable' can simply be added to the curriculum, possibly by taking out something else to create a bit of space.

For one, these topics have an inevitable ill-definedness that require that they be given meaning in a certain place, a certain time in a very dynamic context where new knowledge and insights unfold continuously. Secondly, these topics cannot be well understood unless we consider their connections and interdependencies and our own entanglement in them (Wessels et al., 2022). Thirdly, addressing these issues requires reflecting on ethics and values and taking on a moral position that is based on empathy and care (Noddings, 2010). Fourthly, the search for a more sustainable way of being and living requires reflexivity (Sol et al., 2018), the ability and willingness to pause and think; are we on the right pathway still or do we need to change the course? Are our new insights still helpful or are they blinding us? Are we still able to listen to others or are we locking ourselves up in echo chambers? Clearly, reorienting education towards sustainability calls for a rethinking of what we teach, how we educate, of what we (un)willingly represent—and don't—of what we stand for and stand up for, being explicit of what we value and believe in, but also of what we do, how we behave as individuals, as a group and as a school community. This is no easy task as there is no point denying it: education all over the world is under pressure; staff/pupil ratios remain too high, the freedom to deviate from the curriculum is too limited, the administrative burden is constraining, salaries are too low, professional development is weak or absent altogether in some countries and regions.

Under such circumstances it is quite a challenge to change, especially when policymakers maintain a fixation on testing and examination in an attempt to climb a few places on the *PISA* rankings, and when teachers and school leaders are always being held accountable for what they do by the board, school inspectors and parents. Schools can be seen as a complex adaptive system (Koh & Askell-Williams, 2021) with its own ethos and culture, a system that can be highly resilient, not in the least because of the policy environment and the expectations, also of parents, in which they are entangled. It seems like the key to a sustainable school, one that cannot only sustain itself but can also contribute to the sustainability of people and the planet, can only be realized through a systemic, integrated and critical approach. Such a system overhaul won't be easy and will likely meet resistance. The whole school approach (WSA), central in this book, is one attempt to reorient education towards sustainability, using such a systemic, integrated and critical perspective.

The power of the WSA, as some of the contributors to this book point out, is that sustainability, health and citizenship, among other themes, are regarded as inextricably linked by the qualities and abilities learners need to survive and thrive by co-creating meaningful and fulfilling lives and more hopeful futures. The school itself determines which of these themes or labels appeals most to the imagination, and the identity it is seeking. At the same time, when considering the nature of teaching and learning, a WSA tends to favour more so-called hybrid or blended learning environments and more integrated approaches to teaching and learning. These integrated approaches seek to connect schools with their local communities, blend instructional learning with more open and participatory forms of learning and, utilize Information and Communication Technology (ICT) that helps learners develop sustainability competencies. Within these hybrid learning environments or 'ecologies of learning' (Wals, 2019), different forms of support and scaffolding are utilized, including peer-to-peer learning, mentoring, self-directed learning ICT-supported learning and collaborative forms of inquiry such as citizen science. The WSA is a concept that emphasizes that schools, in synergy, consider a range of aspects in addressing sustainability: curriculum integration, pedagogical and didactical arrangements, professional development of staff, school-community relations, schools' own sustainability practices sometimes referred to as 'walking the talk', and, lastly, school ethos, monitoring, evaluation and assessment, as well as organization and leadership. In the chapter following this introduction, Rosalie Mathie provides an elaborate overview of the WSA, its origins and interpretations.

Internationally such a WSA is gaining traction (see for example European Commission, 2021). UNESCO's strategy for ESD for the next 10 years has identified the WSA to SD as a more integrated systemic approach that can simultaneously lead to school improvement, educational innovation, and a meaningful contribution to sustainable development and global citizenship (UNESCO, 2021). At the UNECE (2022) but also in school practices around the world we see niches unfolding where schools are using a WSA, in different ways, to connect young people meaningfully to complex issues (Wals & Mathie, 2020). Sometimes they are supported by international networks of expertise (e.g. the FEE's EcoSchools programme or UNESCO's ASP Schools Network) and or of peers (e.g. The Network of IB Schools or the International Network of Montessori Schools). Recently we compiled some examples of school practices from around the world that utilize a WSA (Mathie & Wals, 2022).

With this book, we hope to expand our understanding of what a WSA as a concept entails, how the concept links to other areas of educational research, and to elaborate on the practical implementation of a WSA in praxis. In addition, we hope to reach scholars and education researchers who work in the domain of interrelated 'planetary educations' such as health education, sustainability education, development education, global citizenship education and climate education; international, national and regional policymakers working on the educational operationalization of the SDGs in general and SDG 4 and 4.7 specifically; school leaders and school coordinators seeking to implement the WSA in their own schools and teachers with a strong interest and commitment in strengthening sustainability in their own teaching and within their own school.

This edited volume brings together a range of scholars and reflective practitioners from across the globe, albeit mainly from the global North who are both investigating and enacting a WSA in education. The contributions span different levels of the formal educational system, with a predominance of contributions investigating the primary and secondary educational levels. The book has been organized into two sections: 'Principles and Perspectives' and 'Practices' which follow this introductory chapter, and a synthesizing closing chapter. The Principles and Perspectives section contains chapters outlining key characteristics, assumptions and principles that underly a WSA. Perspectives vary from educational philosophy to organizational change and curriculum innovation. Different contextual origins of the WSA in education are also traced including health and well-being, citizenship education, and environment and sustainability. Potential pitfalls and risks are also being presented as lessons learnt from past attempts to change school systems and cultures and the resistances encountered. The Practices section contains case studies of formal education trying to develop and enact a WSA. They show a variety of interpretations but also some of the common and special struggles to change schools in more fundamental ways. The synthesizing closing chapter is written by the editors and contains a critical reflection on both sections, highlighting some 'bold spots' or area's where many seem to be treading as well as some 'blind spots' or missing perspectives that might be critical as well as some 'hotspots' referring to emerging topics and phenomena that likely will affect education in relation to sustainability in the near future.

References

Biesta, G. (2023). Whose school is it anyway? On the insistence of education and the need for the emancipation of the school. In C. A. Säfström & G. Biesta (Eds.), *The new publicness of education democratic possibilities after the critique of neo-liberalism*. Routledge. https://doi.org/10.4324/9781003289067

Brundiers, K., Barth, M., Cebrián, G., Cohen, M., Diaz, L., Doucette-Remington, S., Dripps, W., Habron, G., Harré, N., Jarchow, M., Losch, K., Michel, J., Mochizuki, Y., Rieckmann, M., Parnell, R., Walker, P., & Zint, M. (2021). Key competencies in sustainability in higher education—toward an agreed-upon reference framework. *Sustainability Science, 16*(1), 13–29.

Crutzen, P. J. (2006). The "Anthropocene". In E. Ehlers & T. Krafft (Eds.), *Earth system science in the anthropocene*. Springer. https://doi.org/10.1007/3-540-26590-2_3

European Commission. (2021). *A systemic, whole-school approach to mental health and well-being in schools in the EU—Executive summary*. Publications Office of the European Union. https://data.europa.eu/doi/10.2766/208726

Frumkin, H. (2022). Hope, health, and the climate crisis. *The Journal of Climate Change and Health, 5*. https://doi.org/10.1016/j.joclim.2022.100115

Greig, S., Pike, G., & Selby, D. (1987). *Earthrights: Education as if the planet really mattered*. World Wildlife Fund UK.

Greig, S., Pike, G., & Selby, D. (1989). *Greenprints for changing schools*. World Wide Fund for Nature/Kogan Page.

Jensen, B. B., & Schnack, K. (1997). The action competence approach in environmental education. *Environmental Education Research, 3*(2), 163–178. https://doi.org/10.1080/1350462970030205

Koh, G. A., & Askell-Williams, H. (2021). Sustainable school-improvement in complex adaptive systems: A scoping review. *Review of Education, 9*(1). https://doi.org/10.1002/rev3.3246

Mathie, R. G., & Wals, A. E. J. (2022). Whole school approaches to sustainability: Exemplary practices from around the world. *Wageningen: Education & Learning Sciences/Wageningen University, 109*. https://doi.org/10.18174/572267

Meadows, D. H., et al. (1972). *The limits to growth; a report for the Club of Rome's project on the predicament of mankind*. Universe Books.

Noddings, N. (2010). Complexity in caring and empathy. *Abstracta*, (V).

Nussbaum, M. C. (2012). *Not for profit: Why democracy needs the humanities*. Princeton University Press.

Ojala, M. (2023). Hope and climate-change engagement from a psychological perspective. *Current Opinion in Psychology, 49*. https://doi.org/10.1016/j.copsyc.2022.101514

Pedersen, H. S., Windsor, B. K., Sanders, D., Wals, A., & Franck, O. (Eds.). (2023). *Education for sustainable development in the 'Capitalocene'*. Routledge.

Raworth, K. (2017). *Doughnut economics: Seven ways to think like a 21st century economist*. Chelsea Green Publishing.

Sol, J. M., van der Wal, M., Beers, P. J., & Wals, A. E. J. (2018). Reframing the future: The role of reflexivity in governance networks in sustainability transitions. *Environmental Education Research, 24*(9), 1383–1405. https://doi.org/10.1080/13504622.2017.1402171

UNECE. (2022). Whole school approach conference. International Conference held in Lunteren, The Netherlands. 20 March 2022. https://unece.org/sites/default/files/2022-06/REPORT~1_0.PDF

UNESCO. (2021). *Reimagining our futures together—A new social contract for education*. UNESCO.

United Nations. (1972). *Report of the United Nations conference on the Human Environment, Stockholm, 5–16 June 1972*. United Nations.

Wals, A. E. J. (2019). Sustainability-oriented ecologies of learning. In R. Barnett & N. Jackson (Eds.), *Ecologies for learning and practice: Emerging ideas, sightings, and possibilities* (pp. 61–78). Taylor & Francis.

Wals, A. E. J., & Mathie, R. (2020). Whole school approaches to sustainability: Critical case studies from Europe education for sustainable. In *Education for sustainable development in Asia and Europe - A comparative perspective*. Konrad Adenauer Stiftung. Report accessible via: https://www.kas.de/en/web/politikdialog-asien/panorama/detail/-/content/education-for-sustainable-development

Wessels, K. R., Bakker, C., Wals, A. E. J., & Lengkeek, G. (2022). Rethinking pedagogy in the face of complex societal challenges: Helpful perspectives for teaching the entangled student. *Pedagogy, Culture & Society*. https://doi.org/10.1080/14681366.2022.2108125

Woiwode, C., Schäpke, N., Bina, O., et al. (2021). Inner transformation to sustainability as a deep leverage point: Fostering new avenues for change through dialogue and reflection. *Sustainability Science, 16*, 841–858. https://doi.org/10.1007/s11625-020-00882-y

Part 1

Principles, Perspectives, and Challenges

A Whole School Approach: A Synthesis of Interconnected Policy, Practice, and Research Conceptualisations

2

Rosalie Gwen Mathie

Key Message
If we are to address siloed thinking and competition over what educational innovations are prioritised in school reforms today, finding synergies between different but inter-connected educational initiatives is vital. To authentically integrate sustainability-oriented education, critiquing what a Whole School or Whole Institution Approach means for education generally, not just Education for Sustainable Development, is required. Subsequently, education policy is required that supports boundary-crossing, increases collective capacity-building across and between education institutions, embeds reflexive thinking into every day practice, and increases school-based participatory research opportunities.

2.1 Introduction: Educational Change, Innovation, and School Reform in Times of Global Sustainability Challenges

Transforming societies sustainably is a complex task that calls for an immediate collective effort and action from all sectors of society. It is not enough to teach sustainability-oriented values; one also needs to have the capacity to act and experience those values in practice (Shallcross, 2003). Therefore, to meaningfully experience sustainability-oriented education there needs to be consistency in what we teach in and outside of the classroom: requiring education to be relevant, responsible, and responsive in a world challenged by (un)sustainability (Shallcross & Robinson, 2008; Wals, 2019). Wals (2015, 2019) points out that just like business-as-usual, education-as-usual is no longer an option: arguing that, as the nature of sustainability challenges is *complex, value-laden, ambiguous, emotional, uncertain, and emergent*, and that conventional learning forms, that predominantly focus on knowledge transfer, are insufficient to engage learners with the sustainability challenges we face today. This shift in pedagogical focus is also linked to the progression seen, as Henderson and Tilbury (2004) describe, in terms of moving from purely focusing on Environmental Education (EE), to instead moving towards Education *for* Sustainability (EFS):

> EFS differs from traditional approaches to EE in that it focuses sharply on more complex social issues, such as the links between environmental quality, human equality, human rights and peace and their underpinning politics. This requires citizens to have skills in critical enquiry and systemic thinking to explore the complexity and implications of sustainability. This new educational approach also requires a new pedagogy which sees

R. G. Mathie (✉)
Norwegian University of Life Sciences, Ås, Norway
e-mail: rosalie.mathie@nmbu.no

A. E. J. Wals et al. (eds.), *Whole School Approaches to Sustainability*, Sustainable Development Goals Series, https://doi.org/10.1007/978-3-031-56172-6_2

learners develop skills and competencies for partnerships, participation and action (Henderson & Tilbury, 2004, p8)

As Shallcross (2003) also summarises, the call for more relational, transactional, and ideally transformational forms of learning has multiple, distinct but interconnected, theoretical roots. For example, critical pedagogy (Freire, 1972), emotional and ecological literacy movements (Orr, 1992; Goleman, 1996), and holistic education (Greig et al., 1989). These forms of learning, and their theoretical foundations, all require a reorientating of curriculum. However, a deeper shift is also called for that goes beyond making changes in curriculum content and structures. Such a shift also addresses what holistic education necessitates: a change in '[…] an attitude of mind on the part of teacher and student alike which prioritises and searches out relatedness to the whole' (Greig et al., 1989, p20 as cited in Shallcross, 2003, p138–139). Therefore, education institutions around the world are being asked, or are demanding from themselves, to simultaneously change multiple aspects of their approach to education: the curriculum content, how students are assessed, teacher training, where and how the students are taught, and how schools are organised (Wals & Mathie, 2022). In turn, to meaningfully engage with sustainability challenges, schools are also asked to consider a manifold of learning approaches such as action-orientated and participatory pedagogy (UNESCO, 2020).

Despite this pressing need for changes in education, educational innovation and curriculum reforms also have a long history of failing, highlighting that embedding long-lasting change in schools is difficult to achieve (Hargreaves, 2002; Priestley et al., 2011; Serdyukov, 2017; Whiteside, 2019). As Hargreaves (2002) states 'We live in a world of endless and relentless change' (p189), and yet institutional change is notoriously difficult, especially in the education context. For example, previously, top-down, centrally prescribed educational innovations and policy disseminations have been shown to be ineffective in creating systemic and lasting educational change (Priestley and Sime, 2005). This puts into question, as Priestley et al. (2011) discuss, the sustainability of educational change itself. Consequently, if education, as called for, is to play a central role in realising the aims of sustainability-oriented education, schools must be given the support needed to embed continual and effective processes that foster a culture of change in schools. Localized, contextualized methods that are more readily welcomed by the school community are especially needed. As Priestley et al. (2011) conclude, instead of this onslaught of related but separate reforms, we need to '[…] consider the process of change in a more complex relational manner than is often the case, analysing how these ingredients come together in their particular enactments in specific settings' (p31).

So, how do we support educational innovations and reforms to be more effective? Multiple sustainability-related educational innovations today seek to do this by ensuring all stakeholders are involved in the educational change process: to have top-down support, but also harness bottom-up engagement (Ávila et al., 2017; Shallcross et al., 2006). The United Nations also advocates for education initiatives that stress the importance of multistakeholder engagement, dialogue, and active participation at the school level in order to realise the Sustainable Development Goal (SDG) 2030 Agenda. Ensuring that education is more than a cognitive experience, and that learning moves beyond the traditional classroom setting to proactively engage learners with real-world issues and solutions (Leicht et al., 2018; Scott, 2013; UNESCO, 2017), is also echoed throughout the SDG 2030 Agenda. Discussed in this chapter are three specific education initiatives: Education for Sustainable Development (ESD), Education for Health and Wellbeing, and Global Citizenship Education. These initiatives all emphasise the need to students, and other key stakeholders, to be meaningfully engaged in global, national, and local matters, and highlight how contextually based situated methods are essential for empowering students to participate in 'real-life' locally based change (UNESCO, 2015, 2016, 2017).

While it is complex to see how the global issues represented by the SDGs are interconnected, research shows that holistic and systemic thinking is vital for dealing with this complexity and are required to innovate and integrate sustainable solutions (Leicht et al., 2018; Henderson & Tilbury, 2004). Thus, education that seeks to transform and meaningfully engage societies in being part of creating a sustainable future needs to model this thinking. Recent policy documents, such as UNESCO's (2020) ESD Roadmap, mention that educational innovation and school change required by the SDGs can be achieved through a Whole Institution Approach. More recently, an increase in nations supporting a Whole Institution Approach is also evident in the United Nations Economic Commission for Europe (UNECE) *Framework for the implementation of the United Nations Economic Commission for Europe Strategy for Education for Sustainable Development from 2021 to 2030,* where a Whole Institutional Approach is recognised as a transformative learning tool relevant with the aim to support all institutions becoming *'[...] communities of transformational learning'* (p12). While this chapter focuses on a Whole School Approach within formal school-level education, as UNECE (2022) emphasises the need for a Whole Institution Approach in all institutional settings is important to acknowledge, as this also helps to see why transformative education is both relevant and necessary throughout all aspects of society:

> The "whole institution approach" concept embraces settings of all levels of formal education and institutions providing formal, non-formal and informal education in the public and private sectors. This learning process is fundamental for quality education in all aspects: learning programme, governance, infrastructure, connection to community and society (UNECE, 2022, p12).

The term Whole Government Approach is also used (European Commission, 2022), showing the versatility and breadth of contexts and related conceptualisations concerning a Whole Institution Approach. However, for this study, the term Whole School Approach (WSA) is used due to the scope and context of this enquiry being limited to formal school-level education (primary to upper-secondary).

In recent years WSA's in a formal education setting have particularly gained traction in terms of ESD policy. One of the main reasons for this interest in a WSA is that the framework supports moving beyond learning about sustainability in the classroom and instead enables learners to experience sustainability in action throughout the school: where opportunities and spaces in students' everyday lives to proactively 'live what they learn' are commonplace (UNESCO, 2017). WSA related models also exist in educational change contexts beyond ESD, such as general 'whole-school reform', for example, comprehensive school reforms (Slavin & Madden, 2013), integrating whole-school pedagogic change, for example, formative assessment (Priestley & Sime, 2005), and school interventions, for example, anti-bullying 'whole-school intervention' approaches (Cross et al., 2018). While a WSA has multiple manifestations and origins, in essence, a WSA promotes the earlier mentioned holistic, participatory, and co-engaged approaches needed to move towards more systemic school reform and educational change: an approach where the whole school, and their surrounding community, collectively discover what sustainability means and how becoming 'more sustainable' can be approached and realised in their own context.

2.2 Methods, Scope, Aims, and Objectives

The objective of this study is to lay the foundation for a larger research project co-developing a WSA as a thinking tool in four Norwegian upper-secondary schools. Of note is that a WSA conceptualisation (Wals & Mathie, 2022; Mathie & Wals, 2022) is specifically utilised to co-develop reflexive professional development practices as part of the PhD research. Therefore, Wals and Mathie's (2022) WSA conceptualisation, alongside ESD literature reviewed in a previous study (Mathie, 2019), means there is a pre-defined departure point for this study: Wals and Mathie (2022) offer a broad conceptualisation of a WSA showing '[...] how all areas, levels and stakeholders can be engaged in a WSA' (p3). The flower model depicts six interlinked and movable WSA strands (Fig. 2.1).

Subsequently, the aim set for this literature synthesis study is to critique and gain a deeper understanding of *the principles, problematics, and prospects of a WSA both in and beyond ESD*. The reason for this aim is to broaden the author's own knowledge of a WSA, and to see if and/or how a WSA is understood differently in other contexts within and beyond ESD. To explore different WSA conceptualisations beyond the predefined ESD literature, the following research question was set: *What are the key principles of Whole School Approaches in the context of sustainability-related Educational Innovations?* To answer this, a broader search for WSA conceptualisations in ESD, alongside two distinct but inter-connected sustainability-oriented fields—Education for Health and Wellbeing (EHW) and Global Citizenship Education (GCE), commenced. The reason for choosing to search for WSA conceptualisations in the EHW and GCE fields is due to the larger research project relating to three interdisciplinary topics introduced in the 2020 Norwegian curriculum renewal: Today all schools in Norway are required to facilitate learning in three interdisciplinary topics based on dominant societal challenges: (1) health and life skills, (2) democracy and citizenship, and (3) sustainable development (UDIR, 2020). The premise of selecting EHW and GCE is also due to a broad and holistic understanding of sustainability recognising EHW and GCE as a part of ESD and vice versa.

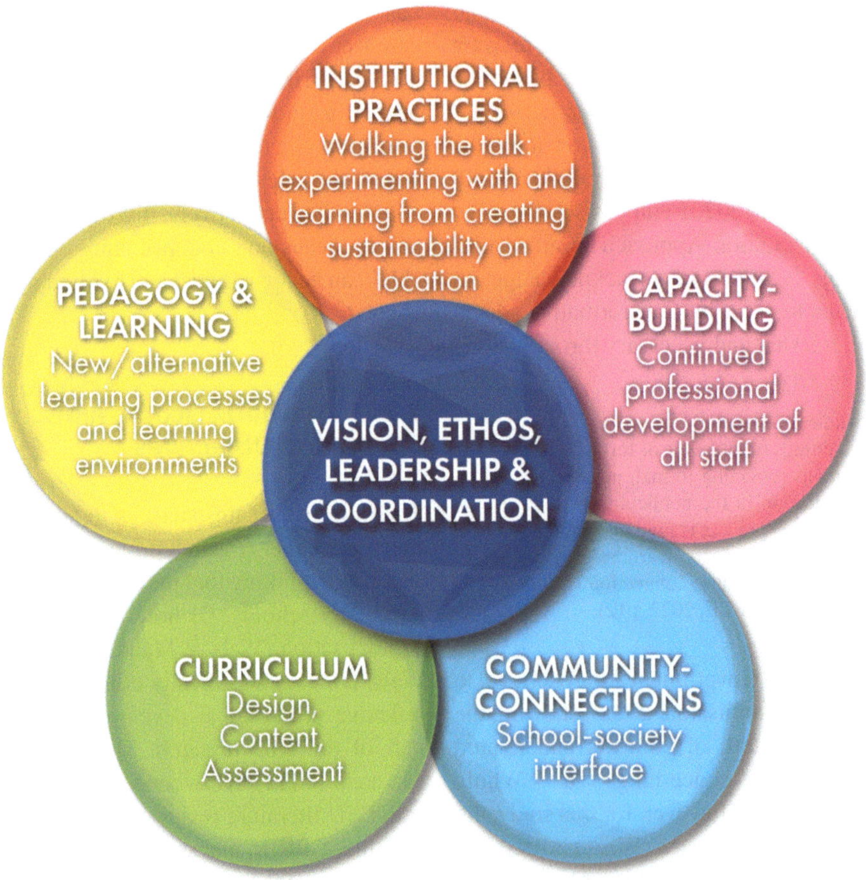

Fig. 2.1 The whole school approach flower model with its six key components. (Reproduced from Wals & Mathie, 2022, p4)

This study presents and compares works of literature found through a non-systematic literature search using the following databases—Eric, Google Scholar, Oria, and Web of Science. The main selection criteria were that the literature included a clear outline and description of each distinctive WSA strand. The literature selected for the synthesis also had to meet at least one of these requirements:

A. *An educational reference or framework clarifying what a WSA is*
B. *Of historical and foundational importance concerning a WSA within the selected field*
C. *Give current examples of how a WSA is being used in practice*

The methods chosen are purposive and explorative, allowing for a broad synthesis of literature from multiple educational initiatives to represent and collate both current and foundational WSA's. As Cook (2019) describes, purposive methods allow for a broad reflection, bringing together different frameworks, philosophies, and research beyond one field. Explorative methods were also chosen as the literature searches at the time revealed limited empirical-based peer-reviewed research concerning a WSA in all ESD, EHW, and GCE contexts. While multiple journal articles identified mention a WSA, very few went into detail about the different strands of a WSA and were thus excluded from this study as they did not meet the main selection criteria set. A lack of peer-reviewed journal articles meant the literature search was broadened to include WSA-related policy documents, curriculum frameworks, and educational resources. Though the explorative nature of the methods has limitations, for example, the literature selected is broad thus not all literature is peer-reviewed, however, the documents selected are considered both relevant and suited to the aims, objectives, and breadth of this study.

Out of just over 150 pragmatically scanned papers, 27 literature documents (from 2003 to 2020) were selected for comparison representing the different but inter-connected normative contexts of ESD, EHW, and GCE. Through the lenses of educational change (Hargreaves, 2002, 2010; Hubers, 2020; Priestley et al., 2011) and educational innovation (Serdyukov, 2017), patterns, variances, and commonalities in the way a WSA is conceptualised were identified, compared, and evaluated. First, an initial round of analysis comparing and interrelating each of the document's WSA frameworks took place. The initial analysis was followed by a second round of abductive thematic analysis using NVivo. This second round of analysis was performed to, (a) to check the rigorousness of the initial analysis and (b) to be able to systematise the analysis results and more clearly visualise the multiple interconnections between the different WSA conceptualisations.

Of note is the limited size and scope of this study. When conducting the WSA and other connected synonyms literature database searches (for publications only written in English), the results relevant to this study predominately originate from westernised countries (Europe, America, Canada, Australia, and New Zealand). Therefore, the choice to limit this study to a westernised perspective of a WSA was taken. Time and accessibility constraints also precluded the inclusion of book publications (a surge of which is seen in the early 2000s), and literature focusing on a WSA in higher, informal, or non-formal education contexts was also excluded. Nor does this study include all WSA-related policy and curriculum framework documents identified, of which there are predominately more than empirical peer-reviewed publications. Meaning, the WSA conceptualisations chosen are not an exhaustive account of a WSA. Instead, for the purpose of gaining a deeper insight of a WSA in various settings, the literature selected acts as a representation of WSA conceptualisations in their respective fields.

2.3 A Synthesis of Whole School Approach Conceptualisations Within Distinct But Connected Education Initiatives

A wealth of knowledge exists about why engagement with societal challenges across all sectors is needed (UNESCO, 2017, 2020). There are also multiple international agendas, such as the SDGs, that detail why education needs to play a central role in resolving these societal challenges. At first glance, it seems ESD (SDG 4.7) is put forward as the initiative required to fulfil the 'educational' requirement of sustainable development. However, when looking at UNESCO's (2017) learning objectives and definition of ESD, interconnections, and overlaps between ESD, GCE, and EHW are found, and all are promoted as central to supporting the SDG 2030 Agenda generally, and specifically in SDG 4—Quality Education. Furthermore, UNESCO policy documents concerning all three education initiatives also promote a WSA (UNESCO, 2015, 2016, 2017). In turn, other UN-based policy documents, such as steering committee meeting papers connected to UNECE (2014, 2022), exist that fully support a WSA. While situated challenges inevitably exist to fully integrate and mainstream a WSA, all 27 conceptualisations synthesised in this study mention in some way that a WSA develops an integrated effort for school-based transformations and a viable way to nurture the multistakeholder partnerships this requires. An overview of the main themes found during the synthesis followed by a brief introduction to a WSA in the contexts of ESD, EHW, and GCE, and an overview of the WSA conceptualisations selected for the synthesis is now presented.

2.3.1 Thematic Structure of the WSA Conceptualisation Synthesis

The following section (Figs. 2.2, 2.3, 2.4, 2.5 and 2.6) gives an overview of the main WSA themes, with their respective sub-themes, developed and coded to analyse the documents, and structure the synthesis. These figures give a broad overview of the different ways the WSA strands are phrased in all 27 WSA conceptualisations and show the number of references found relating to each sub-theme. While all WSA conceptualisations refer to each of the five coded WSA interconnected overarching themes, of note is the varying sizes of the documents analysed. For example, two WSA-related PhD publications are included in the analysis under ESD, which due to their size naturally results in a higher number of references found in these documents. Various sub-themes also include another layer of sub-themes. For a more detailed account of the coding structure, please contact the author directly. The overarching themes are proposed as five interlinking WSA strands phrased to capture the main strands of a WSA from all ESD, EHW, and GCE perspectives. Together, this overview proposes seeing a Whole School Approach to support general quality educational change processes:

2.3.2 Education for Sustainable Development and a Whole School Approach

While WSA's to enacting ESD do exist, today it still seems, as Hargreaves (2008) highlighted over a decade ago, few examples exist of a WSA to ESD being achieved on a system-wide or national basis. Before presenting the ESD-based conceptualisations, of note is the increasing commitment on national levels to supporting WSA's, such as UNECE's implementation strategy for ESD 2021–2030 (UNECE, 2022), which opens up and provides the much-needed infrastructure and guidance required for the genuine mainstreaming of a WSA (or in this context the Whole Institution Approach) in the ESD context. Also of note is the Scottish Government's utilisation of a WSA for their education policy Learning for Sustainability (Scottish Gov., 2012, 2020) Qablan, (2018) also serves as another example in the Scottish context, in particular their policy and educational frameworks aiming to integrate, monitor, and assess a WSA, whereby whole-school and community approaches are

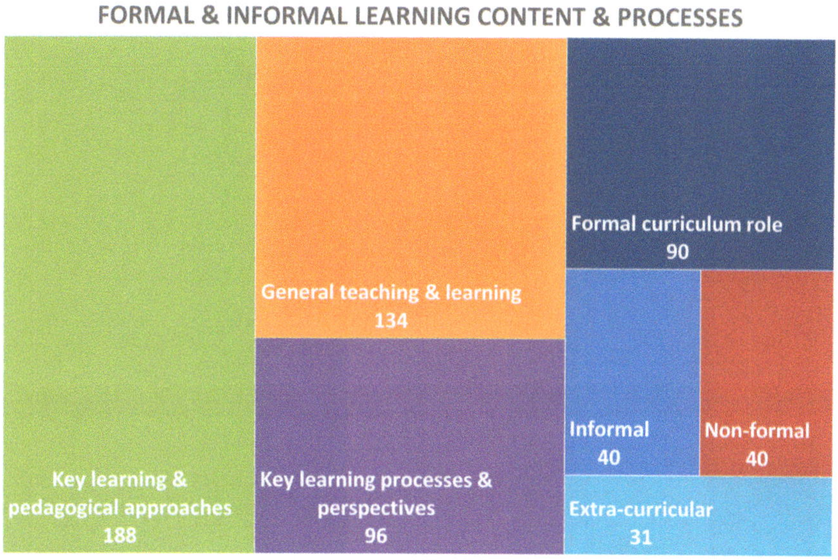

Fig. 2.2 WSA strand 1—formal and informal learning content and processes: overview of references present under each sub-theme

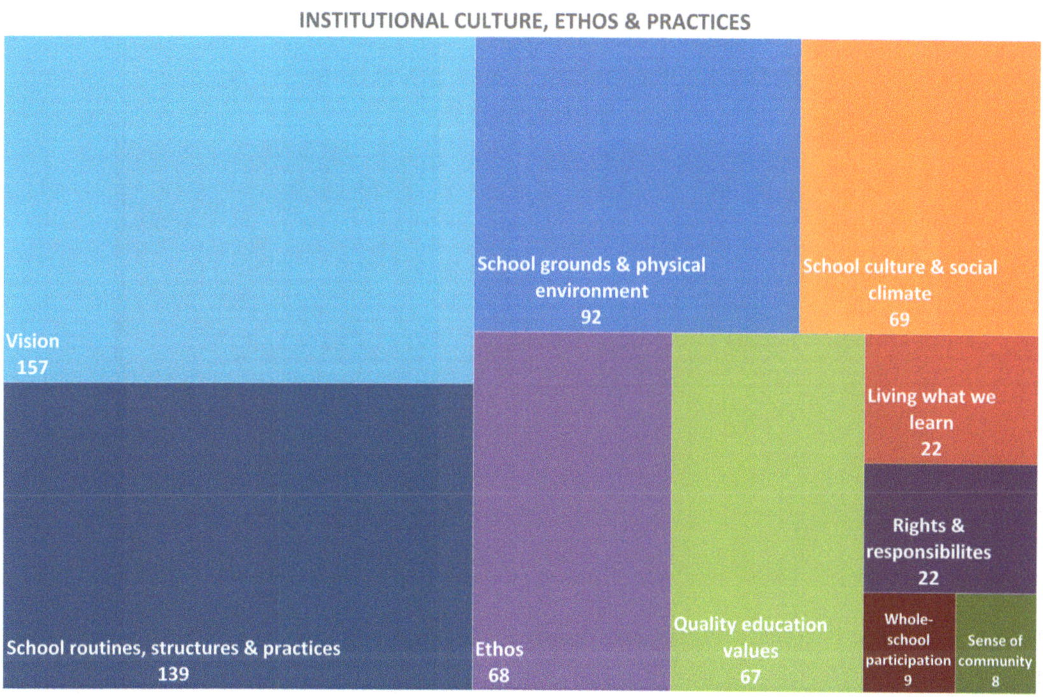

Fig. 2.3 WSA strand 2—institutional culture, ethos, and practices: overview of references present under each sub-theme

promoted on multiple education levels: by weaving together ESD, outdoor learning, and GCE principles (Education Scotland, 2022). While not included in the WSA conceptualisation synthesis, both the origin of the UNECE and Scottish Government WSA contexts are also noteworthy. For example, UNECE's (2014) WSA approach to ESD pre-dates and appears to

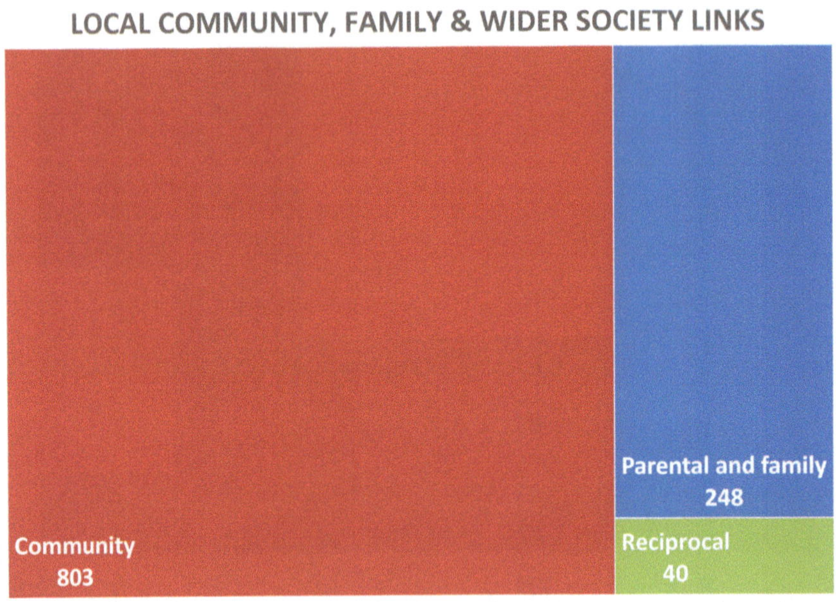

Fig. 2.4 WSA strand 3—local community, family, and wider society links: overview of references present under each sub-theme

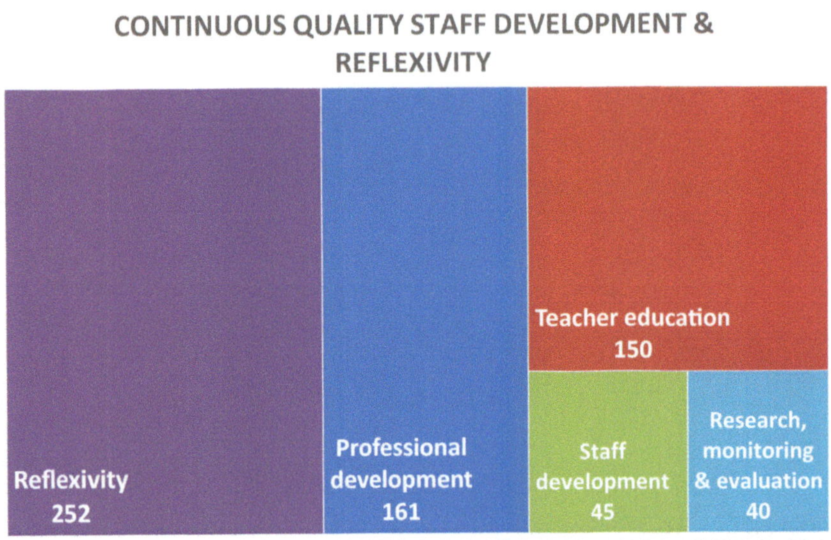

Fig. 2.5 WSA strand 4—continuous quality staff development and reflexivity: overview of references present under each sub-theme

closely influence UNESCO's (2017) policy Whole Institution Approach conceptualisation, and the Scottish WSA model can be traced back to the 4C model 'culture, curriculum, campus and community' (Scottish Gov., 2012) that originates from a higher education context—the Plymouth University Centre for Sustainable Futures (Selby, 2009). However, due to (a) the original literature search engines not identifying these policy documents, and (b) it being beyond the scope of this chapter to include all ESD-based WSA policy conceptualisations, UNESCO (2017) was chosen to represent a ESD-based policy perspective.

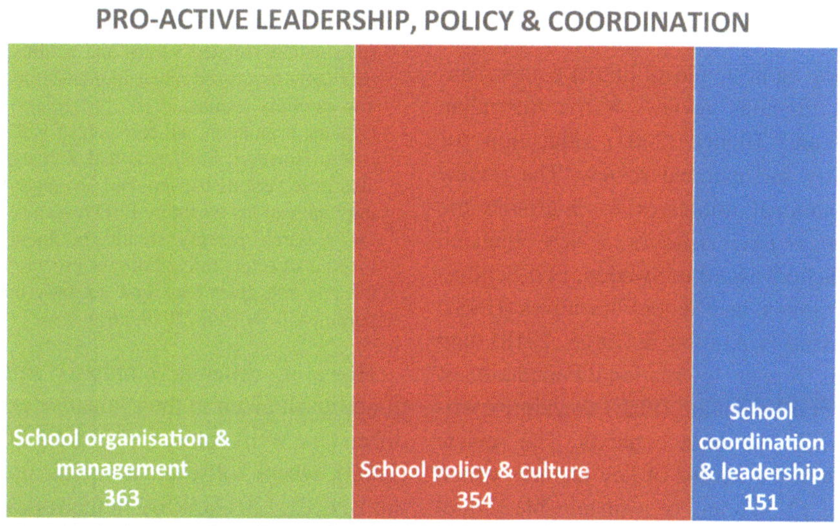

Fig. 2.6 WSA strand 5—pro-active leadership and coordination: overview of references present under each sub-theme

UNESCO defines ESD as empowering '[…] learners to take informed decisions and responsible actions for environmental integrity, economic viability and just society for present and future generations' (UNESCO, 2017, p7). While some contest the term ESD, since the United Nations (UN) Decade of Education for Sustainable Development (DESD), ESD dominates global policy and represents the current semantic progression that education concerning people and planet has taken since the late nineteenth century (Wals & Benavot, 2017). As outlined in the introduction, a WSA is being promoted internationally as an effective way to fulfil the ESD policy requirements. This is in part due to a WSA interconnecting central aspects of ESD policy and theory, such as the ESD-ascribed pedagogies, action and learner-oriented approaches, and transformative learning (UNESCO, 2017, 2020).

UNESCO's (2017) WSA conceptualisation (in this context referred to as a Whole Institution Approach) provides a four-strand model being utilised for example, by UNESCO Associated Schools Project Network (ASPnet), in an international pilot involving schools in 25 countries, to build Climate Ready Schools. Chopin's et al. (2018) study uses UNESCO's WSA model as a lens to evaluate 28 Canadian schools' approaches to integrating climate action into all aspects of the school. The study includes both pilot and reference schools, half of which took part in the ASPnet climate-ready schools' pilot. Both Chopin et al. (2018) and, more recently UNESCO's (2020) ESD roadmap, refer to a WSA as a viable method for schools to mobilise ESD. However, Chopin et al. (2018) highlight that while the ASPnet pilot schools presented a broad range of integration and activities throughout all four WSA components, the reference schools (non-ASPnet schools) in some cases reported higher results. Chopin et al. (2018) also mention that when schools become more aware of the complexity of issues, for example, through professional development seminars connected to the pilot project, schools' self-evaluation and assessment may become more critical and result in less self-perceived progress. Chopin et al. (2018) also discuss the vital role collaborative networks play in a WSA. Through these networks, schools learn from and with each other, which is further supported by having 'diverse climate action actors' (multi-stakeholders). Equally important is for ESD content (in this context, climate change) to be local and relevant, to utilise technology creatively, and to ensure concrete and feasible actions are experienced (Chopin et al., 2018).

One of the earliest, and still most comprehensive, WSA literature reviews is the Australian Henderson and Tilbury (2004) Education *for* Sustainability international review. The review examines national initiatives from around the world, such as *Enviroschools* in New Zealand, the Green School award in Sweden, Eco-Schools from Environment and School Initiatives (ENSI) (also discussed in Affolter & Varga, 2018), and Eco-schools from the International Foundation of Environmental Education (FEE) to gain experience of WSA's in these contexts. The review concludes with an outline of key features which characterise 12 'visions' of a sustainable school (Henderson & Tilbury, 2004, p45) and are used to represent distinctive WSA strands. In addition, Henderson and Tilbury (2004) present five key considerations for Education for Sustainability programs: (1) Relevant (to the school community's needs); (2) Resourced (expertise and financially), (3) Reflective (critical evaluation at all levels); (4) Responsive (flexible school structure); and (5) Reformative (capacity for change). These key WSA characteristics, features, and considerations remain relevant in today's ESD context, for example, in the European Commission's development of a WSA for learning about Environmental Sustainability (Tilbury & Galvin, 2022).

Shallcross (2003) an extensive PhD study on a WSA, Shallcross and Robinson (2008) represent an early conceptualisation of a WSA in the ESD context. Shallcross and Robinson's (2008) WSA model shows key characteristics attributed to each strand. While similar to the Chopin et al. (2018) and UNESCO's (2017) WSA conceptualisation, Shallcross (2003) in particular emphasises reflexive research, monitoring, and evaluation as an integral standalone strand. Central to this is the role of self-evaluation and teacher-led 'action-research' where this WSA conceptualisation shows a cautious stance towards how value-laden ESD can be as summarised in Mathie (2019):

> Shallcross (2003) references Bonnet's (1999) caution that damage can be done if the impact of sustainability-related values and actions, that are inevitably expressed when implementing ESD, is not carefully evaluated. [...] it seems process-oriented evaluation and research of ESD-practice is key. However, Shallcross and Robinson (2008) also raise various barriers and leverages concerning implications for future ESD research, stressing the research process should be inclusive and involve all actors throughout the process to ensure the research gives back and not only takes from participants (Mathie, 2019, p61)

However, different emphasis and the type of approach given to the evaluative processes do differ. One WSA framework similar to Shallcross (2003), where self-evaluation is emphasised, is the UNECE (2014) WSA framework where self-assessment also is shown as a standalone WSA strand, highlighting:

> It [self-assessment] is an essential means to establish a continuous and effective ESD school planning process, which supports continuous enhancement of implementation quality. The self-assessment is a tool only for the school itself, its improvement and development in ESD and not connected or linked to an external reporting mechanism (UNECE, 2014, p12)

Mathar's (2016) WSA conceptualisation is from the German ESD Curriculum Framework publication (Schreiber et al., 2016), and details a mind map that visualises in detail nine strands, with multiple sub-strands, that make up a WSA. The main strands of this mind map utilise ESD and Global Development education as a working field for the whole school. Mathar's (2016) chapter also offers examples of each WSA strand in action from various German schools. The Curriculum Framework, in general terms, also advocates for a holistic approach that integrates ESD as a foundation of all subjects, school management, and extra-curricular activities, and that adopting a WSA is a way for schools to do this (Schreiber et al., 2016). German publications, such as Jucker and Mathar (2016) and Affolter and Varga's (2018) ENSI network 30-year review, also provide WSA-related case studies from around the world.

Mogren's et al. (2019) conceptualisation builds upon Scherp's school improvement model and consists of four WSA-related interlinked dimensions: (1) routines and structures, (2) professional knowledge creation, (3) practical pedagogical work, and (4) holism (p511–512). This conceptualisation takes a holistic view of how a school operates, what functions, what doesn't, and how all these dimensions interact and impact each other (Mogren et al., 2019). Mogren et al. (2019) summarise why operationalising Scherp's model as an empirical research tool and analytical guide is suitable for the ESD context. Reasons are given such as the model being useful to assessing how a holistic vision is integrated in practice; to assess school improvement from a generic perspective; and to visualise the ESD staged responses to sustainability (Scott, 2013) model (Mogren et al., 2019). Mogren's (2019) PhD publication presents four principal quality criteria for transformative ESD in local school organisations which are consequently added to the WSA conceptualisation synthesis: (1) student centred education, (2) collaborative interaction and school development, (3) proactive leadership and long-term perspective, and (4) cooperation with local society. Mogren (2019) also points out the importance of recognising the normativity of ESD: that situating a WSA through a school improvement lens can provide a new way to research a WSA to ESD by addressing the implications of politicisation and normativity concerning ESD policy and practice.

Lastly, Bosevska and Kriewaldt's (2020) WSA conceptualisation presents a recent research case study looking into a whole-school community response to fostering sustainable education. While similar to other WSA conceptualisations in the synthesis, of note here is the reference to Sterling's seven *Operational Dimensions* of a school (2004, as cited in Bosevska and Kriewaldt, 2020, p60). However, Bosevska and Kriewaldt (2020) add the 'historical context' dimension to a WSA, the premise being to capture the case-study school's own historical progression (from learning *for* sustainability to sustainable learning), which played a pivotal part in the school's development, vision, ethos, and focus.

Table 2.1 gives an overview of the nine ESD-related WSA conceptualisations included in the

Table 2.1 Matrix overview of the ESD literature selected for comparison—an overview of references coded to each interlinking WSA strands

	Institutional culture, ethos, and practice	Formal and informal learning content and processes	Local community, family and wider society involvement	Continuous quality staff development and reflexivity	Pro-active leadership, policy and coordination
Bosevska and Kriewaldt (2020)	16	13	16	8	12
Chopin et al. (2018)	19	19	64	4	32
Henderson and Tilbury (2004)	53	35	74	66	77
Mathar (2016)	15	10	12	5	13
Mogren et al. (2019)	31	9	10	12	29
Mogren PhD (2019)	50	24	17	15	69
Shallcross PhD (2003)	112	79	197	161	90
Shallcross and Robin-son (2008)	17	12	24	10	7
UNESCO (2017)	37	23	26	23	34

synthesis and shows the prevalence of references found organised thematically under the five interlinking WSA strands:

2.3.3 Education for Health and Well-Being and a Whole School Approach

UNESCO defines EHW's goal as supporting '[...] the contribution of national education sectors to ending AIDS and promoting better health and well-being for all children and young people. This in turn will contribute to the achievement of the Sustainable Development Goals, particularly those related to education, health, and gender equality' (2016, p8). To realise this goal, UNESCO identifies two priorities: '(1) To ensure that all children and young people benefit from good quality, comprehensive sexuality education that includes HIV education (2) To ensure that all children and young people have access to safe, inclusive, health-promoting learning environments' (UNESCO, 2016, p8). Five guiding principles are mapped out by specifying EHW to be guided by: an international human rights approach; addressing gender and diversity; alignment with national ownership, priorities, plans, and processes; scientific and evidence-based approach; and active participation of young people in policy-making (UNESCO, 2016).

The policy document selected for comparison is a more recent publication from the United Nations Girls Education Initiative (UNGEI) promoting a WSA to Prevent School-Related Gender Based Violence (SRGBV). UNGEI (2019) stipulates eight WSA components that address five SRGBV drivers: (1) normalisation of violence against children through social norms that justify violence; (2) silence around violence against women and girls; (3) rigid gender roles; (4) stereotypes of masculinity and femininity; and (5) inequality and discrimination (UNEGI, 2019). UNEGI (2019) also highlights the need for minimum standards that, out of all the documents reviewed, sets out the most comprehensive WSA monitoring framework for countries to follow. The framework lays out national and school-level indicators for measuring improvement in all eight WSA strands. Participatory methods in both pre-service teaching courses and the classroom are emphasised, as is the need for a holistic approach that 'addresses the drivers and root causes of violence at both the school and the community levels' (UNEGI, 2019, p7). UNEGI (2019) also emphasises involving community and government stakeholders as central to a WSA.

Numerous stand-alone EHW education initiatives utilise a WSA. One example is Save The Children's (STC) ABC anti-bullying practice (Barber et al., 2010). While not selected for the main WSA conceptualisation comparison, it is of note as Barber et al. (2010) provide practical tools for schools to develop annual feedback cycles for the whole school community confronts an issue together. Barber et al. (2010) also highlight that a WSA enables the change process to become a 'normal and sustainable part of the school community' (p3). Another example is the 'Safe to Learn' global initiative (STL, 2020) to end violence against children, who also refer to UNGEI's (2019) minimum standards and monitoring framework. STL (2020) shares similarities with a WSA analytical framework developed by Mathie (2019), whereby even if a WSA is not utilised in schools, a WSA still offers a framework to monitor, evaluate, and benchmark schools' ESD enactment. Studying the effectiveness of frameworks that benchmark and measure a WSA progress, is noteworthy for future research.

Rowe et al. (2007), Rowe and Stewart (2011), and Scott's (2005) WSA conceptualisations are indicative of when a WSA started to be utilised in EHW in connection to the World Health Organization's (WHO) Health Promoting School principles.

Rowe et al. (2007) present an Australian-based WSA regarding school connectedness stressing the importance of processes, structures and wider community connections, democracy, and social capital. Rowe and Stewart (2011) discuss four main strands of a WSA and offer a detailed critique concerning WSA and the

Health Promoting School mechanisms that influence school connectedness.

Scott (2005) presents a New Zealand–based foundational WSA conceptualisation concerning behaviour management that concludes that for an effective WSA, a strong collective vision that the whole-school community commits to is essential. Moreover, 'If all students are to experience a sense of belonging and inclusion in our schools there must be a shift in the way that schools respond to severe and challenging behaviour' (Scott, 2005, p33). Scott's (2005) findings also point out key leverages and barriers impacting the effectiveness of a WSA and the importance of continuous monitoring and evaluation within this context.

Rowling's (2009) WSA conceptualisation presents 14 core characteristics that support the capacity building of schools to implement WSA for mental health promotion. Unlike ESD, issues of collegiality and consistency between staff, the need for students to uphold high standards, and the need for consistency in disciplinary sanctions are highlighted. Experiences from over a decade of MindMatters implementation in Australia, and other research connected to this, form the basis of this study and conclude two key issues: Leadership and Professional Learning are considered central for WSA-related capacity building. Rowling also discusses that a need for increased educational research and teacher professional development (like Shallcross, 2003) is vital for the future progression of school mental health promotion.

A combination of Coordinated School Health (CSH) and the coordinated school health's Whole School, Whole Community, Whole Child (WSCC) model in Lewallen et al. (2015) is also added for comparison. Added also for comparison is Hunt et al. (2015) paper concerning the WSCC model. Hunt et al. (2015) point out that research concerning how to make the model 'actionable' is limited and subsequently propose a ten-step systematic process to support closing this WSCC theory-practice gap.

Lastly, Cross et al.'s, (2018) WSA conceptualisation again concerns bullying behaviour, this time evaluating the Friendly Schools Project (FSP), an Australian-based intervention that '[...] aimed to reduce bullying and aggression among 3000 students who had recently transitioned to secondary school'. This quantitative study concludes, '[...] a whole-school intervention approach shows promise in the prevention of bullying and negative outcomes' (Cross et al., 2018, p509).

Table 2.2 gives an overview of the nine EHW-related WSA conceptualisations included in the synthesis and shows the references found organised thematically under the five interlinking WSA strands:

2.3.4 Whole School Approaches to Global Citizenship Education and a Whole School Approach

UNESCO defines GCE through three conceptual dimensions that serve to form GCE goals, learning objectives, and competencies: (1) *Cognitive*—concerning issues on local-global levels, interconnectedness, and interdependency; (2) *Socio-emotional*—concerning, common humanity, shared values, respect for differences and diversity; (3) *Behavioural*—concerning effectiveness and responsibility on all local-global levels for a more peaceful and sustainable world (UNESCO, 2015, p15). These conceptual dimensions are also shared with ESD. UNESCO (2015) likewise offers a broad overview and GCE history that interconnects the initiative to human rights education, peace education, gender equality, and ESD-shared perspectives such as transformative learning, critical thinking, lifelong learning, and action-orientated approaches. In this GCE context, a WSA is promoted as an approach to school-wide delivery of GCE by providing opportunities for transforming '[...] curriculum content, the learning environment, and teaching and assessment practices' (UNESCO, 2015, p48). The use of participatory learning methods across subjects and community engagement are listed as some examples of a WSA in practice (UNESCO, 2015). This GCE synthesis, possibly due to the nature of GCE's global con-

Table 2.2 Matrix overview of the EHW literature selected for comparison—an overview of references coded to each interlinking WSA strands

	Institutional culture, ethos, and practice	Formal and informal learning content and processes	Local community, family and wider society involvement	Continuous quality staff development and reflexivity	Pro-active leadership, policy and coordination
Cross (2018)	4	2	16	7	10
Hunt et al. (2015)	7	10	17	5	16
Lewallen (2015)	9	14	25	6	22
Rowe and Stewart (2011)	3	12	22	1	5
Rowe et al. (2007)	13	10	20	2	14
Rowling (2009)	4	2	10	8	18
Scott (2005)	7	3	9	4	10
UNESCO (2016)	17	8	14	7	28
UNGEI (2019)	11	6	65	26	62

text, includes a higher number of WSA-related international curriculum frameworks and school guides. This is because the literature searches found multiple WSA-related curriculum and education frameworks as opposed to academic studies concerning national- or school-based empirical research.

In efforts to support GCE policy in Europe, the Council of Europe Education Department (COE, Ed.) developed a Reference Framework of Competences for Democratic Culture (RFCDC) that utilises a WSA. The RFCDC provides practical guidelines for enacting a WSA—a reflexive cyclical process involving situation analysis, planning, and executing action plans, reflexive evaluation, and sharing lessons learned (COE, Ed., 2018). In the RFCDC WSA conceptualisation, integrating democratic and human rights principles, along with the safety and well-being of students, is central. The framework also discusses (as in Henderson & Tilbury, 2004) the need to address the *hidden curriculum*, meaning all that is not explicitly taught but still learnt beyond the formal lesson content. The COE, Ed, (2018) also emphasises the importance of informal, nonformal, and formal education. As with other WSA conceptualisations, the COE, Ed. states, 'A whole-school approach implies the active involvement and commitment of all stakeholders in a school' (COE, Ed., 2018, p91).

Barrett's (2020) review of the RFCDC also mentions that it is well known that a WSA is an effective way to teach, for example, human rights and equality, and highlights how the framework gives practical examples of a WSA application in the context of building a democratic culture. Barret also gives an overview of the benefits this approach can have:

> [...] increased learner responsibility, self-efficacy, self-esteem and intrinsic motivation; improvements in learning and academic standards; better understanding by learners of the nature of rights and responsibilities; learners participating actively in decision making [...]; learners being empowered to undertake action in defence of human rights; and learners acquiring more positive attitudes towards diversity (Barret, 2020, p12).

The RFCDC also mentions sustainability in different contexts, such as creating a *sustainable* democratic and inclusive society; the need for long-term support to have a *sustainable* impact; and the relevance of other 'transversal topics' that interconnect and can overlap with a Democratic Culture such as *sustainable* development (COE, Ed., 2018).

While Henck's (2017, 2018) WSA conceptualisation offers a comparatively brief overview, they represent a well-established WSA framework used to inspire learners and transform schools within the GCE context, the America-based Association for Childhood Education International Global Schools First (GSF) programme. The GSF programme offers support for schools concerning GCE by utilising a WSA. Here, similarities in the ESD-based UNESCO (2017) WSA model are found, yet informal learning experiences and arenas are clearly emphasised in this GCE context.

Hunt and King (2015) discuss a WSA in the context of the UK's Global Learning Programmes (GLP) (2013–2017), which utilised a WSA to deliver a school-wide citizenship scheme. Hunt and King (2015) also provide concrete WSA didactic approaches, case-study examples, and a strong focus on measurement tools for assessing the impact of global learning.

Another example from GCE is Anne Beauvallet's (2016) England, Wales, Scotland, and Northern Ireland comparative study documenting educational approaches to Citizenship Education. Here, England is highlighted as the only nation not utilising a WSA to enact GCE. One of the main WSA case-study examples in Beauvallet is a Scottish Government GCE resource document published by Learning and Teaching Scotland (2011). From reviewing Beauvallet (2016) and in more depth Learning and Teaching Scotland (2011), four WSA strands are determined: (1) active participatory activities, (2) school ethos, (3) organisational structures, and (4) community links. Also, of note in Beauvallet (2016) is the Northern Ireland case study mentioning whole-school events and extra-curricular activities as core to the WSA in this context.

Van Driel's et al. (2016) WSA conceptualisation derives from a European review about fostering tolerance in children (in terms of social tensions and unrest concerning issues that arise due to ethnic and religious diversity). Here, a WSA is documented extensively in terms of defining a WSA and concludes that WSA's '[…] with strong and dynamic ties to the local community have great potential for promoting cohesion. They create a sustainable positive school atmosphere, as well as a stronger sense of belonging' (Van Driel et al., 2016, p9).

Oxfam's WSA conceptualisation (2015) is another example of a practical guide for schools concerning GCE. Within the guide, a WSA framework is presented as 'a stimulus for whole school planning' and a 'starting small but thinking big' attitude is advocated (Oxfam, 2015, p14). In addition, participation is emphasised in this practical guide, stressing the importance of learner-led participation through curricular and extra-curricular activities (Oxfam, 2015). Table 2.3 gives an overview of the nine GCE-related WSA conceptualisations included in the synthesis, showing the prevalence of references found organised thematically under the five interlinking WSA strands.

2.4 Discussion: What are the Main Principles, Processes, and Strands of a Whole School Approach in and Beyond Education for Sustainable Development?

All 27 WSA conceptualisations in this chapter showcase how a WSA is relevant for and can be adapted to support schools in enacting specific school initiatives. While a more extensive systematic literature review would provide additional rigour, an indication of central WSA principles in a 'general quality education' context is found. Fig. 2.7 depicts the synthesis findings through a composite model outlining central WSA principles when viewed in this broader *context*. WSA principles are divided into the following keys: (a) overarching processes, (b) underpinning perspectives; and (c) five aspects of an educational institution that a WSA requires for

Table 2.3 Matrix overview of the GCE literature selected for comparison—an overview of references coded to each interlinking WSA strands

	Institutional culture, ethos, and practice	Formal and informal learning content and processes	Local community, family and wider society involvement	Continuous quality staff development and reflexivity	Pro-active leadership, policy and coordination
Beauvallet (2016)	4	10	10	1	4
COE, Ed. (2018)	25	81	67	70	47
Henck (2017)	3	3	4	3	3
Henck (2018)	3	7	5	2	5
Hunt and King (2015)	17	27	28	20	19
Learning & Teaching Scotland (2011)	12	10	22	12	6
OXFAM (2015)	11	16	13	19	9
UNESCO (2015)	12	41	56	8	18
Van Driel et al. (2016)	22	54	96	58	49

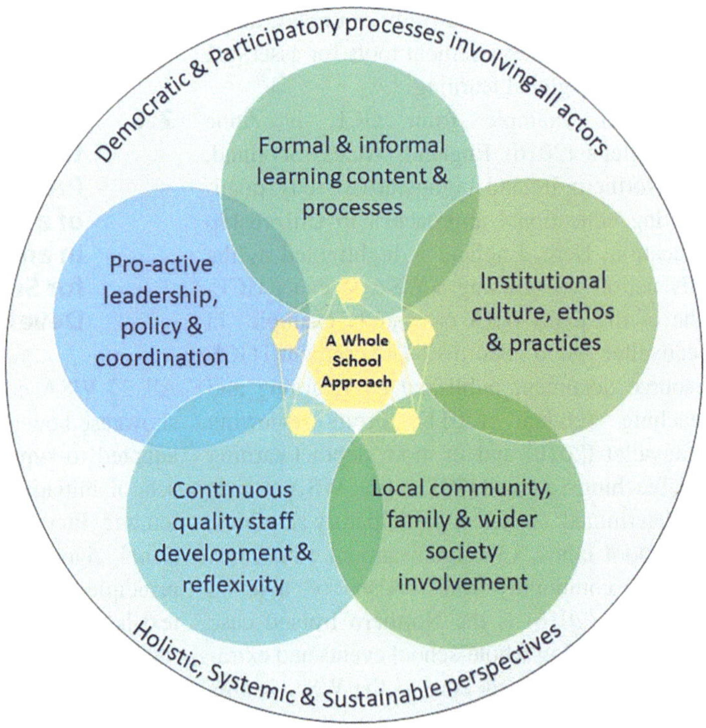

Fig. 2.7 Overarching principles, processes, and strands of a Whole School Approach to Support Educational Change Processes

engagement and collaboration from within the whole-school community.

2.4.1 Holistic, Systemic, and Sustainable Perspective

Holistic, Systemic, and *Sustainable* perspectives have been chosen as they provide a normative direction to work towards when enacting a WSA and help unravel different mechanisms at play within a WSA. The majority of WSA conceptualisations reviewed refer specifically to a holistic and systemic perspective as a part of this approach; however, this terminology is mentioned more frequently in the ESD conceptualisations. In a broader context, sustainability as a concept has multiple meanings and dual purposes. As Hargreaves (2002) discusses, in educational change sustainability can have five interconnected dimensions: '(1) Improvement that sustains learning—not merely change that alters schooling; (2) improvement that endures over time; (3) Improvement that can be supported by available or achievable resources; (4) improvement that doesn't impact negatively on the surrounding environment of other schools and systems; (5) improvement that promotes ecological diversity and capacity throughout educational and community environment' (p193). So, in this WSA underlying *'perspective'* context, sustainability is understood in terms of environmental, societal, and economic sustainability, and also to determine whether a specific initiative, procedure, or aspect being implemented can be sustained. Subsequently, sustainability has strong normative values and is both adaptable and relevant for *sustaining* educational change and innovation over time. In Mogren (2019) there is a need to recognise the normative and politicisation attached to ESD; this is something to consider further. Hargreaves (2010) also summarises that, if educational reforms are to succeed (they have mainly failed so far) sustainability in a school reform context needs to encompass more than just maintainability:

> A deeper philosophy and practice of sustainability returns to and embraces the environmental roots of the concept. It emphasises, celebrates, and advocates for the learning in depth about one's place in and contribution to the world [...] Our future has to be sustainable. So must our schools be (Havelock, 1971; Hargreaves, 2010, p293).

Therefore, a broad sustainability perspective, with its multiple meanings, is a critical part of a WSA, especially if it is to be utilised to integrate continual educational change processes instead of supporting one-off initiatives. Of note also is the prevalence throughout all WSA conceptualisations to broader social and societal contexts (Table 2.4) that very much link to the holistic and systemic values. This highlights why weighting the importance of environmental, social, political, and economic contexts equally is needed, and how developing an awareness of holistic and systemic perspectives can support this.

2.4.2 Democratic and Participatory Processes Involving All Actors

Democratic and participatory processes are, to some extent, part of all the WSA conceptualisations compared. Although the extent varies (Table 2.5), and different definitions are given, WSA conceptualisations within ESD, EHW, and GCE often refer to democratic and participatory processes as critical for integrating a WSA. Therefore, the choice was made to emphasise these as key overarching processes to be present, not just within the curriculum or one specific strand, but throughout all aspects of a WSA.

The decision to include democratic and participatory as overarching processes is also due to questions arising throughout the synthesis analysis regarding how and why promoting a WSA for general quality educational change processes, instead of for specific 'whole-school' initiatives, is necessary. Therefore, these processes are illustrated as an overarching process because to have a participatory and democratic approach, all school and community members, including parents, staff members, students, and teachers, are required to be involved, not just the students.

Table 2.4 Matrix overview of the holistic, systemic, and sustainable themes and sub-themes referenced under ESD-, EHW-, and GCE-related WSA conceptualisations

	EHW conceptualisations	ESD conceptualisations	GCE conceptualisations
Systemic	2	34	4
Holistic	10	130	20
Sustainability	8	321	17
Global and local	2	20	30
Relational	4	13	2
Social and societal	129	311	241
Economic	26	154	58
Environmental	0	504	24
Political	6	79	70
Cultural	6	109	95
Historical	0	9	12
Ethical	2	133	45

Table 2.5 Matrix overview of the Democratic and Participatory themes and sub-themes referenced under ESD-, EHW-, and GCE-related WSA conceptualisations

	EHW conceptualisations	ESD conceptualisations	GCE conceptualisations
Democratic	4	39	53
Participatory	20	73	29
Integrative	2	16	1
Agency	7	87	7
Praxis (reflection and action)	9	143	114

2.4.3 Five Strands of a Whole School Approach to General Quality Education

While some WSA conceptualisations give more details than others, all the different variations of what a WSA entails fit into, or under sub-categories of, the following five overarching strands: (1) Formal and informal learning content and processes, (2) Institutional culture, ethos, and practices, (3) Local community, family, and wider society links, (4) Continuous quality staff development and reflexivity, and (5) Pro-active leadership and coordination. These five overarching strands comprise the core aspects of an institution's community, both in terms of place and representing the key stakeholders that a WSA requires to engage in the process. The overlapping circles in Fig. 2.7 illustrate the importance of educational change processes throughout the school and for each aspect to interconnect. Most WSA conceptualisations a highlight the importance of interconnections between the strands and that the learning, messages, and the formal and hidden curriculum are to be echoed in practice throughout the school. For example, Rowe et al. (2007, p524) highlight the importance of school connectedness for enacting a WSA, 'defined as the cohesiveness between diverse groups in the school community, including students, families, school staff and the wider community'. This relational side of building sustainable societies is also why institutional culture, ethos, and practices are all identified as part of this WSA strand.

UNESCO's (2020) ESD 2030 roadmap describes that change is needed in how societies and individuals think and act. Hence, education also needs to change if we are to create a sustainable and peaceful world. However, as previously mentioned changing well-established institutions is a massive undertaking in which past efforts show curriculum reforms often fail, and long-lasting embedded change is hard to achieve

(Hargreaves, 2008). Enabling 'change' to be accepted and embraced in schools is proposed as possible through enacting a WSA, yet the examples synthesised often focus on enabling change to support a specific initiative. It is clear multiple WSA conceptualisations presented in this study highlight that a WSA can enact and operationalise a variety of education initiatives. Subsequently, a WSA could also act as a link between different but inter-connected educations and topics (such as climate, health, democracy, justice, biodiversity, and citizenship) through a common united goal for school improvements, educational innovation, and sustainability-oriented transformations. Equally, when WSA is utilised for just one specific education initiative, such as bullying, citizenship, or climate change, it runs the risk of doing the opposite of its intentions by unintentionally fostering siloed thinking and creating competition over what is prioritised in terms of educational innovations at a school.

2.4.4 Tensions, Problematics, and Dilemmas

While the theory and the practice-based WSA examples synthesised in this chapter are compelling, this type of holistic approach is not without its challenges. Synthesising the 27 WSA conceptualisations, alongside examining the additionally assembled WSA-based literature sources, two dominant and recurring barriers to enacting a WSA are found as commonplace: (1) resources and capacity and (2) commitment and interdependency from all stakeholders. This is because a WSA requires commitment and effort on every level, along with a willingness to restructure and rethink institutional power dynamics in order to forge new ways of working together: As Hunt (2015) also touches upon, this rethinking and restructuring require all stakeholders to have the capacity (competency, time, resources), readiness, and opportunity to engage.

While it is evident the knowledge and willingness to integrate a WSA are present on many levels, for example, UNECE (2022) shows 56 UNECE nations committing to and recognising a Whole Institution Approach as a high-priority issue, it is clear capacity and commitment issues are evident and impact every level. Having the capacity and necessary policy structures to fully realise a WSA is not always present. What Shallcross et al. (2006) mentioned nearly 20 years ago seems to remain a challenge for enacting a WSA, that not all schools will have the capacity to fully enact a WSA without additional support due at large to limited resources that often connect to broader socio and economic struggles. In Mathie and Wals (2022) a similar conclusion is drawn, that creating healthy policy environments (Fig. 2.8) is essential for a WSA to move from the margins into the mainstream. What is clear however, as UNECE (2022) illustrates, is that nations committing to prioritising a WSA in national policy and strategy is increasing.

After analysing all 27 WSA conceptualisations, an issue concerning labels and names has also arisen. Many conceptualisations that exist under a WSA 'label' appear to share similar theoretical underpinnings. However, do they have a shared understanding of the terminology? In turn, it is also clear that numerous other models exist, while they may not explicitly be called a WSA, that also share similar theoretical underpinnings and objectives, many of which have come about in a local grassroots manner suitable for the context promoted within. It is beyond the scope of this chapter to seek to answer questions concerning these themes, however, some questions for future consideration are proposed:

- What is it about a WSA that makes it stand apart from other educational change frameworks?
- Does a WSA need to have a globally recognised name, or is it appropriate for a WSA to have multiple names?
- Does a commonly recognised 'label' help schools, or does it pigeonhole the 'methods'

Fig. 2.8 Healthy policies for enabling a Whole School Approach. (Reproduced from Mathie & Wals, 2022, p102)

too much so the approach is not situated and contextually appropriate enough?

- What opposes a WSA?
- Would the holistic, systemic language not be welcomed in some cultural settings?
- What other 'labels' or methods exist that can also support schools to arrive at the same outcome, but in a different manner?

Questions like this arise, albeit hypothetically vague, because a WSA represents a set of values, approaches, methods, and ways of thinking that are crossing boundaries and being implemented worldwide in different ways.

2.4.5 The Role of Reflexivity

The role of reflexivity and critical reflection to build up the capacity of all school community-related staff and actors are dominant themes in ESD, EHW, and GCE. When analysing the barriers mentioned that a WSA faces, reflexive monitoring and evaluation (RME) practices are highlighted as a potential solution. For example, if RME practices become embedded into a school's organisational structures and everyday practices, this could provide a catalyst for capacity building, awareness raising, and collaboration between key stakeholders. Increasing and promoting a culture of change is also an essential part of this.

While the task of integrating reflexive education practices is well known for being arduous and tremendously challenging (Shallcross et al., 2006), when comparing all the WSA conceptualisations in this study, a WSA appears to have the potential to enable reflexive processes becoming more commonplace. In turn, RME practices seem pivotal to integrating a WSA as they can support combatting operational challenges, such as the difficult task seen in sustainability-oriented education: '[...] to create an integrated and systemic response that addresses sustainability issues meaningfully, consistently and effectively' (Wals & Benavot, 2017, p409).

Another important WSA aspect, also connected to RME, is determining achievable indicators of progress and methodologies for researching, monitoring, and evaluating a WSA. It is also common knowledge that educational innovations and reforms face monitoring and evaluation issues. As do global agendas, such as the SDGs, highlight that current monitoring and evaluation methods are insufficient (Leicht et al., 2018). Therefore, detailed WSA assessment frameworks, in particular UNEGI's (2019) monitoring and evaluation framework, Hunt and King's (2015) auditing tools, and the Scottish government's reflexive auditing tools (Education Scotland, 2020), all show realistic processes and existing frameworks others can build upon. Moreover, these frameworks call for schools to actively participate in developing RME practices at the school level. A WSA, especially with RME as a primary focus, could help provide accessible and meaningful monitoring and evaluation systems that support schools in assessing themselves. In turn, this could lead to more active participation, and support schools to directly engage with leading their own school's journey towards sustainability: A journey where sustainability issues are not viewed as static problems to solve, but seen through a reflexive lens that recognise the ever-changing and uncertain nature this process brings (Wals et al., 2009).

2.5 Conclusions

This chapter provides an overview of different but interconnected conceptualisations of a WSA in sustainability-related educational innovations. It is evident all 27 WSA conceptualisations synthesised call for action-orientated participatory pedagogical approaches to innovate and change the way mainstream education is currently conducted. While not an exhaustive literature review, through purposively collating a broad variety of WSA-related conceptualisations within ESD, EHW, and GCE, a deeper understanding of how a WSA can support educational change, both in and beyond ESD, is realised. From this, three main principles and points of interest have arisen (Fig. 2.7): *(a) democratic and participatory processes are overarching tools throughout all aspects of a WSA; (b) holistic, systemic, and sustainable perspectives are central underpinning theories; and c. all aspects of an institution's organisation that a WSA seeks to engage and involve can be situated under five broad interlinking strands*. Looking at a WSA beyond ESD and through a GCE and EHW context also highlights the importance of emphasising the social and societal aspects of a WSA: To support the forging of reciprocal relationships and collaborations within all levels of society towards the common goal of strengthening social cohesion. The purpose of viewing a WSA in this broad way helps to stress that to meaningfully integrate a WSA it needs to be less about one agenda, innovation, or initiative, and as a means to encourage schools to use a WSA as a thinking tool for integrating general quality educational change processes, whilst still grounded in the values of a peaceful, inclusive and sustainable future for people and planet.

The lenses of educational change, reform, and innovation show us that schools are often overwhelmed by the amount of change and new agendas thrust upon them. With the knowledge that top-down curriculum reform often falls short of the theoretical visions when implemented in practice, schools require a way of approaching educational innovation where neither top-down nor bottom-up, individualistic, or one-off projects dominate. The examples from ESD, EHW, and GCE all show promise that a WSA can support schools in developing this multistakeholder approach that encourages collaboration and learning together in all arenas. However, more school-based empirical research is required to determine practical methods for supporting schools to utilise a WSA as a thinking tool in their own situated context—to support their own, and their surrounding communities, journey with relevant educational innovations, and engagement with sustainability-oriented transitions. Future considerations for WSA research also include understanding better the effectiveness of different WSA monitoring and evaluation frameworks.

To conclude, this chapter proposes framing a WSA as a continuous reflexive thinking tool to enable schools to integrate educational change processes into their everyday practice, thus supporting the type of school reform and action required today. With the proper support and open-minded experimental attitude, a WSA is relevant for supporting multiple forms of quality educational change and innovation. A WSA also has the potential to be adopted internally by schools as a reflexive analytical tool to proactively monitor, evaluate, and operationalise their own, and their surrounding communities, journey towards a more sustainable and just future. However, when a WSA is utilised to integrate only singular education initiatives, such as bullying, citizenship, or climate change, it still runs the risk of further fostering siloed thinking and creating competition over what is prioritised in terms of educational innovation at a school. Instead, a broad and inclusive understanding of a WSA is essential as this has the potential to trigger continual educational innovation processes, more reflexive schools, and create synergies between different but inter-connected sustainability-oriented educations and topics.

Acknowledgements I would like to recognise and thank my supervisors Arjen Wals and Astrid Sinnes for their ongoing insights and support during the writing of this paper and my PhD journey. Thank you also to the peer reviewers for their critique and constructive feedback, and for bringing even more knowledge and insight needed to understand what authentically integrating a WSA entails. Lastly, I also give thanks to my sustainability-education peers and critical friends—Ane Eir Torsdottir, Beth Roberts, Brita Phuthi and Stephanie Leite—for their ongoing support in proofreading and thought provoking conversations!

References

Affolter, C., & Varga, A. (2018). *Environment and school initiatives: Lessons from the ENSI network-past, present and future*. Environment and School Initiatives-ENSI.

Ávila, L. V., Leal Filho, W., Brandli, L., Macgregor, C. J., Molthan-Hill, P., Özuyar, P. G., & Moreira, R. M. (2017). Barriers to innovation and sustainability at universities around the world. *Journal of Cleaner Production, 164*, 1268–1278.

Barber, J., Owens, E., & Brown, A. (2010). *All together now! A whole school approach to anti-bullying practice*. Save the Children.

Barrett, M. (2020). The Council of Europe's Reference Framework of Competences for Democratic Culture: Policy context, content and impact. *London Review of Education, 18*, 1–18. https://doi.org/10.18546/LRE.18.1.01

Beauvallet, A. (2016). Four nations going their own ways? Citizenship education in the United Kingdom. *Revue française de civilisation britannique, 21*(1), 1–15. https://doi.org/10.4000/rfcb.727

Bonnet, M. (1999). Education for Sustainable Development: a coherent philosophy for environmental education. *Cambridge Journal of Education, 29*(3), 313–324.

Bosevska, J., & Kriewaldt, J. (2020). Fostering a whole-school approach to sustainability: Learning from one school's journey towards sustainable education. *International Research in Geographical and Environmental Education, 29*(1), 55–73.

Chopin, N., Hargis, K., & McKenzie, M. (2018). *Building climate-ready schools in Canada: Towards identifying good practices in climate change education*. Sustainability and Education Policy Network, University of Saskatchewan.

Cook, D. A. (2019). Systematic and non-systematic reviews: Choosing an approach. In *Healthcare simulation research: A practical guide* (pp. 55–60).

Council of Europe. Education Department (COE, Ed.). (2018). *Reference framework of competences for democratic culture—Volume 3. Guidance for implementation: Whole-school approach (Vol. 3)*. Council of Europe.

Cross, D., Shaw, T., Epstein, M., Pearce, N., Barnes, A., Burns, S., & Runions, K. (2018). Impact of the friendly schools whole-school intervention on transition to secondary school and adolescent bullying behaviour. *European Journal of Education, 53*(4), 495–513.

Education Scotland. (2020). *Whole school and community approach to learning for sustainability (LfS) Self-evaluation and improvement framework*. https://education.gov.scot/improvement/self-evaluation/whole-school-and-community-approach-to-learning-for-sustainability-lfs-self-evaluation-and-improvement-framework/

Education Scotland. (2022). *A summary of learning for sustainability resources*. National Improvement Hub. https://education.gov.scot/improvement/learning-resources/a-summary-of-learning-for-sustainability-resources

European Commission. (2022). *EU whole-of- government approach: The EU comprehensive EU approach towards implementing the UN's 2030 Agenda for Sustainable Development*. https://ec.europa.eu/info/strategy/international-strategies/sustainable-development-goals/eu-holistic-approach-sustainable-development_en

Freire, P. (1972). *Pedagogy of the oppressed*. Penguin Books.

Goleman, D. (1996). *Emotional intelligence- Why it can matter more than IQ*. Bloomsbury.

Greig, S., Pike, G., & Selby, D. (1989). *Greenprints for changing schools*. WWF and Kogan Page.

Hargreaves, A. (2002). Sustainability of educational change: The role of social geographies. *Journal of Educational Change, 3*(3–4), 189–214.

Hargreaves, L. G. (2008). The whole-school approach to eduation for sustainable development: From pilot projects to systemic change. *Policy and Practice-A FDevelopment Education Review, 6*, 69–74. https://www.developmenteducationreview.com/issue/issue-6/whole-school-approach-eduation-sustainable-development-pilot-projects-systemic-change

Hargreaves, A. (2010). Sustainable educational reform. In P. Peterson, E. Baker, & B. McGaw (Eds.), *International encyclopedia of education (Third Edition)* (pp. 289–294). Elsevier.

Havelock, R. G. (1971). The utilization of educational research and development. *British Journal of Educational Technology, 2*, 84–97.

Henck, A. (2017). ACEI's global schools first: A whole-school assessment. *Childhood Education, 93*(6), 514–515.

Henck, A. (2018). Looking beyond the classroom: Integrating global citizenship education throughout your whole school. *Childhood Education, 94*(4), 75–77.

Henderson, K., & Tilbury, D. (2004). *Whole-school approaches to sustainability: An international review of sustainable school programs*. Australian Research Institute in Education for Sustainability: Australian Government. http://aries.mq.edu.au/projects/whole_school/files/international_review.pdf

Hubers, M. D. (2020). Paving the way for sustainable educational change: Reconceptualising what it means to make educational changes that last. *Teaching and Teacher Education, 93*(103083), 1–14.

Hunt, F., & King, R. P. (2015). *Supporting whole school approaches to global learning: Focusing learning and mapping impact*. UCL Development Education Research Centre. Research Paper No.13 for the Global Learning Programme. https://www.ucl.ac.uk/ioe/sites/ioe/files/franhunt2015supportingwholeschoolapproachestogloballearning.pdf

Hunt, P., Barrios, L., Telljohann, S. K., & Mazyck, D. (2015). A whole school approach: Collaborative development of school health policies, processes, and practices. *Journal of School Health, 85*(11), 802–809.

Jucker, R., & Mathar, R. (Eds.). (2016). *Schooling for sustainable development in Europe*. Springer International. https://link.springer.com/content/pdf/10.1007/978-3-319-09549-3.pdf

Learning and Teaching Scotland. (2011). Developing global citizens withing curriculum of excellence. *Learning Teaching and Scotland*. ISBN: 978-184399-183-0.

Leicht, A., Heiss, J., & Byun, W. J. (2018). *Issues and trends in education for sustainable development* (Vol. 5). UNESCO Publishing. https://unesdoc.unesco.org/ark:/48223/pf0000261954

Lewallen, T. C., Hunt, H., Potts-Datema, W., Zaza, S., & Giles, W. (2015). The whole school, whole community, whole child model: A new approach for improving educational attainment and healthy development for students. *Journal of School Health, 85*(11), 729–739.

Mathar, R. (2016). Chapter 5. Global development education/ESD - a task for the whole school. In T. Stukenberg (Ed.), *Curriculum framework, education for sustainable development - A contribution to the Global Action Programme. Education for sustainable development (Translation of German edition)* (pp. 401–419). Engagement Global gGmbH.

Mathie, R. G. (2019). *Education for sustainable development in Norway: Calling for a whole institution approach* [Master Thesis, Faculty of Landscap and Society, Norwegian University of Life Sciences]. https://nmbu.brage.unit.no/nmbu-xmlui/handle/11250/2638496

Mathie, R. G., & Wals, A. E. J. (2022). *Whole school approaches to sustainability: Exemplary practices from around the world* (pp. 1–109). Education and Learning Sciences/Wageningen University. https://www.wur/wholeschoolapproach

Mogren, A. (2019). *Guiding principles of transformative education for sustainable development in Local School Organisations: Investigating whole school approaches through a school improvement lens* [Doctoral Dissertaion, Department of Environmental and Life Sciences. Karlstad University]. https://kau.diva-portal.org/smash/record.jsf?pid=diva2%3A1368940&dswid=-487

Mogren, A., Gericke, N., & Scherp, H.-Å. (2019). Whole school approaches to education for sustainable development: A model that links to school improvement. *Environmental Education Research, 25*(4), 508–531.

Norwegian Directorate for Education and Training (UDIR). (2020). *Core curriculum: Interdisciplinary topics.* https://www.udir.no/lk20/overordnet--del/prinsipper-for-laring-utvikling-og-danning/tverrfaglige-temaer/?lang=eng

Orr, D. W. (1992). *Ecological literacy; education and the transition to a postmodern world.* Albany State University of New York Press.

Oxfam. (2015). *Education for global citizenship: A guide for schools*. Oxfam.

Priestley, M., & Sime, D. (2005). Formative assessment for all: A whole-school approach to pedagogic change. *The Curriculum Journal, 16*(4), 475–492.

Priestley, M., Miller, K., Barrett, L., & Wallace, C. (2011). Teacher learning communities and educational change in Scotland: The highland experience. *British Educational Research Journal, 37*(2), 265–284.

Qablan, A. (2018). Building capacities of educators and trainers. In A. Leicht, J. Heiss, & W. J. Byun (Eds.), *Issues and trends in education for sustainable development* (pp. 133–156). UNESCO Publishing.

Rowe, F., & Stewart, D. (2011). Promoting connectedness through whole-school approaches: Key elements and pathways of influence. *Health Education, 111*, 49–65.

Rowe, F., Stewart, D., & Patterson, C. (2007). Promoting school connectedness through whole school approaches. *Health Education, 107*(6), 524–542.

Rowling, L. (2009). Strengthening "school" in school mental health promotion. *Health Education, 109*, 357–268.

Schreiber, J.-R., Siege, H., & Stukenberg, T. (2016). *Curriculum framework: Education for sustainable development: A contribution to the global action programme education for sustainable development.* Engagement Global.

Scott, B. (2005). Getting to the heart of the matter: Examining the efficacy of a whole-school approach to behaviour management. *Kairaranga, 6*(1), 29–34.

Scott, W. (2013). Developing the sustainable school: Thinking the issues through. *Curriculum Journal, 24*(2), 181–205.

Scottish Gov. (2012). *Learning for sustainability—Report of the One Planet Schools Ministerial Advisory Group.* https://education.gov.scot/improvement/Documents/One-planet-schools-report-learning-for-sustainability.pdf

Scottish Gov. (2020). *The educational outcomes of learning for sustainability: A brief review of literature*. https://www.gov.scot/publications/educational-outcomes-learning-sustainability-brief-review-literature/pages/6/

Selby, D. (2009). Towards the sustainability university: The centre for sustainable futures, University of Plymouth. *Journal of Education for Sustainable Development, 3*(1), 103–106.

Serdyukov, P. (2017). Innovation in education: What works, what doesn't, and what to do about it? *Journal of Research in Innovative Teaching and Learning, 10*(1), 4–33.

Shallcross, A. G. (2003). *Education as second nature: Deep ecology and school development through whole institution approaches to sustainability education* [Doctoral dissertation, Institute of Education Crew abd Alsager Faculty, Manchester Metropolitan University] (pp. 1–382).

Shallcross, T., & Robinson, J. (2008). Sustainability education, whole school approaches, and communities of action. In *Participation and learning* (pp. 299–320). Springer.

Shallcross, T., Robinson, J., Pace, P., & Wals, A. E. J. (2006). *Creating sustainable environments in our schools*. Trentham Publishers.

Slavin, R. E., & Madden, N. A. (2013). Success for all at 27: New developments in whole-school reform. *Journal of Education for Students Placed at Risk, 18*(3–4).

Sterling, S. (2004). Higher education, sustainability, and the role of systemic learning. In P. B. Corcoran & A. Wals (Eds.), *Higher education and the challenge of sustainability: Problematics, promise, and practice* (pp. 47–70). Kluwer Academic Publishers.

STL. (2020). *Global programmatic framework and benchmarking tool: From call to action to programme responses*. Retrieved from https://www.end-violence.org/safe-to-learn: https://www.end-violence.org/safe-to-learn

Tilbury, D., & Galvin, C. (2022). *European Commission input paper: A whole school approach to learning for environmental sustainability*. Expert briefing paper in support of the first meeting of the EU Working Group Schools: Learning for sustainability. https://education.ec.europa.eu/document/input-paper-a-whole-school-approach-to-learning-for-environmental-sustainability

UNECE. (2014). *Information paper 4 working group on 'ESD School Planning': Outcomes*. United Nations Economic Commission for Europe steering committee on education for sustainable development 9th meeting, 3 and 4 April 2014. https://unece.org/fileadmin/DAM/env/Information_document_4_school_planning_02.pdf

UNESCO. (2015). *Global citizenship education: Topics and learning objectives*. https://unesdoc.unesco.org/ark:/48223/pf0000232993

UNESCO. (2016). *UNESCO strategy on education for health and well-being - contributing to the sustainable development goals.* https://unesdoc.unesco.org/ark:/48223/pf0000246453

UNESCO. (2017). *Education for sustainable development goals: Learning objectives.* https://unesdoc.unesco.org/ark:/48223/pf0000247444

UNESCO. (2020). *Education for sustainable development: A roadmap.* https://unesdoc.unesco.org/ark:/48223/pf0000374802

United Nations Economic Commission for Europe (UNECE). (2022). *Framework for the implementa-*

tion of the United Nations Economic Commission for Europe strategy for education for sustainable development from 2021 to 2030. https://unece.org/sites/default/files/2022-05/ece_cep_ac.13_2022_3_e.pdf

United Nations Girls' Education Initative (UNGEI). (2019). *School-Related Gender-Based Violence (SRGBV): A whole school approach to prevent School-Related Gender-Based Violence - minimum standards and monitoring framework*. https://www.ungei.org/publication/whole-school-approach-prevent-school-related-gender-based-violence-1

Van Driel, B., Darmody, M., & Kerzil, J. (2016). *Education policies and practices to foster tolerance, respect for diversity and civic responsibility in children and young people in the EU*. Neset Ii Report. EU– Executive summary, Publications Office. https://doi.org/10.2766/797305.

Wals, A. E. J. (2015). Social learning-oriented capacity-building for critical transitions towards sustainability. In R. Jucker & R. Mathar (Eds.), *Schooling for sustainable development in Europe* (pp. 87–107). Springer.

Wals, A. E. J. (2019). *Sustainability-oriented ecologies of learning as a response to systemic global dysfunction*. In R. Barnett & N. Jackson (Eds.), *Learning ecologies: Sightings, possibilities, and emerging practices* (pp. 61–78). Taylor and Francis.

Wals, A. E. J., & Benavot, A. (2017). Can we meet the sustainability challenges? The role of education and lifelong learning. *European Journal of Education, 52*(4), 404–413.

Wals, A., & Mathie, R. G. (2022). Whole school responses to climate urgency and related sustainability challenges. In *Encyclopedia of Educational Innovation* (pp. 1–8). Springer.

Wals, A. E. J., van der Hoeven, N., & Blanken, H. (2009). *The acoustics of social learning: Designing learning processes that contribute to a more sustainable world* (pp. 1–34). Wageningen Academic Publishers.

Whiteside, T. (2019). *The sociology of educational innovation: Contemporary sociology of the School Routledge Library Editions* (Vol. 58). Routledge, Taylor and Francis. (Original work published 1978, by Methuen & Co Ltd).

3 Insights, Challenges, and Prospects on Whole School Approach to Sustainability in the UN Economic Commission for Europe (UNECE) Region

Aravella Zachariou, Stella Hadjiachilleos, Chrysanthi-Kadji Beltran, and Roel van Raaij

Key Message
This chapter provides valuable insights about the ways forward with WSA in ESD. It encourages individual and collective reflection through examples from the UNECE countries' sustainable school vision, in order to impact the curricula and their quality, tools, materials, and collaboration between stakeholders within and outside the school. The diversity within UNECE brings out a variety of good examples where sustainable schools provide meaningful learning, and the transformative context within which learners develop the capacities needed to mitigate sustainability challenges.

A. Zachariou (✉)
Unit of Education for Environment and Sustainable Development Cyprus Pedagogical Institute, Frederick University, Nicosia, Cyprus
e-mail: aravella@cytanet.com.cy; pre.za@frederick.ac.cy

S. Hadjiachilleos
Unit of Education for Environment and Sustainable Development Cyprus Pedagogical Institute, Nicosia, Cyprus
e-mail: Hadjiachilleos.st@cyearn.pi.ac.cy

C.-K. Beltran
Frederick University, Nicosia, Cyprus
e-mail: pre.kch@frederick.ac.cy

R. van Raaij
Ministry of Agriculture, Nature and Food Quality, The Hague, The Netherlands
e-mail: r.m.vanraaij@minlnv.nl

3.1 Introduction

The whole school approach (WSA), in the context of education for sustainable development, can be traced back to the 1990s with the recognition of environmental issues' inextricable links to a multitude of socioeconomic and political issues (Henderson & Tilbury, 2004) and reflection of its holistic-systemic view into educational reforms (Wals & Mathie, 2022). Education for sustainable development (ESD) is not only viewed as a body of information or knowledge but also as the aim and framework of education (Gleason et al., 2020). This means that sustainability is addressed through a range of complex and diverse aspects in schools' operation, such as curriculum and pedagogy, school governance, sustainable consumption, connection with the community, the schools' infrastructure, and landscaping. The day-to-day school practices consider sustainability, also in the nonformal or hidden curricula (Ferreira et al., 2006).

Education for sustainable development is a key priority for creating more just, prosperous, and sustainable societies. In many regions, member states jointly initiated regional plans, strategies, and policies aiming to integrate ESD into their national and regional context more effectively. Such an example of regional collaboration is the Mediterranean Strategy on ESD. The strat-

A. E. J. Wals et al. (eds.), *Whole School Approaches to Sustainability*, Sustainable Development Goals Series, https://doi.org/10.1007/978-3-031-56172-6_3

egy serves as a flexible framework, for motivating and guiding governments in the Mediterranean region and to develop and implement policies and practices that incorporate Sustainable Development into Education and learning. Countries of the Mediterranean are also encouraged to develop and incorporate Education for Sustainable Development (ESD) into their formal, nonformal, and informal educational systems (MIO/ESCDE, 2014 p.16). The most concrete, long-term, and coherent framework of regional collaboration, though, is the UNECE ESD Strategy which constitutes a complete, holistic and coherent framework. It aims to support the 56 States of the United Nations Economic Commission for Europe (UNECE) region to implement ESD in the most effective way, based on their needs and priorities and taking into consideration: (a) the policy, regulatory, and operational frameworks that support ESD; (b) the promotion of sustainable development through formal, nonformal, and informal learning; (c) the equipping of educators with the competence to include sustainable development in their teaching; (d) the provision and accessibility of adequate tools and materials for ESD; (e) the promotion of research on and development of ESD; and (f) the strengthening of cooperation on ESD at all levels within the UNECE region (UNECE, 2005).

This chapter focuses on exploring how the WSA in the UNECE region has contributed to the infusion of ESD in the schools' vision, ethos, leadership, and coordination, in what we teach, how and where we learn, whom we learn from, and whom we learn with. In the introduction, we analyze what the WSA is and discuss its importance within the UNECE Strategy for ESD.

3.1.1 Defining the Whole School Approach

The WSA refers to a multilevel integration of sustainability within the school unit, meaning that sustainability is addressed through the curriculum (what is learnt) and drives school management, operations, and outreach. By engaging students in school decisions, community projects, and global initiatives, the WSA provides meaningful learning, connected to real-life conditions, transformation through action and practical outcomes that benefit both school and community (Tilbury & Galvin, 2022). Within this approach, the priorities shift from including ESD issues into a few selected subjects or interventions, to applying ESD as a foundation for all educational institutions' functions, including management and budget, education, professional development, campus design, resource use, community relations, etc. (Abrahamsson et al., 1999). The WSA provides a framework for a "whole system redesign" and it is associated with the term "sustainable school" (Tilbury & Galvin, 2022), which envisions sustainability to be integrated into education through holistic, systemic, and reflexive efforts, collectively introduced by those affected:

- *Holistically* indicates the multiple perspectives and the integrated and relational way in which sustainability is introduced.
- *Systemic* refers to education as a system in which sustainability permeates all of its aspects: curriculum, pedagogy and learning, professional development, school–community relationships, school practices, ethos, vision, and leadership.
- The *collective-reflexive* embraces the importance of including multiple voices in participative co-developed processes that lead to change and continuous learning (Wals & Mathie, 2022).

The growing acknowledgement of the value of WSA in supporting society's transition to sustainability leads to an intense dialogue on its transformative potential. Sterling (2004a, b, 2010a, b), for example, argued that educational organizations are systems that need to be aligned with sustainability in order to support an education-driven change. The whole school approach includes the involvement of all school's actors and expert stakeholders and brings learning closer to wider social issues (Mogensen & Schnack, 2010). In the same direction, the

Environment and Schools Initiative (ENSI) expanded the concept of WSA for ESD, beyond the school grounds focus, to become much more inclusive and holistic (in Ferreira et al., 2006) and support a local sustainable future (Summers & Childs, 2007).

Progressively, the term whole institutional approach (WIA) was prompted by policymakers, in order to also address institutions and stakeholders beyond schools (UNECE, 2014; UNESCO, 2014). The term WIA has been principally adopted by higher education institutions, whereas for primary and secondary education settings, WSA is more commonly used (Mogren, 2019). It requires the reorientation of every institution's strategy and culture toward sustainability (Gleason et al., 2020).

So, how do the terms "whole school approach" and "whole institution approach" differ? Often the two terms are used interchangeably. In general, the term WSA refers mainly to school education and "addressing the needs of learners, staff and the wider community, not only within the curriculum, but across the whole-school and learning environment. It implies collective and collaborative action in and by a school community, to improve student learning, behaviour and wellbeing, and the conditions that support these" (UNESCO IBE, 2012).

The term WIA is addressed in the new implementation framework of the UNECE Strategy on ESD 2021–2030. As stated ESD through WIA goes beyond schooling and it is relevant to any institution and organization that seeks to mainstream sustainability into all aspects of its operation. It requires reviewing and revising the institution's internal operations and external relationships, and becomes an essential component in any strategy aligned with the SDGs (UNECE, 2022). In this chapter, we use the term WSA, being the term most widely used. The term WIA is used in the text, where appropriate

The WSA/WIA is viewed as a continual reflexive process for sustainability through which a school or an institution encourages ESD within and across all domains of school life or an institution's operations, where governance (vision, ethos, leadership, and coordination), pedagogy, and learning (e.g., cross-curricular projects; assessments; action learning); institutional practices/infrastructure, capacity building, community connections, research, and innovation are highlighted as important aspects of it (see Wals & Mathie, 2022; Verhelst et al., 2022; NGA and NAAE, 2022; Hargis et al., 2021; UNECE, 2014).

The importance of WSA and the need for a coherent framework for organizing its dimensions are documented by literature, through several WSA models, devised by researchers and organizations, to support its implementation by schools and institutions: the OECD model (Mathar, 2013); the UNESCO model (UNESCO, 2014; p. 89); the NGA and NAAE model (NGA and NAAE, 2022); the conceptual framework of an ESD effective school (Verhelst et al., 2022); and the "flower model" (Wals & Mathie, 2022). The models converge to the dimensions and aspects of WSA, they differ though, in the way they are developed. This is important for the implementation of WSA on ESD because institutions are able to choose the model that best suits their profile, plans, and aspirations. Also, the variety of WSA models proves that its application is neither linear nor static, but constitutes a dynamically evolving process, which can be applied in various ways and at multiple levels in each organization.

The aim of this chapter is not to analyze the various WSA models, nevertheless, we consider it necessary to briefly present the "flower model" by Wals and Mathie (2022), which we chose for the examination of the WSA implementation in the UNECE region. This model intends to trigger and guide a multistakeholder dialogue about infusing sustainability in school settings in the best possible ways. The model exists in a number of slight variations and, in any case, it presents in an organized way the most important aspects of WSA. The principal model consists of six interrelating elements that illustrate how all levels, areas, and stakeholders can be engaged in a WSA. The central element in the flower is the vision, which reflects the ethos, leadership, and coordination needed for a WSA, and operates as the connecting point for the six components: curriculum, pedagogy and didactics, management, operations, professional development, and school environment. The "flower model" was chosen because we consider that it presents the WSA dimensions in a complete, concise, and unified manner under a cyclical form that highlights the sequence, connections and continuity that must permeate every organization. Additionally, we consider that it is closer to the UNECE ESD Strategy and the way WSA is captured through it.

3.1.2 Whole School Approach in the UNECE Region

Fifty-six member states belong to the UNECE region. These are grouped into four sub-regions: Eastern Europe, the Caucasus, and Central Asia (EECCA); West Europe and America (EU-West); South Eastern Europe (SEE); and Western Asia (WA). The UNECE ESD Steering Committee is the body responsible for giving guidance and strategic directions for the implementation of the UNECE Strategy for ESD to the member states. The UNECE Strategy for ESD stressed the importance of the WSA since the early beginning of its implementation. It serves as a roadmap for ESD, addresses ESD axes holistically, and ensures its effective implementation (Zachariou & Pace, 2022) by the member states of the UNECE. The strategy clearly states that "an educational institution as a whole [...], should follow the principles of SD" (UNECE, 2005, par. 29) highlighting the importance of integrating ESD through a whole school/institution approach.

The UNECE Strategy for ESD is fully aligned with other regional and international frameworks (WHO, 2019; UNESCO, 2020; EC, 2022a, b), but differs in the ways it seeks to provide countries and organizations with the tools they need to support the implementation of the main priority areas of the UNECE Strategy on ESD. Such a tool is the development of the UNECE school planning framework, a crucial component of a WSA, which can serve as a guide for developing national and local approaches to ESD school planning (UNECE, 2013). The UNECE ESD School Plan implementation framework goes beyond simply teaching about sustainability in schools. It provides meaningful, flexible, and feasible policy recommendations for integrating WSA more effectively, by identifying and analyzing through practical examples the core dimensions of WSA (UNECE, 2014). It addresses, explains, and describes the generic areas of a WSA/WIA as a concrete framework that can be applied in a variety of institutions, organizations, sectors, and contexts, which are implementing ESD and the SDGs (UNECE, 2012).

The third evaluation report findings of the UNECE Strategy on ESD indicated that nearly 71% of the member states promoted the WSA/WIA. Nevertheless, most of them continue to focus on enhancing curricula with ESD, through projects and voluntary schemes for ESD recognition and certification of schools (UNECE, 2016, p. 7). These findings were an important lever for mobilizing the process of enhancing and strengthening WSA in the third implementation framework (2015–2018). WSA became one of the priority areas in the UNECE region being acknowledged as an important vehicle for achieving holistic and structural implementation of ESD in formal education (primary and secondary), vocational and higher education, and in enhancing formal and nonformal education links (UNECE, 2015, p. 5).

Despite the advances, the progress report on the implementation of the UNECE Strategy for ESD across the ECE region (2015–2018) states that WSA mainly concerns primary and secondary schools rather than other formal, nonformal, and informal education and other institutions (Hadjiachilleos & Zachariou, 2022). Zachariou and Hadjiachilleos (2022) comment that WSA in the UNECE region is viewed as a distinct aspect of the ESD Strategy implementation rather than as an issue, horizontally integrated into each of its parameters (e.g., policies, legislations, teacher competency development, research, development, etc.). Similarly, European Commission policy recommendations noted that WIA to sustainability is not always sufficiently present (Tilbury & Galvin, 2022).

The importance of establishing WSA in the educational systems of the UNECE region countries is emphatically stated as one of the key priority strands, in the new framework for the implementation of the UNECE Strategy for ESD from 2021 to 2030 end evolves from WSA to WIA, reaching beyond schooling in order to address every institution as a community of transformational learning for ESD. WSA/WIA in the new framework of the UNECE ESD Strategy is not just "a cerebral educational exercise" but it is the actual and core part of the organizations' operations. It envisions ESD as a "component integrated across all educational and training programs, together with the sustainable management of the institution and its interface with the local community and system-wide interventions, where learners will acquire a "whole system" view, develop systemic thinking, and become able and willing to address problems and act accordingly" (UNECE, 2022, p. 12–13).

3.2 Methodology

Because of the UNECE regions (a) wide geographical representation and the large number of countries participating in its processes and mechanisms and (b) new developments and challenges set for ESD2030 with WSA being in the core of the UNECE Strategy for ESD activities, the UNECE region is considered as a good example for presenting how WSA is implemented in a regional level. In this chapter, we use the implementation of the strategy for ESD by the 56 member-states of the UNECE region, as a case study. We seek to critically reflect on how WSA is integrated through the UNECE Strategy for ESD implementation and discuss current practices and the steps forward.

Thirty-four National Implementation Reports (NIRs) (out of the 56 member-states) were submitted at the end of the fourth implementation phase of the UNECE ESD Strategy for the period 2015–2018. These 34 NIRS correspond to 8 EECCA region countries, 21 from the EU-West, 4 from the SEE, and 1 from the WA region. The NIRs reported on 48 qualitative and quantitative indicators, that reflected the 6 objectives (regulatory frameworks, ESD in formal, nonformal education, the teachers' capacity on ESD, ESD education tools, ESD research, collaboration on ESD) of the UNECE Strategy for ESD implementation. The WSA in the UNECE ESD Strategy was specifically addressed through indicator (2.3), which includes three sub-indicators (2.3.1. "adopted by institutions," 2.3.2 "incentives," and 2.3.3 "ESD Indicators"). Given the fact that elements of WSA permeate the set of indicators concerning the UNECE ESD report format, the examination of WSA, solely through this specific indicator, would be very limiting. We would like to clarify that the first number in parentheses corresponds to the UNECE Strategy for ESD Issue, the second number to the indicator, and the third number to the sub-indicator. An overview of coding corresponding to the UNECE ESD Strategy can be found in ECE/CEP/AC.13/2014/5. Phase III: Format for reporting on the implementation of the UNECE ESD Strategy.

This chapter addresses all the indicators of the UNECE ESD Strategy and uses the "flower model" aspects (Wals & Mathie, 2022) as the basis for the analysis of all the NIRs submitted. This approach can provide a better insight into how WSA is viewed and addressed by the member states in the UNECE region and reveals possible challenges met by the countries and opportunities to promote WSA to ESD. It also helps highlight the connections and correlations of the UNECE Strategy for ESD with the dimensions of WSA and allows us to identify gaps and steps that need to be undertaken for improving UNECE ESD Strategy in this particular strand.

Table 3.1 presents an overview of the indicators and sub-indicators referring to each of the "flower model" petals. The NIR template, requested examples from each country's national context, for each of the indicators and sub-indicators included in the report. These results should be interpreted with caution, due to the disparity of sizes of the UNECE member states. Additionally, they are not exhaustive and may omit other examples of WSA in ESD.

In the data analysis, we cross examine the UNECE Strategy for ESD NIR's data with each

Table 3.1 Cross-examination of NIRs and the flower model

WIA Parameters	Issue 1: Policy, regulatory and operational frameworks	Issue 2: Formal, informal, nonformal learning	Issue 3: Educator competences	Issue 4: Tools and materials	Issue 5: Research and development	Issue 6: Cooperation on ESD In ECE region	Issue 7: Conservation of knowledge of indigenous people
What do we learn?	1.2.2. Policy documents, e.g., curricula 1.2.4. Raising public awareness 1.2.6. Multistakeholder cooperation on policies 1.2.7. Funding	2.1. SD key themes in formal education (curriculum, learning outcomes, e.g., skills/attitudes/values) 2.4. Quality assessment systems 2.5. ESD methods/instruments for nonformal & informal learning to assess changes in knowledge, attitude, and practice			5.1. Research on ESD is promoted (e.g., on content and methods—5.1.1., in strategy outcomes—5.1.2, innovation and capacity building—5.2.1)	6.1.1. Public authorities cooperate/support international networks on ESD 6.1.2. Participation in international networks 6.1.3. Government takes steps to promote ESD in forums outside ECE region	Traditional knowledge, knowledge of indigenous people (rationalization, conservation)
What do we teach?	1.2.2. Policy documents, e.g., curricula 1.2.3. Policy documents referring to nonformal/informal education	2.2. Strategies on ESD are clearly identified 2.4. Quality assessment systems 2.5. ESD methods/instruments for nonformal & informal learning to assess changes		4.1. Production 4.2. Quality control mechanisms			Traditional knowledge, knowledge of indigenous people (rationalization, conservation)
How do we learn?	1.1.4. National implementation plan 1.1.5. Synergies	2.6. ESD implementation as a multistakeholder process					Nonformal context

Where do we learn?		2.5. ESD methods/ instruments for nonformal & informal learning to assess changes					
Who do we learn with?		2.6. ESD implementation as a multistakeholder process	3.1. ESD is included in the training of educators 3.2. Opportunities for educators to cooperate on ESD		5.2.1. Research is promoted on innovation and capacity building in ESD practice 5.3. Dissemination of research results		Intrergenerational cooperations Local communities Indigenous people

of the petals of the "flower model" in order to reveal how and to what extent WSA is integrated into ESD across the UNECE region. Table 3.1 presents an overview of how indicators and sub-indicators of the UNECE ESD Strategy's issues are categorized under each of the five WSA parameters or "petals" of Wals and Matthie's (2022) Flower Model. Following this framework, results refer to quantitative and qualitative data for each of the parameters of WSA as reported under the specific UNECE Strategy for ESD issues, indicators, and sub-indicators.

3.3 Results

3.3.1 WSA and Curriculum: What Do We Teach?

Curricula were one of the WSA aspects most frequently addressed through the NIR indicators and sub-indicators (Table 3.1). In order to examine how ESD curricula are relevant to WSA through the UNECE Strategy for ESD, prerequisite measures were examined (ESD Strategy Issue 1) with specific reference to policy documents and their indicators and sub-indicators as stated in the parentheses: (1.2.2.), raising public awareness (1.2.4), multistakeholder cooperation on policies (1.2.6), and funding (1.2.7). In terms of formal, nonformal, and informal learning (Issue 2), measures examined included SD key themes in formal education (curriculum, learning outcomes, e.g., skills/ attitudes/ values) (2.1), quality assessment systems (2.4), and ESD methods/instruments for nonformal and informal learning to assess changes in knowledge, attitude, and practice (2.5). Curriculum was also viewed in the framework of research and development, specifically the promotion of research on the content of ESD (5.1), on strategy outcomes (5.1.2), and innovation and capacity building (5.2.1). ESD is incorporated in the national curricula of 24 out of 32 reporting countries. Figure 3.1 depicts the extent to which ESD is embedded in curricula across the International Standard Classification of Education (ISCED) levels (UNESCO, 2012). ISCED is the official framework used to facilitate international comparisons of education systems, developed by UNESCO. ISCED is a universal categorization of education levels, eliminating discrepancies emerging from differences among the countries' national education systems.

Figure 3.1 presents frequencies and percentages of countries (out of 34 submitted reports), referring to whether ESD is included in national curricula, standards, ordinances, and requirements in all levels of formal education, as understood in each country's education system. Caution should be taken when interpreting this question as it referred to all levels of education and each country responded on whether or not ESD is included in national curricula, standards, and ordinances without further justification. Considering that ESD and its level of implementation are differentiated from country to country, the quality of this integration across policies cannot be further discussed since countries were not required to provide in-depth information as to specific frameworks in which this takes place. Nevertheless, as presented in Fig. 3.1, it should be noted that 32 out of 34 countries report that ESD is integrated across curricula, standards, ordinances, and other requirements, primarily throughout school education (ISCED 0 through ISCED 3). For example, in Iceland, there is a National Curriculum Guide for all levels of school education including specific reference to sustainability as a fundamental pillar of education. In Serbia, the Law of Fundamentals of the Education System provided the framework for the compilation of The Rulebook of Foundation Education Programs, specifically referring to the kindergarten as a place of democratic and inclusive practice under which sustainability is a key strand. The lowest rates of incorporating ESD in national curricula were observed for preuniversity education levels (ISCED 4 and 5). Often, even though ESD is implicitly addressed in the curricula of preuniversity education through various actions, it seems that there is a lot to be done to achieve its full integration at this level. Indicatively, Romania points out a lack of a specific discipline for ESD at preuniversity education, even though key themes of SD are included in core or optional at this level.

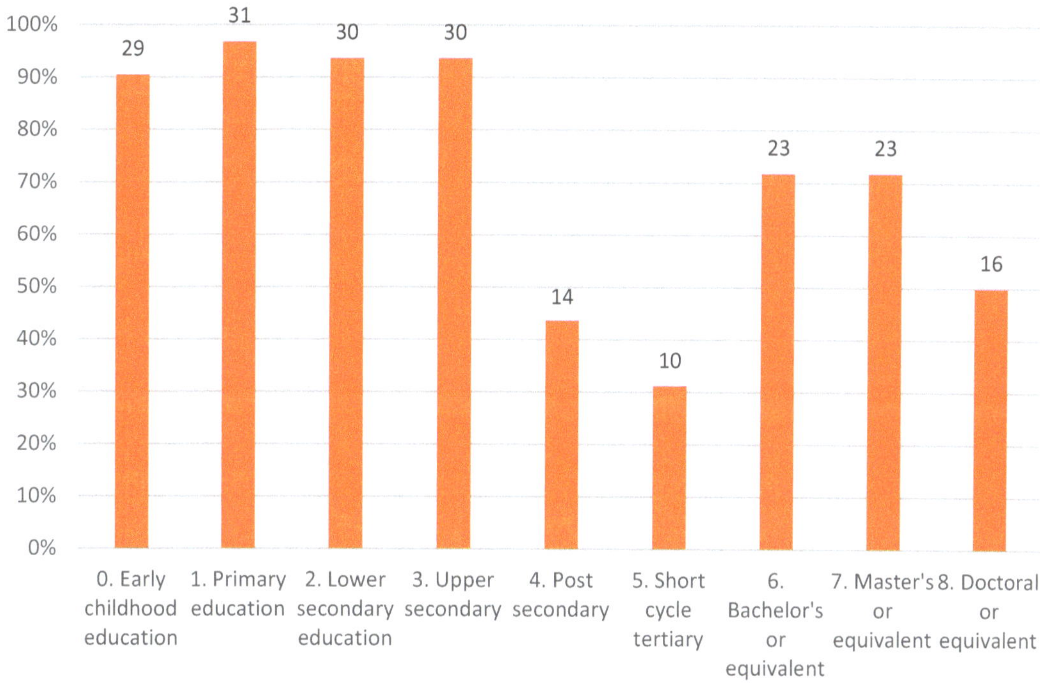

Fig. 3.1 ESD is addressed in national curricula/standards/ordinances/requirements as understood in each country's education system: Distribution across ISCED levels (out of 32 countries). (Hadjiachilleos & Zachariou, 2022, p. 36)

Throughout the NIRs, it was evident that Sustainable Development (SD) key themes (2.1.1.) are established within curricula as >85% of the countries indicate that they have SD key themes integrated across various education levels but to a different extent, depending on the education level. Indicatively, in kindergarten and primary education, environmental protection and environmental health are the most highly integrated themes. SD key themes are more emphasized in lower secondary compared to the other education levels, especially biological and landscape diversity and environmental protection. In short-term tertiary education, cultural diversity was the SD key theme most extensively addressed, whereas in higher education, emphasis was placed on human rights, ecological principles, and oceans.

Figure 3.2 shows that the key themes of corporate social responsibility, culture's contribution to SD, economic growth, and poverty alleviation were the themes least addressed by the countries at all education levels. This might be attributed, amongst other reasons, to the complexity of the themes, which may hamper their integration at the lower ISCED levels, particularly in early childhood and primary education levels. The way in which sustainable development themes are defined depends on the context of each country and thus more emphasis could be placed on environmental (e.g., Slovakia, Serbia), social (e.g., Estonia, Turkey, Belarus), and economic issues (e.g., Romania, Tajikistan) or to a combination of the three (e.g., Belgium, Finland, Hungary, Latvia, Croatia, Switzerland, Cyprus, Greece, Kyrgyzstan, and Russian Federation).

ESD key-theme integration level also varies according to education level. Specifically, themes integrated into school education pertain more to the environmental aspects of ESD such as environmental protection, biology and landscape diversity, and environmental health. The least addressed themes in school education refer to social responsibility, economic growth, and good jobs. Nevertheless, as suggested in the Serbian NIR, sociocultural and economic aspects of ESD

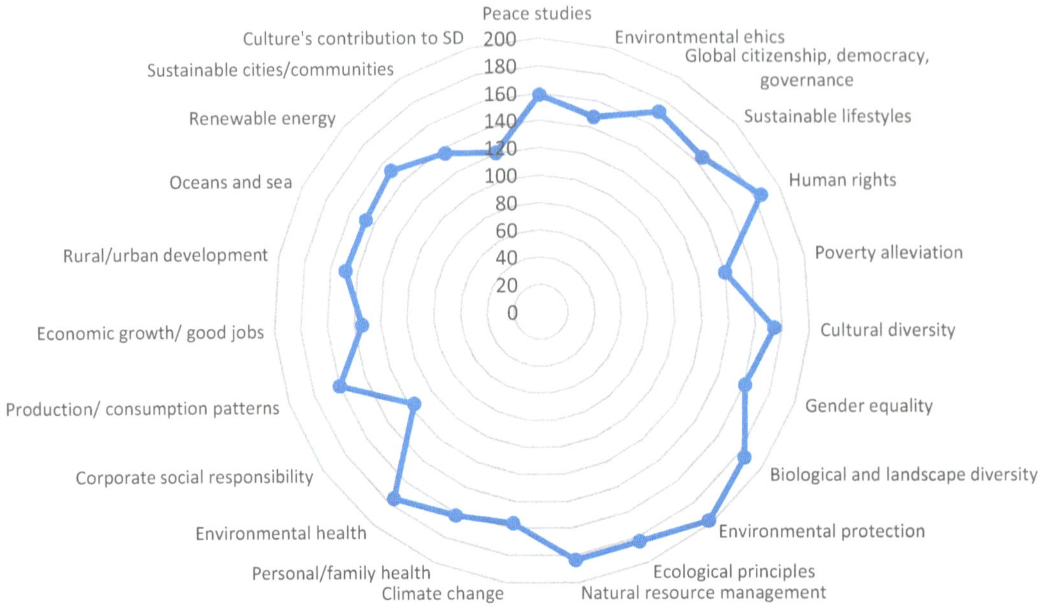

Fig. 3.2 Key theme integration across education levels (Hadjiachilleos & Zachariou, 2022, p. 55) *Scale 0–200 refers to the sum of the number of ticks indicated by member states for each of the key themes as provided in Appendix I(a) in the report (ECE/CEP/AC.13/2014/5) template

are often in place but they are not always recognized as SD. Examples of key-theme integration in Technical Vocational Education and Training (TVET) provided by Romania, where part of the "Local development curriculum" which is adopted to local specificities in SD. In Montenegro, 25 modularized curricula for TVET ensure the attainment of key SD competences. ESD key themes with strong social and economic elements are more extensively addressed at post-secondary and in higher education levels. For example, in Estonia, the Estonian Refugee Council, Tartu University, and local Non-Governmental Organizations (NGOs) offered several example activities as a response to the current refugee crisis.

3.3.2 WSA and Pedagogy/Didactics: How Do We Learn?

In formal, nonformal, and informal education, pedagogy and didactics were viewed through the pedagogical methods utilized (2.1.) and through implementing ESD as a multistakeholder process (sub-indicator 2.6). Finally, conserving traditional knowledge and knowledge of indigenous people (Issue 7) was mainly understood through utilizing nonformal settings.

NIR results indicated that various teaching and learning methods are addressed explicitly in curricula by 26 (76.5%) countries. How these methods are used depends on the education level. For kindergarten and primary levels, teaching and learning methods used are more connected to concrete life experiences, such as role-playing, excursions, and outdoor learning. In secondary education, the methods used are rather oriented toward problem-solving, simulations, and Information Communication Technology (ICT), which, in fact is one of the most extensively used methods across preuniversity and university education. Surveys, simulations, and problem-solving are also extensively integrated at these education levels.

To a large extent, the methods used at the formal and nonformal levels depend on the teacher. Schools, education institutions, and teachers in Finland, for example, have autonomy regarding the learning methods they use (Finland NIR). In the Netherlands, under the "freedom of Education" law, the national curriculum only describes the "what" (content) in highly abstract terminology. The "how" and "when" are the responsibilities of individual schools (Netherlands NIR).

Some countries report specific teaching methods in the curricula, the majority of which promote learner-centered learning (e.g., Finland, the Netherlands, Hungary, Latvia, Romania, Estonia, and Slovenia). As presented in Fig. 3.3, the teaching and learning methods most widely used to promote SD throughout all ISCED levels are simulations, role-playing and games, ICT, excursions and outdoor learning, discussions, and problem solving. The least utilized teaching methods to promote SD are workplace experience, good practice analyses, philosophical inquiry, and conceptual and perceptual mapping (Hadjiachilleos & Zachariou, 2022). Regarding the pedagogies and actions related to hands-on learning concerning the conservation of traditional knowledge (Issue 7), a number of countries (e.g., Belarus, Croatia, Cyprus, Greece, Hungary, Kirgizstan, and Russia) indicate that they are bringing students in connection with local populations with programs that concern traditional crafts and local products.

3.3.3 WSA and Building Management/Operations: Where Do We Learn?

This parameter was only examined under Issue 2: ESD in formal, nonformal, and informal learning, specifically in relation to whether instruments for formal, informal, and nonformal learning are in place to assess changes in knowledge, attitudes, and skills. Coombs et al. (1973) (as cited in Smith, 2002) distinguished the three forms of education by stating that (a) *formal edu-*

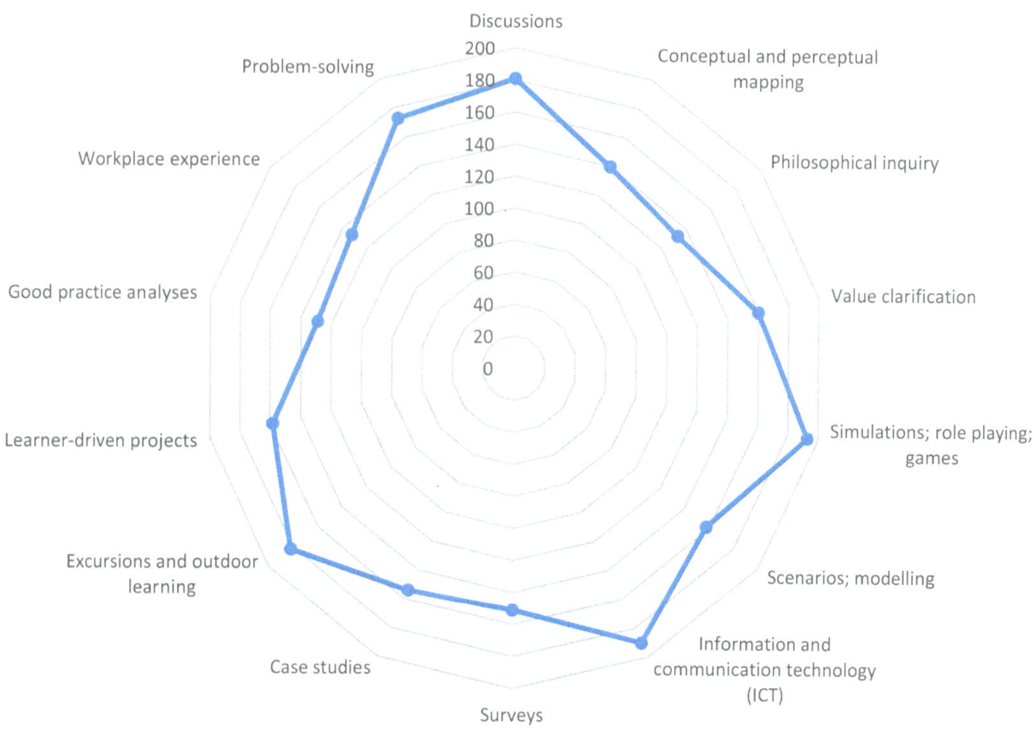

Fig. 3.3 Teaching–learning methods integration across ISCED levels (Hadjiachilleos & Zachariou, 2022, p. 77) *Scale 0–200 refers to the sum of the number or ticks provided regarding teaching and learning methods across all ISCED levels as required in Appendix I(c) in the report (ECE/CEP/AC.13/2014/5) template

cation is hierarchically structured, chronologically graded "education system," running from primary school through the university and, including, in addition to general academic studies, a variety of specialized programs and institutions for full-time technical and professional training, (b) *informal education* is the truly lifelong process whereby every individual acquires attitudes, values, skills, and knowledge from daily experience and the educative influences and resources in his or her environment—from family and neighbors, from work and play, from the market place, the library and the mass media, and (c) *nonformal education* is any organized educational activity outside the established formal system—whether operating separately or as an important feature of some broader activity—that is intended to serve identifiable learning clienteles and learning objectives.

The significance of nonformal and informal learning in ESD is emphasized throughout the UNECE Strategy for ESD and has been reflected in the previous reporting cycles (UNECE, 2015, para. 32). Thus, member states have long recognized that addressing ESD solely through formal education is insufficient, and have emphasized nonformal and informal learning as a means of orienting societies toward sustainable development. Therefore, even though there was an attempt in the strategy to examine infrastructure in terms of the places/buildings where ESD is delivered throughout the various education levels, examining school buildings and the way in which they are utilized to promote sustainability and ESD was overlooked.

Few countries provided examples of sustainable school infrastructure (e.g., Cyprus, Finland, Greece, Hungary, Slovakia, Slovenia, and the Netherlands). Indicatively, green spaces or gardens in schoolyards have been particularly emphasized as part of WSA efforts in Cyprus and Slovenia. In Cyprus, green school spaces are integrated into the schools' Sustainable Environmental Education Policy, whereas in Slovenia they are promoted through the Eco-Schools program at all levels of education. The Faculty of Education in Ljubljana is the first eco-friendly faculty in Slovenia, which fully implements WSA through addressing energy and water saving, local food and effective food management, and maintaining vegetable gardens.

3.3.4 WSA and Professional Development: Whom Do We Learn From?

A key factor in successfully promoting ESD implementation is the educator, indicating the importance of teacher professional development and initial education. This parameter also refers to stakeholders such as the local community, traditional communities, and local artisans, which can be considered as facilitating the aspirations of ESD in terms of learning outcomes and becoming learning arenas for teachers, students, and other school actors.

"Whom do we learn from" explored educator ESD competences in the training of educators (3.1) and opportunities for educators to cooperate on ESD (3.2). Additionally, we reviewed research and development on ESD issues, specifically research on innovation and capacity building in ESD practice (5.2.1) and ways in which research results are disseminated (5.3) and feed ESD professional development. Intergenerational cooperation, local communities, and indigenous people (Issue 7) were viewed as factors promoting professional development.

Countries reported specifically on whether national policies on ESD support teacher professional development. Malta for example explained that their ministry provides educators with clear and updated guidelines for the implementation of ESD across the curriculum through a formal structure that coordinates ESD implementation in educational institutions, provides financial support, and provides adequate continuous professional development opportunities to educators and school leaders.

Countries were also requested to provide data as to how teacher professional development is delivered to preservice teachers, in-service teachers, and school leaders. An overview is presented in Fig. 3.4; most member states report that ESD

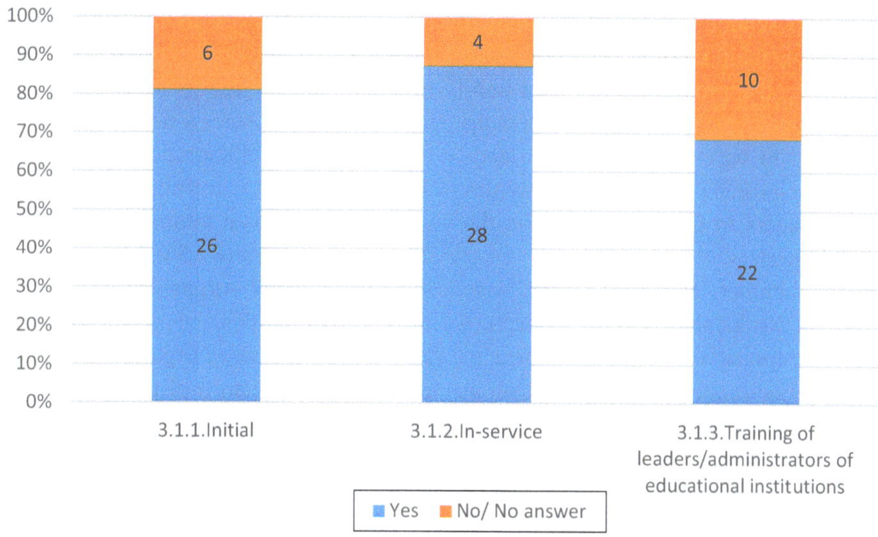

Fig. 3.4 Training of educators (Hadjiachilleos & Zachariou, 2022, p. 108)

is part of initial training (26 countries, 81%) and part of in-service training (28 countries, 87.5%), with 22 member states (68%) also addressing ESD competences in training programs for education leaders and administrators. These outcomes should be interpreted with caution, especially on teachers' professional development which is often offered by tertiary-level institutions that have a high level of autonomy and reporting on this area was often fragmented.

Approaches to in-service training in ESD competences vary from country to country and are offered as part of teacher professional development education in 28 member states (87.5%). It is important to mention that ESD is included in teachers' training but often takes place on a voluntary basis. These optional opportunities are provided by extended networks on ESD actions, established in several countries. For example, Germany, Hungary, Ireland, and Slovenia have embraced the green kindergarten network. In Germany, 29 model networks supported by the foundation "Haus der Kleinen Forscher" (Little Scientists' House) provide pedagogical resources and a continuous professional development program. Training workshops on the integration of ESD in early childhood education are offered to caring/teaching staff, reaching 30,000 early childhood education institutions. In Hungary, the green kindergarten network and the eco-school network are efficient representatives of the WSA and also provide teacher training, reaching more than 20% of the overall population of Hungarian teachers. In Austria SD and ESD lectures, courses and seminars are offered, among others, through summer academies, or specific courses offered by local universities. Despite their optional nature, an impressively high number of in-service educators have attended such courses. In Israel, the Ministries of Environmental Protection and Education train more than 2700 teachers per year on how to incorporate ESD in their teaching.

In addition, mandatory in-service ESD training was reported by several member states (e.g. Bulgaria, Romania, Slovenia, Montenegro, Cyprus, Belarus, and Tajikistan) indicating the importance placed on developing in-service teachers' ESD competences. ESD competences in the training of education leaders and administrators have increased significantly as reported during phase IV compared to reporting on phase III (UNECE, 2015) from just above 50% of reporting member states offering such education opportunities in the period 2011–2015 to 68.7% (22 countries) in the period 2015–2018.

In Cyprus, mandatory and optional ESD courses are offered not only to teachers but also to principals and school leaders on an annual basis, focusing on their schools becoming sustainable. In Georgia, approximately 80% of public school principals have received leadership academy training to become instructional leaders. Various Official ESD Awards and activities by other stakeholders in Germany focus on the training of ESD leaders.

Twenty-six member states (81.2%) have affirmed that networks/platforms of educators and/or leaders/administrators are established in their country. For example, the Education Coalition NGO in Switzerland is a network of approximately 30 NGOs, youth associations, environment associations, health, international cooperation, and human rights organizations, aiming at integrating and anchoring ESD in civil society and politics.

Governmental support for ESD networks and platforms comes mainly in the form of coordination and/or financing. Specifically, 22 (68.7%) member states report that such support exists for specific initiatives, such as Austria's Sustainability Award for Higher Education, Finland's SD certificate for schools, and Slovakia's Ministry of Environment funding for education institutions. Concrete governmental structures to provide funding, equip educators with ESD competences, support ESD Strategy implementation, and guide and/or coordinate such efforts seem to be missing.

3.3.5 WSA and School Environment: Whom Do We Learn With?

Sustainable development is attracting an increasing number of people, including entrepreneurs and governments. The transition from an unsustainable to a sustainable world is a learning process for everyone. The school environment and its connection to society are critical in the WSA. The school environment has been examined through the NIRs, in relation to ESD competences included in educators' training (3.1), ESD cooperation opportunities (3.2), and research and development (5.2 and 5.3). Finally, whom we learn with was a parameter examined through the countries' initiatives to promote ESD through intergenerational cooperation, local communities, and indigenous people (Issue 7).

The strategy places a strong emphasis on multistakeholder collaboration for ESD. Subindicator 2.6.1 examines whether ESD implementation is a multistakeholder process in 28 member states (87.5%). Local governments, nongovernmental organizations (NGOs), the media, communities, the private sector, organized labor, and faith-based organizations are involved in the implementation of the ESD Strategy. The actions in which stakeholders participate in ESD Strategy implementation—such as raising public awareness, training, quality education, etc.—are presented in Fig. 3.5.

Raising public awareness is the ESD aspect stakeholders are mostly engaged in, whereas they are least engaged in reorienting education toward sustainable development. Reorienting education refers to empowering learners to be engaged in lifelong learning processes that will allow them to make informed decisions, develop creative problem-solving skills, scientific and social literacy, and participate in individual and cooperative action for a sustainable future (Bowers, 1995). Stakeholders such as community-based organizations, NGOs, the media, and local governments are the ones mostly involved in raising public awareness and social learning. Local governments have the most multifaceted role in promoting ESD implementation, specifically, training, quality education, and public awareness. This can be attributed to their authority to establish national policies and collaborations and also to funding opportunities they can provide in the national context.

3.4 Discussion

Despite the fact that it is not possible to extensively analyze and discuss each dimension of the WSA, within this chapter, the results allow us to summarize an overall picture of the WSA implementation in the UNECE region. What is noted from the synopsis of the results is that there is a clear orientation toward WSA. However, frag-

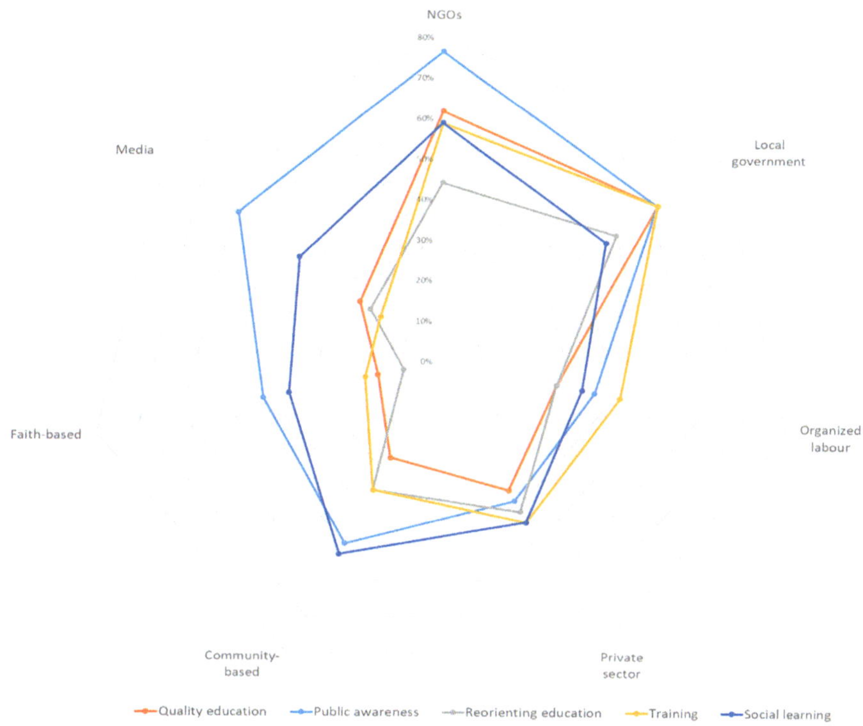

Fig. 3.5 Multistakeholder cooperation (Hadjiachilleos & Zachariou, 2022, p. 97)

mented data provided by countries reflect a partial view of *how* WSA is implemented in ESD in the UNECE region. As demonstrated through the data analysis, the strategy in itself does not place equal value on each of the "petals" of the WSA to ESD "flower model" taking into consideration that the focus is on curricula and the school environment, while the school infrastructure is neglected. Below, the results of the WSA as realized through the UNECE ESD Strategy are discussed, organized under the aspects-petals of the WSA to ESD "flower model."

3.4.1 What Do We Learn? (Curricula)

It was evident through the NIRs that WSA to ESD in the UNECE region is well integrated across curricula, especially in school education levels. Depending on the national context, key themes examined emphasize either the environmental, social, cultural, or economic aspects of SD issues. This is a promising fact as the well-defined context and pedagogy of the curricula are important prerequisites of quality in education (Kadji et al., 2017; Kadji & Zachariou, 2022). Maintaining the integrity of life support systems (environmental sustainability), equity between ethnic and social groups as well as between generations (social sustainability), the nourishment and sharing of attitudes and values that represent diverse ways of viewing the world (cultural sustainability), and using resources to provide necessary and desirable products and services for the next generation without compromising their ability to do the same (economic sustainability) (Wals, 2014) are all valued within curricula but understood differently across UNECE countries. The different degree of emphasis placed on each of the SD parameters through curricula is indicative of how factors such as education level and different priorities determine which SD key themes will be emphasized.

In tertiary education, despite the considerable efforts to incorporate ESD in university curricula, there is a lot to be done regarding *how* Higher

Education Institutions (HEIs) address ESD through WIA. The 2022 European Strategy for Universities supports the higher education sector in implementing WIA for the green transition and sustainable development, through highlighting the development of skills, competences, and technological innovation. The high level of the HEIs' autonomy can be used as a factor in promoting innovation through WIA in ESD.

In order to successfully implement WIA, the tertiary level has to highlight entrepreneurship (Hopkinson & James, 2013). Countries have made considerable efforts to connect school learning with the labor market, mainly through vocational education and tertiary education (e.g., through connecting what is taught in TVET with market needs in the area or through creating quality frameworks documenting competences acquired through ESD in HEIs).

3.4.2 Whom Do We Learn From? (Professional Development)

In the UNECE region, teacher training on ESD implementation is offered by multiple stakeholders on an optional basis. Mandatory training is gaining attention in several countries revealing the growing acknowledgement of the importance of ESD. In the UNECE region, governmental support for professional development is usually in the form of coordination and/or financing, whereas teachers' initial training is mainly offered through HEIs.

University teacher education is also an important milestone in teachers' professional development. According to Albareda-Tiara et al. (2019), active teaching/learning strategies in ESD embedded in curricula of initial teacher education highly contribute to future teachers' sustainability competences. In order to examine this, a clearer picture of ESD integration in university education is required. The authors also claim that sustainability competences are needed to orient teachers toward sustainability and empower them to teach the way they live. This assumption places tremendous importance on teacher training for effective ESD implementation in general. Under this framework, teachers are required to make appropriate connections among various aspects of sustainability. That is why teacher professional development should be ongoing.

In addition, in the UNECE region, there is increased attention on educating school leaders for effective ESD implementation, as they can facilitate strategy implementation and dissemination. Reorienting a school to ESD requires specific leadership practices (Kadji-Beltran et al., 2012) such as empowering teachers, encouraging challenging current approaches, exploring alternative curricula and pedagogies, and collaborating with others to support the school vision. In addition, through WSA in ESD addressing school leaders' competences improves overall coordination among stakeholders and reduces resistance to change. External stakeholders and networks have also been found to provide learning opportunities and authentic experiences for all learners in school.

3.4.3 How Do We Learn? (Pedagogy/Didactics)

UNECE countries reported a variety of pedagogical approaches in addressing ESD both in formal and nonformal contexts, some being more frequently used than others. Encouraging experimentation with multiple and alternative teaching techniques is a dynamic element of WSA in ESD, which can be encouraged through well-informed and progressive school leadership (Zachariou et al., 2013). Furthermore, alternative pedagogies that evolve around problem solving should be encouraged more, through national education systems, to promote the necessary paradigm shift and foster sustainability competences (Gokoo-Ramdoo & Rumjaun, 2017; Kadji-Beltran & Zachariou, 2022).

Strong experiences and appropriate pedagogies that foster critical thinking, reflection, and sustainability competences, in general, can have a transformative impact on learners. Meaningful experiential learning in nature and communities, as well as connecting places and people, are therefore important learning spaces that support shifting from transmissive to transformative forms of education (Bell, 2016). As our data analysis has revealed, the UNECE countries address pedagogical methods integration in ESD in different ways, with some countries providing full autonomy for teachers to select and use appropriate teaching and learning methods and others embedding these methods in curricula. Student-centered pedagogical approaches are highly emphasized in all education levels. Nevertheless, in school education, especially in kindergarten and in primary education, methods build more on students' concrete learning experiences in formal and nonformal settings (e.g., field work, excursions, and role-playing), whereas in secondary and tertiary education methods gradually orient toward more abstract mental processes, engaging critical thinking (e.g., problem solving and ICT).

3.4.4 Where Do We Learn? (Building Management/Operations)

Sustainable infrastructure plays an important role in the green economy transition, as it sits at the heart of the sustainable development goals (SDGs) (UNEP, 2021). The need to transition to green infrastructure investments is clearly stated through the new policy text of the UNECE Committee of Environmental Policy "Greening the economy in the pan-European region: working toward sustainable infrastructure." Green infrastructure investments are becoming urgent because of the ongoing climate crisis and the COVID-19 pandemic, which press the governments in the region to achieve collective tangible progress for the green economy beyond individual actions, especially with regard to sustainable infrastructure development (CEP, 2022, p. 4).

For schools, the creation of sustainable infrastructure is very important because it contributes to a healthy school, increases the self-esteem of students, affects their psychological mood positively, and ultimately contributes to the dissemination of knowledge as well as their creativity (Tapia-Fonllem et al., 2020). In addition, sustainable school infrastructures contribute to the protection of the environment or act as a measure of adaptation to climate change (Barrett et al., 2019). It is no coincidence that the European Commission addresses its commitment to provide member states with new financial resources to make school buildings and operations more sustainable (EC, 2019, p. 19). Despite acknowledging the need to transition to green infrastructures and sustainable school buildings, this dimension is the least addressed aspect of WSA in the current UNECE ESD Strategy. This can be justified by a number of factors such as the capacity and resources available and mainly because sustainable infrastructure is complex in nature (CEP, 2022). The results of the present study indicate that in the UNECE region, more attention should be paid to school infrastructure, schools being the place where ESD is primarily delivered. School buildings should be a sustainability paradigm: inspire students to live and act sustainably, provide experiences hard to obtain elsewhere, and develop a sense of pride in their sustainable school. The improvement of school building's infrastructure is an issue that needs to be addressed by governments, school leadership, parents' associations, local communities, strengthening the school's connections, and benefitting at the same time the community (Espinet & Zachariou, 2014).

3.4.5 Whom Do We Learn With? (School Environment)

Apart from schools, stakeholders mostly involved in the UNECE Strategy for ESD implementation are NGOs, the media, local governments, and community-based organizations. These are highly engaged in public awareness actions and generate a special type of learning that aims to connect people effectively and actively in achieving a better quality of life in their own context (Espinet & Zachariou, 2014). This collaboration creates learning and action communities and leads to collaborative change. Bringing together heterogeneous perspectives is crucial for fully understanding sustainability challenges (Beyers & Leventon, 2019).

The status of an institution, such as for example public authorities or faith-based organizations, NGOs, or the media, calls for a different type of engagement in ESD such as, for example, quality education. Still, quality education is the least addressed aspect of multistakeholder collaboration, which reveals a gap that needs to be addressed in the future. The primary goal of participatory ESD is to educate people so that they can participate as active citizens in democratic processes concerning sustainable development issues. At the same time, external factors such as the requirements of employment may help to provide additional incentives to promote multistakeholder collaboration in ESD (Matthiews & Vargas, 2016), especially in higher education.

3.5 Ways Forward

There are many ways to move forward with WSA in ESD. To start with, the UNECE countries need to reflect—individually and collectively—on the vision of the sustainable school they aspire to. Indicatively, starting with curricula, there needs to be a thorough evaluation or mapping of their content across all levels of education in the UNECE countries from an ESD perspective in order to find out the status of ESD content examined (Hadjiachilleos & Zachariou, 2022). Quality control mechanisms for ESD curricula (e.g., guidelines, evaluation mechanisms etc.), for tools and materials, and the kinds of collaboration established between various stakeholders outside the school, need to be set. In this way, more informed decisions on ESD implementation can be taken, promoting competences and increasing youth employability (Lambini et al., 2021). Pedagogies and curricula could be viewed from the perspective of the SDGs and "reach beyond the institution by engaging in sustainable and inclusive social change in line with the civic mission of the institution" (Gregersen-Hermans, 2021, p. 461).

Taking into consideration the complexity of sustainability issues and the need to move to more radical models of learning, which will change our frame of mind to more sustainable behaviors, WSA seems to be the only way to move forward since it "captures" unified and concretely the vision as well the challenge of the institutions to evolve from education organizations to communities of learning for ESD. McGregor (2019) pointed out that mainstream ESD should radically transform to "promote a way for people to walk a different path, especially considering the environmental challenges staring them in the face." To achieve this radical transformation, deep, meaningful learning should be put forward and best educational practices should be integrated synergistically into a unified whole. This idea could be put forward through WSA, which encourages transformative learning settings (Ostergaard, 2021). Such settings help students to develop the capacities required to mitigate the challenges facing the environment in the long term and require internal and external synergies of the school, which are mainstreamed through WSA.

The fragmentation in the way WSA in ESD has been viewed in the current UNECE Strategy for ESD progress report (Hadjiachilleos & Zachariou, 2022) and the orientation toward transformational learning seem to be well addressed through the new framework for the implementation of the UNECE Strategy for ESD2021–2030 (UNECE, 2022). WSA approach is considered one of the key strands and in connection with the rest of the key strands (Quality Education and ESD, Digital Education, ICT and ESD, and Entrepreneurship, Employment, Innovation and ESD). The new strategic framework for the implementation of the UNECE Strategy for ESD 2021–2030 aspires to improve educational systems and other mechanisms that enable people to learn, live, and work in a more prosperous, just, creative, healthy, and sustainable planet and region (UNECE, 2022). Additionally, the new strategic framework seeks to empower WSA/WIA across the spectrum of organizations within formal, nonformal, and informal education and beyond school education. Importantly, it addresses the need for emphasizing the impact of school infrastructure and buildings, on how ESD is realized. Barrett et al. (2019)

provide evidence documenting the impact of pedagogy and school buildings on the overall quality of education. Undoubtedly, addressing school infrastructure and its role in ESD implementation require planning, ambition, long-term perspectives, and, of course, funding (Barrett et al., 2019). That is why addressing school infrastructure through the new strategic framework of ESD Strategy in the UNECE region is vital. School infrastructure can facilitate the opening of the school to the community and the formation of multistakeholder collaborations (Mula et al., 2022).

WSA in ESD is currently realized in various ways through the UNECE region. Countries tend to integrate WSA differently because of the different educational priorities, particularities, and variety of educational systems as well as the disparity of size, cultural heritage, and geographic position of the countries in the UNECE region. Having that in mind, in view of the new framework for the implementation of the UNECE Strategy for ESD2021–2030 currently in progress, a lot is expected concerning how policymakers, school leaders, practitioners, students, local communities, and other stakeholders will be involved and oriented toward sustainable development in the future, through well established and validated WSAs. The UNECE can make invaluable contributions to how WSA is implemented in ESD due to the "polyphony" of countries and also due to the opportunities for international cooperation, capacity, expertise, and exchange of good practices that they can promote.

References

Abrahamsson, B., Berg, G., & Wallin, E. (1999). Organisations - och läroplansperspektiv. [Organisation and curriculum perspective]. *Pedagogisk Forskning i Sverige, 4*(2), 145–161. https://open.lnu.se/index.php/PFS/article/view/1065/916

Albareda-Tiana, S., García-González, E., Rocio-Jiménez, F., & Solis-Espallargas, C. (2019). Implementing pedagogical approaches for ESD in initial teacher training at Spanish Universities. *Sustainability, 11*(18), 4927. https://www.mdpi.com/2071-1050/11/18/4927/htm

Barrett, P., Treves, A., Smis, T., Ambasz, D., & Ustinova, M. (2019). *The impact of school infrastructure on learning: A synthesis of the evidence*. World Bank Group. https://doi.org/10.1596/978-1-4648-1378-8

Bell, D. (2016). Twenty-first century education: Transformative education for sustainability and responsible citizenship. *Journal of Teacher Education for Sustainability, 18*(1), 48–56.

Beyers, F., & Leventon, J. (2019). Learning spaces in multi-stakeholder initiatives: The German partnership for sustainable textiles as a platform for dialogue and learning. *Earth System Governance, 9*, 100–113.

Bowers, C. A. (1995). Educating for an ecologically sustainable culture: Rethinking moral education, creativity, intelligence, and other moral orthodoxies. New York, SUNY Press.

CEP. (2022). *Greening the economy in the pan-European region: Working towards sustainable infrastructure*. ECE/NICOSIA.CONF/2022/4. https://unece.org/sites/default/files/2022-08/ECE_NICOSIA.CONF_2022_4_E_1.pdf

Coombs, P. H., Prosser, R. C. & Mansoor, A. (1973). New paths to learning for rural children and youth. *UNICEF, International Council for Educational Development*.

EC. (2019). *The European green deal*. https://eur-lex.europa.eu/resource.html?uri=cellar:b828d165-1c22-11ea-8c1f-01aa75ed71a1.0002.02/DOC_1&format=PDF

EC. (2022a). *Proposal for a council recommendation on learning for environmental sustainability*. 2022/0004(NLE). https://data.consilium.europa.eu/doc/document/ST-9242-2022-INIT/en/pdf

EC. (2022b). *Communication from the commission on a European strategy for universities*. https://education.ec.europa.eu/sites/default/files/2022-01/communication-european-strategy-for-universities-graphic-version.pdf

Espinet, M., & Zachariou, A. (2014). *Key stones on school community collaboration for sustainable development*. Lifelong Learning Program: CoDeS. https://www.ensi.org/global/downloads/Publications/369/CoDeS-Key%20Stones.pdf

Ferreira, J., Ryan, L., & Tilbury, D. (2006). *Whole-school approaches to sustainability: A review of models for professional development in pre-service teacher education*. Australian Government Department of the Environment and Heritage and the Australian Research Institute in Education for Sustainability (ARIES).

Gleason, R., Kirillov, P., Koryakina, N., Ermakov, S., & Ermakov, D. (2020). Whole-institution approach in education for sustainable development: Theory and practice. *Scholarly Notes of Transbaikal State University, 15*, 36–43. https://doi.org/10.21209/2658-7114-2020-15-4-36-43

Gokool-Ramdoo, S., & Rumjaun, A. B. (2017). Education for sustainable development: Connecting the dots for sustainability. *Journal of Learning for Development, 4*(1), 72–89. https://files.eric.ed.gov/fulltext/EJ1141544.pdf

Gregersen-Hermans, J. (2021). Towards a curriculum for the future: Synthesizing education for sustainable development and internationalisation of the curriculum. *Journal of Studies in International Education, 25*(4), 461–481. https://doi.org/10.1177/10283153211031033

Hadjiachilleos, S., & Zachariou, A. (2022). *Implementation of the UNECE Strategy for ESD across the ECE Region (2015-2018)*. ECE/CEP/196. UNECE. https://unece.org/sites/default/files/2022-09/Implementation%20of%20the%20UNECE%20Strategy_web_final_05.09.2022.pdf

Hargis, K., McKenzie, M., & Le Vert-Chiasson, I. (2021). Chapter 3. A whole institution approach to climate change education preparing school systems to be climate proactive. In R. Iyengar & C. Kwauk (Eds.), *Curriculum and learning for climate action. Toward an SDG 4.7 roadmap for systems change* (pp. 43–66). IBE on Curriculum, Learning, and Assessment.

Henderson, K., & Tilbury, D. (2004). *Whole-school approaches to sustainability: An international review of sustainable school programs*. Report prepared by the Australian Research Institute in Education for Sustainability (ARIES) for The Department of the Environment and Heritage, Australian Government.

Hopkinson, P., & James, P. (2013). Whole institutional change towards sustainable universities: Bradford's ecoversity initiative. In S. Sterling, L. Maxey, & H. Luna (Eds.), *The sustainable university: Progress and prospects* (1st ed.). Routledge. https://doi.org/10.4324/9780203101780

Kadji-Beltran, C., & Zachariou, A. (2022). ESD competences for deep quality in education. In P. Vare, N. Lausselet, & M. Rieckmann (Eds.), *Competences in education for sustainable development: Critical perspectives* (Sustainable Development Goals Series). Springer.

Kadji-Beltran, C., Zachariou, A., & Stevenson, R. (2012). Leading sustainable schools: Exploring the role of primary school principals. *Environmental Education Research, 19*(3), 1–21. https://doi.org/10.1080/13504622.2012.692770

Kadji-Beltran, C., Christodoulou, N., Zachariou, A., Lindemann-Matthies, P., Barker, S., & Kadis, C. (2017). An ESD pathway to quality education in the Cyprus primary education context. *Environmental Education Research, 23*(7), 1015–1031. https://doi.org/10.1080/13504622.2016.1249459

Lambini, C. K., Goeschl, A., Wäsch, M., & Wittau, M. (2021). Achieving the sustainable development goals through company staff vocational training—The case of the Federal Institute for Vocational Education and Training (BIBB) INEBB Project. *Education Science, 11*(4), 179. https://doi.org/10.3390/educsci11040179

Mathar, R. (2013). The concept of the whole school approach—a platform for school development with focus on sustainable development. In ESD Expert Nework (Eds.), *Schools for sustainability—A resource toolkit for teacher training*. https://esd-expert.net/files/ESD-Expert/pdf/Concept-Paper-Mathar.pdf

Matthews, H., & Vargas, V. (2016). Institutional support for ESD in the curriculum: A case study of the School of Science and the environment, MMU. *Learning and Teaching in Action, 13*(1), 97–114. ISSN 1477-1241.

McGregor, S. (2019). David Selby's radical approach to sustainability education. *The Journal of Sustainability Education, 21*. http://www.susted.com/wordpress/wp-content/uploads/2020/01/McGregorJSEDecember2019generalissuepdf.pdf

MIO/ESCDE. (2014). Mediterranean Strategy on Education for Sustainable Development (MSESD) https://ufmsecretariat.org/wp-content/uploads/2014/05/Mediterranean-Strategy-on-Education-for-sustainable-development-.pdf

Mogensen, F., & Schnack, K. (2010). The action competence approach and the 'new' discourses of education for sustainable development, competence and quality criteria. *Environmental Education Research, 16*(1), 59–74.

Mogren, A. (2019). *Guiding principles of transformative education for sustainable development in Local School Organisations. Investigating whole school approaches through a school improvement lens* [Doctoral Thesis, Karlstad University Studies]. https://www.diva-portal.org/smash/get/diva2:1368940/FULLTEXT01.pdf

Mulà, I., Cebrian, G., & Junyent, M. (2022). Lessons learned and future research directions in educating for sustainability competences. In P. Vare, N. Lausselet, & M. Rieckmann (Eds.), *Competences in education for sustainable development: Critical perspectives* (Sustainable Development Goals Series). Springer. https://doi.org/10.1007/978-3-030-91055-6

NGA and NAAE. (2022). Environmental sustainability: A whole school approach. In *A guide for governing boards*. National Governance Association. https://www.nga.org.uk/getmedia/42a15743-7a17-4dfb-b15a-804e5f26b5a2/nga-environmental-sustainability-20211104.pdf

Ostergaard, T. (2021). Alone in the sustainable wilderness; transforming sustainable competences and didactics in a design for change education. *Design and Technology Education, 26*(3), 190–212. https://files.eric.ed.gov/fulltext/EJ1323602.pdf

Smith, M. K. (2002). Informal, non-formal and formal education: A brief overview of different approaches. In *The encyclopedia of pedagogy and informal education*. https://infed.org/mobi/informal-non-formal-and-formal-education-a-brief-overview-of-some-different-approaches/

Sterling, S. (2004a). An analysis of the development of sustainability education internationally: Evolution, interpretation and transformative potential. In J. Blewitt & C. Cullingford (Eds.), *The sustainability curriculum: The challenge for higher education* (pp. 43–62). Earthscan Publications.

Sterling, S. (2004b). Higher education, sustainability, and the role of systemic learning. In P. B. Corcoran & A. Wals (Eds.), *Higher education and the challenge of sustainability* (pp. 49–70). Springer Direct.

Sterling, S. (2010a). Learning for resilience, or the resilient learner? Towards a necessary reconciliation in a paradigm of sustainable education. *Environmental Education Research, 16*(5–6), 511–528.

Sterling, S. (2010b). Living in the earth towards an education for our time. *Journal of Education for Sustainable Development, 4*(2), 213–218. https://doi.org/10.1177/097340821000400208

Summers, M., & Childs, A. (2007). Student science teachers' conceptions of sustainable development: An empirical study of three postgraduate training cohorts. *Research in Science and Technological Education, 25*(3), 307–327. https://doi.org/10.1080/02635140701535067

Tapia-Fonllem, C., Fraijo-Sing, B., Corral-Verdugo, V., Gerza-Teran, G., & Moreno-Barahona, M. (2020). School environments and elementary school children's well-being in northwestern Mexico. *Frontiers in Psychology, 11*(510), 1–8. https://doi.org/10.3389/fpsyg.2020.00510

Tilbury, D., & Galvin, C. (2022). *Input paper: A whole school approach to learning for environmental sustainability*. Expert briefing paper in support of the first meeting of the EU Working Group Schools: Learning for Sustainability. European Commission. https://education.ec.europa.eu/sites/default/files/2022-02/input-paper-whole-school-approach-sustainability.pdf

UNECE. (2005). *UNECE strategy for education for sustainable development*. Adopted at the High-level meeting of Environment and Education Ministries (Vilnius, 17–18 March 2005). CEP/AC.13/2005/3/Rev.1. http://www.unece.org/fileadmin/DAM/env/documents/2005/cep/ac.13/cep.ac.13.2005.3.rev.1.e.pdf

UNECE. (2012). *Report of the United Nations economic commission for Europe steering committee on education for sustainable development on its seventh meeting*. ECE/CEP/AC.13/2012/2. https://unece.org/fileadmin/DAM/env/esd/7thMeetSC/Official_Docs/ece.cep.ac.13.2012.2e.pdf

UNECE. (2013). *Report of the United Nations economic commission for Europe steering committee on education for sustainable development on its eighth meeting*. ECE/CEP/AC.13/2013/2. https://unece.org/fileadmin/DAM/env/documents/2013/ece/cep/ac.13/ece.cep.ac.13.2013.2.e.pdf

UNECE. (2014). *Information paper 4 working group on 'ESD school planning': Outcomes*. https://unece.org/fileadmin/DAM/env/Information_document_4_school_planning_02.pdf

UNECE. (2015). *Draft future implementation framework*. ECE/CEP/AC.13/2015/4. https://unece.org/fileadmin/DAM/env/esd/10thMeetSC/Documents/ece.cep.ac.13.2015.4e.pdf

UNECE. (2016). *Learning from each other: Achievements, challenges and ways forward third evaluation report of the UNECE strategy for education for sustainable development*. ECE/CEP/AC.13/2016/3. https://unece.org/fileadmin/DAM/env/esd/11thMeetSC/Documents/ece.cep.ac.13.2016.3e_Advance_Copy.pdf

UNECE. (2022). *Framework for the implementation of the United Nations Economic Commission for Europe strategy for education for sustainable development from 2021 to 2030*. ECE/CEP/AC.13/2022/3. https://unece.org/sites/default/files/2022-05/ece_cep_ac.13_2022_3_e.pdf

UNEP. (2021). *International good practice principals for sustainable infrastructure integrated, systems-level approaches for policy makers*. First Edition. https://www.unep.org/resources/publication/international-good-practice-principles-sustainable-infrastructure

UNESCO. (2014). *Roadmap for implementing the global action programme on education for sustainable development*. UNESCO. https://unesdoc.unesco.org/ark:/48223/pf0000230514

UNESCO. (2020). *Multi-stakeholder approaches to education for sustainable development in local communities: Towards achieving the sustainable development goals in Asia*. https://www.rcenetwork.org/portal/sites/default/files/Multi-stakeholder%20approaches%20to%20ESD.pdf

UNESCO IBE (International Bureau of Education). (2012). *General education quality analysis/diagnosis framework (GEQAF)*. http://193.242.192.196/en/general-education-system-quality-analysisdiagnosis-framework-geqaf

Verhelst, D., Vanhoof, J., & Van Petegem, P. (2022). Development and validation of the education for sustainable development school organisation questionnaire. *Environmental Education Research, 28*(2), 241–259.

Wals, A. E. G. (2014). Sustainability in higher education in the context of the UN DESD: A review of learning and institutionalization processes. *Journal of Cleaner Production, 62*, 8–15. https://doi.org/10.1016/j.jclepro.2013.06.007

Wals, A. E. J., & Mathie, R. G. (2022). Whole school responses to climate urgency and related sustainability challenges. In M. A. Peters & R. Heraud (Eds.), *Encyclopedia of educational innovation*. Springer. https://doi.org/10.1007/978-981-13-2262-4_263-1

WHO. (2019). *Integrating environment and health considerations into healthy schools' settings. Draft background document*. WHO.

Zachariou, A., & Hadjiachilleos, S. (2022). *Whole school approach and education for sustainable development in the UNECE region: Findings, challenges and steps forward*. Presented at the 11th WEEC Building Bridges, Prague, 14–18 March 2022.

Zachariou, A., & Pace, P. (2022). *Setting up an expert group in UNECE, and a working program for 2022-2025 for WSA expertise*. Presented in the International Conference Implementation of the WSA/WIA, Lunderen, 28–30 March 2022.

Zachariou, A., Kadji-Beltran, C., & Manolis, K., (2013). School principals' professional development in the framework of sustainable schools: A matter of refocusing. *Professional Development in Education, Taylor and Francis, UK, 39*(5), 712–731.

4 How to Institutionalize a Whole School Approach to ESD

Niklas Gericke, Anette Forssten Seiser, Anna Mogren, Teresa Berglund, and Daniel Olsson

Key Message
The theoretical framework and empirical examples presented in this chapter stem from school improvement research and can be used by *schools* that want to work practically according to a whole school approach (WSA) toward education for sustainable development (ESD) through *policymakers* who want to establish guidelines enabling the implementation of WSA to ESD and through *researchers* who want to investigate and analyze the process of institutionalizing WSA to ESD.

4.1 Introduction

The concept of *whole school approach* (WSA) has become a central terminology used in the policy on sustainability education or education for sustainable development (ESD).[1] The reason for this is that regular education within the boundaries of school subjects has been urged to address the need to educate citizens who can meet the sustainability challenges of the twenty-first century (see, e.g., Lotz-Sisitka et al., 2015). According to Wals and Mathie (2022), schools are challenged by the problem of how to respond to emerging societal issues such as runaway climate change. So far, more education has not reversed the negative decline of the environment. Consequently, a WSA to ESD has been suggested to redirect education toward a more sustainable society (Henderson & Tilbury, 2004; Mogren, 2019). Since learning cognitively and decontextualized about sustainability issues and how to solve them has proven not to translate into changed values and actions toward sustainability outside school, learning in authentic settings is suggested to stimulate students' affective dimensions of emotions, attitudes, and values (Eilam & Trop, 2010). Consequently, affective dimensions are often given equal importance as the cognitive knowledge dimension in ESD (Gericke & Torbjörnsson, 2022b). Therefore, the WSA has been suggested as the next step in sustainability education because it imposes an inclusive perspective on teaching and learning, including all

[1] In this chapter, we do not distinguish between the terms *sustainability education, education for sustainability*, and *education for sustainable development*. We consider them here to be synonyms and an educational approach with the overall aim to facilitate development of sustainability in the wider society as expressed in the title of this book *Whole School Approaches to Sustainability*.

N. Gericke (✉) · T. Berglund · D. Olsson
Department of Environmental and Life Sciences, Karlstad University, Karlstad, Sweden
e-mail: niklas.gericke@kau.se

A. F. Seiser · A. Mogren
Department of Educational Studies, Karlstad University, Karlstad, Sweden

A. E. J. Wals et al. (eds.), *Whole School Approaches to Sustainability*, Sustainable Development Goals Series, https://doi.org/10.1007/978-3-031-56172-6_4

stakeholders in the school, and opens education to the society outside the school.

There is rich diversity in the ways the WSA has been described in the literature. However, a common guiding principle for the WSA is the integration of three lines of action: environmental management ("greening") of the school, the establishment of ongoing partnerships with the broader local community to address issues of social-environmental sustainability, and incorporation of sustainability in the curriculum (Gericke, 2022). Hence, a WSA demands the efforts of the whole school community and external stakeholders to expose students to real sustainability issues. Research has, however, continuously shown that this is a very difficult goal to achieve (Holst, 2023; Laurie et al., 2016; Mathar, 2015; Mogren & Gericke 2017a, b; Müller et al., 2021). Therefore, the most central issue as we see it in practice and research on the WSA is how to implement the approach or, to put it differently, how to reform a school to work according to a WSA.

WSA to ESD has mainly been practice-based and lacked a foundation in educational theory on how a school works as an organization. Our research group has been inspired by and has applied methods and theories from the field of school improvement research to study the WSA toward sustainability. In this chapter, we build on those experiences and take a novel process perspective from developed theories within *school improvement research* to suggest how a WSA to ESD can be implemented in schools over time. The theory and its implications are also accompanied and compared with empirical findings from our studies to further elaborate on practical implications for local school reforms on a WSA to ESD. The school improvement processes, its drivers and constraints toward ESD, are discussed longitudinally based on school improvement theories mainly resting on the works of Blossing et al. (2015) and Fullan (2007). Hence, the focus of this chapter will be to propose and critically discuss the school improvement processes leading to a WSA.

4.2 What Is a Whole School Approach?

A WSA to ESD can be defined as the involvement of a whole school community in efforts to promote sustainable development, in addition to the orientation of the teaching, learning, and curriculum toward its promotion (Henderson & Tilbury, 2004; McKeown & Hopkins, 2007). The concept of WSA was originally used in research about schools' ability to incorporate perspectives of health, well-being, and anti-bullying (Rowe & Stewart, 2009; Wyn et al., 2000). The goal has been to engage all departments of a school in common efforts at all levels to improve targeted characteristics, for example, the well-being of staff and pupils. The WSA approach has been suggested as a method to draw on the experiences of external stakeholders and use their expertise on urgent matters for society. External stakeholders can then support education toward societal solutions within certain areas of interest (Wyn et al., 2000). Students have in this way been engaged in relevant development tasks in the wider society while still in school. This has been found to be a successful approach in health education, according to a few studies (see Rowe & Stewart, 2009). The WSA has subsequently been adopted and proposed for use in ESD and sustainability education and is established especially in Australia and New Zealand (see, e.g., Eames et al., 2010; Henderson & Tilbury, 2004). In summary, WSA was originally developed to create shorter links and more extensive cooperation between educational institutions and local societies, attempting to create opportunities to learn from each other and affect each other's agendas.

However, regardless of whether WSA has been suggested or applied in health education or sustainability education, there are very few empirical studies on how to accomplish this and even fewer studies using WSA as a theoretical lens. In a recent chapter on implementing ESD using a WSA, Gericke (2022) gives an historical overview of the literature on WSA relating to ESD for the interested reader to learn more about previous developments of WSA. What stands out in that review is that several models for working according to WSA for ESD have been proposed (see Henderson & Tilbury, 2004; Mathar, 2015; Mogren et al., 2019; Shallcross & Robinson, 2008), and recently the "flower model" by Wals and Mathie (2022) was added to the list. Common to these models is that they point out the many complex features or key components within the school organization that should be addressed to establish connections to the wider community in teaching and learning. For example, Shallcross and Robinson (2008) identify formal curricula, social and organizational aspects, institutional practice, and research and evaluation as crucial to establishing community links, while Wals and Mathie (2022) recognize pedagogy and learning, institutional practice, capacity building, curriculum and visionary leadership, and coordination as crucial components. Based on this conformity in the literature, our conclusion is that to establish a WSA to ESD in a school, there needs to be a reciprocal process, simultaneously reforming and renewing the way the school is working in alignment with the organizational mechanisms in the local school organization.

In traditional educational reforms, school developers are often asked to work in alignment with the schools' existing organization and culture (Cohen & Mehta, 2017), while the ESD discourse has proposed more radical suggestions to reform schools according to the WSA. For example, Sterling (2004) suggests a "whole system redesign," and Lotz-Sisitka et al. (2015 p73) argue that "to break with maladaptive resilience of unsustainable systems it is essential to strengthen transgressive learning and disruptive capacity-building." Due to the urgency of the sustainability issues in nature and society as we are transgressing our planetary boundaries (Steffen et al., 2015), there are of course powerful incentives to radically change and reform schools. However, we would argue that this is difficult to achieve at the local school level because schools in many countries are under the jurisdiction of national law and other regulatory documents that are only possible to influence at a higher political level than the local school organization. This situation is supported by research and practice showing that redirecting schools to the WSA for sustainability is difficult as there are few good examples (Laurie et al., 2016; Müller et al., 2021). In our research group, we have therefore worked according to a more conservative line of action and collaborated with local school organizations to support the reform within the existing educational systems. In so doing, we have looked at and taken inspiration from existing educational sciences with the specific aim of contributing to local school reforms, that is, school improvement research. We argue that much of what has been learnt in school improvement research is also valid in sustainability education and is fundamental when addressing how WSA for ESD is to be established. In the next section, we elaborate on our previous empirical research before going into a more explicit suggestion on how sustainability education can be implemented through a WSA.

4.3 Aligning Inner School Organization with Authentic Sustainability Problems

What then do we empirically know about implementing WSA from a school improvement perspective? In previous and ongoing research within our research group, we have shown that the internal organizational structures of the school (Mogren et al., 2019) and the contextual factors influencing the school (Gericke & Torbjörnsson, 2022a, b) need to be considered in the local reform agenda to accomplish a WSA to ESD.

Mogren (2019) introduced a way to contribute to this gap by linking school improvement research to ESD research. The link between the two provides a way to understand how a WSA to ESD is established based on school improvement research tools and theories, where theoretical models from the field of school improvement embracing the whole school system can be used to investigate education and contribute further understanding of the anchoring process of ESD in school organizations. In Mogren's (2019) overall thesis project, school leaders and teachers from ten of the most recognized ESD schools at the upper secondary level in Sweden, supposedly working according a WSA, participated in a combined interview and questionnaire study. The aim of the project was to contribute new knowledge on ways that school organizations can implement ESD while aligning with a WSA. In this overall project, the school leaders were interviewed and asked to identify the quality criteria for implementing ESD (Mogren & Gericke, 2017a). Clustering the schools' practices according to ESD implementation strategies, based on these criteria, resulted in four main principles for ESD implementation according to WSA: collaborative interaction and school development, student-centered education, cooperation with local society, and proactive leadership and continuity (Mogren & Gericke, 2017b). Moreover, the results showed that the schools focused either on the internal part of the organization, that is, the dimension of routines and structures, or on the external community links, that is, professional knowledge creation. A conclusion from these studies was that a main problem in implementing WSA for the participating schools was to align the internal organizational aspects with an educational approach emphasizing authentic ESD learning by using community links.

The finding that schools trying to implement ESD and work according to a WSA had difficulties in aligning the school organization with the wider society is very problematic, as the literature claims that this is the most important goal of a WSA to ESD (Henderson & Tilbury, 2004; Shallcross & Robinson, 2008; Wals & Mathie, 2022). One reason for this problem could be that the WSA–ESD relationship looks different in the case of sustainability education compared with more conventional topics often addressed in local school reforms, such as increasing interest in science or improving learning outcomes in mathematics (Gericke & Torbjörnsson, 2022a). In the latter cases, what is to be achieved is static, that is, change in attitudes or change in learning outcomes. In ESD, however, the goal is much more complex and fluid, for example, to provide students with an action competence to deal with complex sustainability problems (Olsson et al., 2020). In such problems, issues of economy, society, and environment are entangled, and students are supposed to learn how to decide and act using various kinds of authentic knowledge and values depending on the problem at hand (Berglund & Gericke, 2018, 2022). Hence, it is problematic when schools often run the same ESD themes every year in the whole school without assessing their relevance to students and current society demands and, therefore, miss out on the relevance of education to the outer society, as well as making sustainability education relevant to the students (Henderson & Tilbury, 2004). In the schools we studied, staff and students often collaborated on ESD without connecting to society, and they were often top-down steered by the school leader. Therefore, ESD was superficial and not implemented in the schools' regular educational program (Mogren & Gericke, 2019). To conclude, the issue of aligning the school organization with the outer society seems to be crucial to schools wanting to work according to WSA.

Returning to the ten investigated WSA to ESD active schools, we found that the school leaders' responses in nine schools could easily be linked to an internally or externally oriented way of applying ESD in their school organizations (Mogren & Gericke, 2017b). Only one school managed to combine the two strategies by integrating a focus on student-centered education with cooperation with local society. In this school, connections with organizations and stakeholders in the society were a responsibility for the students themselves in their education. A major aspect of this school's historical development to practice a WSA to ESD revealed by the study was that it had undergone something of a transition from proactive leadership established by its previous school leader toward a more student-centered education-oriented style of leadership by the current school leader (Mogren & Gericke, 2017b). Hence, the importance of school leadership to work locally with a WSA to ESD cannot be overemphasized (Mogren & Gericke, 2019).

Interestingly, these studies at the school leader level were later confirmed by a survey study with the teachers at the same schools. In the teacher study (Mogren et al., 2019), a survey instrument was used covering a model of four theoretical dimensions of the school organization: (1) whether the school had an articulated common holistic vision of their aims concerning student outcomes, and the pedagogic methods and perspectives that should be applied to realize the vision (the holistic idea dimension); (2) how the school upheld the inner school organization (the routines and structures dimension); (3) how the school accommodated the organization in relation to changes in the local community (the professional knowledge creation dimension); and (4) how the teachers and students should create learning platforms together, handling situations in the learning interaction in a manner promoting both the students' learning of new knowledge and the teachers' teaching ability (the teaching and learning dimension). The teacher study confirmed that the schools in which the staff managed to integrate and align the inner school organization and culture with the needs of the outer society, according to the teachers, also increased the quality of teaching and learning in their schools (Mogren et al., 2019). The studies above, based on school improvement research, are good empirical examples of the need to align the inner school organization with the outer society, which therefore seems to be crucial for schools working according to a WSA to ESD.

The findings in our previous studies are also in line with the findings of Leo and Wickenberg (2013) in their case studies on ESD-active schools. Recently, Verhelst et al. (2020, 2023) made a review on educational management in relation to ESD. Based on that review, they developed a conceptual framework for an ESD-effective school organization in terms of what organizational characteristics are of importance to implement ESD effectively within the school organization (Verhelst et al., 2020, 2023). Although this framework does not address WSA to ESD, it still identifies supportive relations to outer society as an important aspect of an ESD-effective school organization, which further strengthens the argument that ESD needs to be related to outer society when addressed from a WSA perspective. Studies of ESD school organizations and WSA models give theoretical and empirical evidence for involving the whole school in relation to its surrounding community, but they say little about how to accomplish this in practice. The studies discussed in this and previous sections give no practical insights into how to implement WSA to ESD while aligning it with the development process of the local school organization. What we know is that this process takes time and have many hurdles to overcome, as observed by Müller et al. (2021 p12) in their review of the field: "Introducing ESD in schools is a marathon, not a sprint." In the next section, we outline this long process, and show how empirically derived theories in school improvement research, mainly based on the work of Blossing et al. (2015) and Fullan (2007), can describe how this process can be achieved over time, and what pitfalls to look out for during the process.

4.4 Capacity Building as a Tool to Conceptualize Implementation of WSA to ESD

Previous research in the field of ESD lacks theoretical and practical tools on how to steer and implement a WSA in local school organizations. To find such tools, we turned to the field of school improvement. From a school improvement perspective, the concept of *capacity building* is often used. A useful metaphor for capacity building in schools—as societal institutions—is different forms of social capital (Hargreaves & Fullan, 2012). Capital in social and educational contexts refers to a quality or an asset that can be exchanged for any desirable educational condition or a desired change in the school. In the case of ESD or sustainability education research, we translate this into the implementation of ESD as a teaching approach and, as a consequence of this, the development of students' action competence (Gericke & Torbjörnsson, 2022b). We are thus interested in developing the capacity of local school organizations to reform their way of working in line with WSA to ESD.

School improvement refers to school-wide improvement processes that include the whole school organization and is "a distinct approach to educational change that aims to enhance students' learning outcomes as well as strengthening the school's capacity for managing change" (Hopkins, 2001 p 139). Knowledge about school improvement and institutionalization processes has accumulated since the middle of the twentieth century (see Miles et al., 1987), and today there is strong empirical evidence that even when schools are motivated to improve, many struggle to implement change or fail to sustain it (Forssten Seiser & Blossing, 2020).

School organizations are complex and dynamic, and even more so when WSA and ESD are involved, as argued above. Therefore, school improvement efforts cannot be assumed as an activity in a linear chain of causes and effects; rather, it needs to be comprehended in a larger context (Blossing et al., 2015). Hence, it is important to take a broad view of the school as an organization and acknowledge that to "succeed" (make a change), one must grasp the complexity of various factors that influence a school's capacity to improve. It is about making the complex a little less complicated and to reach a deeper understanding. Researchers such as Harris (2001) and Hopkins (2001) argue that a *school's improvement capacity* is crucial when it comes to achieving school improvement. Developing schools' self-renewing ability requires commitment, skills, and knowledge of both school leaders and the teachers. Not seldom, it also includes other professional functions in the educational system as both internal and external parts of the system shape and are shaped by each other. In this chapter, we use the four themes identified by Blossing et al. (2015) for developing or enhancing a school's capacity to improve:

- Improvement history of the school
- Infrastructure of the school organization
- Improvement processes
- Improvement roles

To understand these four themes, we address each of them in turn. Theoretically, they are possible to separate, but in practice they are intertwined. Starting with the first theme *improvement history*, every organization has a distinct history, meaning that over time organizations develop rules and routines that are relatively stable and persistent and not seldom developed into norms and social systems in the form of institutions, in this case as schools (Giddens, 1984). Consequently, activities and ideas are often no longer questioned or debated in the institution; instead, they become more or less taken for granted. Therefore, we need historical information that can help us understand why previous improvement efforts have succeeded or failed. Consequently, it is important to keep a school's improvement history alive as this provides an awareness of what actions have been enabling and constraining in former school development efforts. This kind of knowledge is important in planning and designing improvement efforts so that supportive action can be enhanced and constraining actions may be avoided or changed. In our projects with schools to implement WSA to ESD, we encountered this problem in practice. In an implementation study over 3 years, the WSA to ESD was more difficult to establish in schools with bad experiences of previous improvement projects, which is why the same design criteria and process for implementing WSA can turn out differently in schools depending on their improvement history (Gericke & Torbjörnsson, 2022a). This shows that a school's historical experiences influence how much capacity for school improvement a particular school has accumulated and has in store for the school reform at hand.

The second theme in a school's capacity to improve is *the infrastructure*. This pays attention to the whole picture of the school as the infrastructure of the school organization captures the daily social life of students, teachers, and principals concerning rules and routines as well as more cultural and unconscious aspects. The infrastructure is built on a foundation of eight specific systems that can be used for analyzing schools' social life that together constitutes the infrastructure model: *the grouping system, the communication system, the goal management system*, *the power and responsibility system*, *the decision-making system*, *the norm system*, *the reward system*, and *the evaluation system* (Blossing et al., 2015). Exploring these systems renders knowledge about how teachers and students are grouped in the school, how (or if) certain goals are discussed and managed, how the power to make decisions as well as responsibility to execute them is distributed, how teachers perceive and signal "good teaching," what behavior is rewarded or punished, and how information about the quality in the organization is evaluated. This kind of information forms an infrastructure in which the social life in a local school is performed. Moreover, it plays a significant role within school development, which then can be used as a tool in planning and leading the school improvement process. Fullan (2007) stresses the need for an infrastructure that supports the implementation of change as it is necessary that it nurtures and supports ongoing efforts, not least in terms of time enhancing collective learning and collaboration. For example, the frequency of meetings and work in teacher teams is crucial for school improvement success (Smooch & Drach-Zahvy, 2007), which signals that a lack of meetings will become a big obstacle if this is not provided within the school's infrastructure.

The third theme, *improvement processes*, focuses on different phases in a school improvement process. The first subprocess is called *initiation* (Blossing et al., 2015). This is the stage when new ideas are proposed, commitment is sought, and the local school organization hopefully begins to change. Initiation activities are more typical at the beginning but must also occur throughout the entire process. The next subprocess is *implementation*. This stage involves putting new ideas into practice. This is the subprocess that is likely to require the most efforts as it is at this point that everyone is supposed to adapt to the new idea (the reform—such as a WSA to ESD), which also includes new ways of acting (teaching) (Fullan, 2007). The last subprocess is *institutionalization*, which involves incorporating the "new" idea into the school's internal infrastructure so the "new idea" no longer is new. The three subprocesses mentioned are closely related to each other and often overlapping. But it is theoretically and practically important to distinguish the specific features of each subprocess.

The improvement processes are closely linked to the school's improvement history as the latter may show that a particular school has developed a talent for performing a rigorous initiation phase, introducing new ideas for the teachers. But the history may also reveal that the initiation phase was perceived as "the whole improvement process" resulting in an improvement process that slowly stopped and finally vanished because the process was seen as completed already after a semester. A whole school improvement process often lasts from 5 to 8 years before it can be assessed as institutionalized (Blossing et al., 2015).

To prevent the school improvement process to cease in the intensive and demanding implementation phase, the use of *improvement roles* is recommended (Blossing et al., 2015). This is the fourth and last theme of school improvement capacity and deals with the roles it can take on in a school improvement work and how these roles vary in importance during the whole process. The different roles are to be understood as extended leadership for change. A successful initiation is often lead by skilled *visionaries*, who serve a central function of communicating the "new" and of creating a fundamental understanding among the staff. Midthassel and Bru (2001) found that the degree of relevance attributed by the teachers was the main motivating factor to their involvement. This emphasizes how important it is for participants to perceive the improvement work as relevant. To survive the implementation phase, when concrete action is required and defense mechanisms arise, it is necessary to have *inventors*. Inventors are teachers who communicate how to put "the new" into practice. In our previous studies on locally implementing WSA via professional development efforts, we found that it is necessary to continuously negotiate a shared vision and build trust around the WSA during both initiation and implementation phases with the teachers to make the reform survive (Gericke & Torbjörnsson, 2022b). Listening to the inventors are the *early appliers*. These are brave teachers who go ahead and try out the inventors' innovations. There will also be a need for *drivers* especially in the implementation phase. The drivers' role is to speed up the practical work when some teachers are ignoring the improvement efforts and just want to be left alone so they can keep doing what they have been long doing. The *goal keepers* remind people of the aim of the new and the *inspectors* gather information and propose action to keep the improvement process going on the right track. Finally, the *preservers* play an important role in reminding what already works and making sure that these things do not vanish among the new idea. These roles should not be perceived as personal qualities but as professional roles or functions, and it is important that they are all represented in a school development process.

In this section, we have addressed the four themes identified by Blossing et al. (2015) for developing or enhancing a school's capacity to improve in an intended direction. Hence, by considering the themes and their underlying systems, subprocesses, and roles that are described in this section, it is possible to develop a local school reform program with the aim of establishing a WSA to ESD. In the following section, we take a chronological perspective and describe how an improvement process can develop over time. The time factor is of outmost importance when aiming for a WSA to ESD (Müller et al., 2021).

4.5 A School Improvement Process Leading to Establishing WSA to ESD

Each school improvement process has its own history, starting with its initiation and, in successful cases, ending with its institutionalization. As described above, institutionalization is a process of building in structures that facilitate educational changes in a lasting way so they can continue as stable routines of a local school's daily life. When the "new" (in this case WSA to ESD) is accepted by the users as something normal, it is incorporated into the local school organization (infrastructure) as a natural pattern or structure. This process is complex and dynamic, and the question of how institutionalization happens is not simple. Therefore, we cannot be prescriptive, but we provide a description from a theoretical perspective of how changes in school organizations can become stabilized and lasting in practice.

Based on the empirical results from our 3-year-long WSA to ESD project (Boeve-de Pauw et al., 2022; Forssten Seiser et al., 2023; Gericke & Torbjörnsson 2022a, b), we have included some recommendations in the descriptions of how we worked to drive a WSA for ESD process forward during the school improvement phases. We outline the recommendations under the respective school improvement phase. We would also like to emphasize that these phases do not follow each other in a straight line but rather overlap with each other. Included in each stage are some examples of the function of different improvement roles. We also demonstrate how the school infrastructure can be used as a tool for leading the improvement process. Finally, a well-known and well-documented school improvement history is understood as an absolute prerequisite in planning, designing, and leading the process.

4.5.1 Initiation

In terms of time, activities in the initiation phase can last from 0.5 to 1 year, depending on the context (Blossing et al., 2015). The aim in this phase is to create curiosity, interest, and knowledge of the WSA to ESD. A way of doing this can be to arrange specific WSA to ESD conferences where the whole staff is invited. Another is to engage experts giving lectures on WSA and ESD. Arranging study visits to ESD-effective schools and organizing workshops around the subject are other suitable activities in this initiation phase. This will require a coherent infrastructure that enables and nurtures these kinds of activities. A supportive norm system in the form of a positive language is crucial as well as a decision-making system providing resources for activities to create interest and engagement in WSA to ESD, and so on.

In the initiation phase, the role of the visionary is essential. A school leader may take on this role but also an engaged teacher, or even better both. If the school improvement capacity is low, and perhaps also the general knowledge of WSA to ESD, a suitable strategy could be to engage an external agent in form of an "ESD expert" for creating engagement and developing a deeper understanding. However, this strategy implies that the external change agent will be slowly removed during the improvement process and that this function should consciously be handed over to the teachers and the school leaders.

At this stage of the process, there will be progressive teachers who very soon want to try to put ESD into practice. It is of great importance to identify these teachers who then can function as inventors and early appliers. It is important that bottom-up initiatives are encouraged and supported as these teachers are likely to become the school's internal ESD experts and facilitators during the ensuing process. An important strategy in the initiation process is to connect these efforts to the school infrastructure and more specifically to the reward system. It is important to facilitate communication between the participants in the WSA to create and negotiate a common vision for the WSA to ESD. All activities that promote engagement, participation, and the development of shared responsibility are advantageous as the next phase in the process is critical and often conflict filled.

4.5.2 Implementation

In this second phase of the WSA to ESD process, the initial learning is meant to be transformed into practical work. This phase is most critical as it often means conducting teaching differently within new organizational settings (infrastructure), also connected to stakeholders in the wider society. This phase needs persistence, as research has shown that the implementation phase can last up to 7 years (Blossing et al., 2015).

In this phase, it is no longer enough that only progressive teachers have an ESD approach in their teaching as the aim is that every teacher should start to work in line with this approach. This explains why the implementation phase is often conflict-filled, as some teachers are not yet prepared to change how they teach. This means that there could be a resistance toward ESD, and working collegially and in cross-curricular themes might seem threatening to some of the teachers. Often, it is at this point that the progression starts to shift back and forth, and a united and committed ESD leadership is a prerequisite for continued progression. In our empirical studies, we found that ESD implementation does not follow a linear development but evolves in waves over time as resistance is overcome (Boeve-de Pauw et al., 2022). This is the time for the drivers to step forward. Teachers and school leaders acting as drives are necessary when defense mechanisms begin to develop. The driver's function is to ensure that everyone is capable of working according to ESD and encouraging skeptical teachers to start (or at least try) to work according to the WSA to ESD. The goal keepers are the ones reminding everyone of the schools shared vision of WSA to ESD. The function of the drivers is an example of how to enhance the goal system within the school's infrastructure. From a leading perspective, it is important to stick to the goals and to implement what was learned in the initiating phase. This involves, for example, reorganizing the school grouping system by organizing teachers in cross-curricular teams. Cooperative activities need to be promoted, and regular professional development in the form of seminars, conferences, and similar activities must continue. The aim is to also develop a collaborative culture and to challenge individual norm systems.

Being a school leader or teacher with the role as a facilitator in a local school reform toward WSA to ESD is an exposed position, as it often means challenging colleagues. This means that school resources are needed for assigning several facilitators so they as a community can handle the situations and critique that might arise. The development of an ESD facilitator group meeting regularly and providing guidance to each other on how to act is a wise strategy. A sign to heed in this phase is that teachers might start to argue within the school's norm system that "ESD is not anything new as this is the way they have always worked." This could be an indication that ESD is turning into "window dressing," meaning that things that may seem like ESD (such as decorating walls with SDGs) are not necessarily the same as conducting a WSA to ESD. Therefore, it is important that the facilitators and the school leaders meet regularly for reflection and analysis. We endorse school leaders to be operative and engaged in the actual work that takes place in the local school reform. A democratic decision-making system is also a wise strategy for establishing trustful relations, as this ensures that individuals know that they can influence the reform at hand.

4.5.3 Institutionalization

If the teachers and school leaders overcome the resistance in the implementation phase and work persistently according to a WSA to ESD for 3–7 years, the improvement process is likely to move into the institutionalization phase (Blossing et al., 2015). ESD as a WSA is assessed as institutionalized when it becomes a norm and routine in the daily work of the school. In other words, ESD is now included in the school's infrastructure and has developed into routines described as "how we do things around here." This is when the school more or less works in a way that is prescribed by the different WSA to ESD models for sustainability education (Mogren et al., 2019; Shallcross & Robinson, 2008; Wals & Mathie, 2022).

From a school leading perspective, there is no time to relax even if WSA to ESD has reached the institutionalization phase, as local schools are constantly exposed to external demands. Leading activities promoting regular professional development, collegial collaboration, and cross-curricular activities still need to be in focus—as new projects are always a competitive factor—and exert external pressure on capacity building to work in line with WSA to ESD (Gericke & Torbjörnsson, 2022a). Teachers' voluntariness to participate or not in a school improvement effort such as WSA to ESD is often justified, although not clearly expressed, by professional arguments. In many countries, it is often part of teachers' professional skills to determine how to teach and what content to focus, and to decide whether they consider it worthwhile to participate (Eilam & Trop, 2010). In cases where participation is mandatory, the engagement is often low, and there is a lack of shared responsibility. School improvement research shows how both voluntary and mandatory participation can promote or hinder school improvement. Blossing and Ertesvåg (2011) recommend that the voluntary nature be included later in the school improvement process to give the teachers the opportunity first to experience the process and current content before deciding to participate or not.

4.6 Concluding Remarks

We have in this chapter discussed that to achieve success in a WSA to ESD, schools need the capacity to improve and a shared leadership with the ability to integrate the present and the past while keeping a focus on the future. To conclude, we have shown a way to institutionalize WSA to ESD in school organizations based on school improvement theories. Institutionalization is here seen as a process of establishing organizational structures that facilitate educational changes in a lasting way so they can continue as stable routines of a local school's daily life, subsequent to having participated in a local school reform (Forssten Seiser et al., 2023). We would argue that a local school reform aiming to implement WSA to ESD has a greater possibility to succeed if it attends to the recommendations given in this chapter on how to build the capacity to improve and follows these empirically identified capacity-building guidelines. However, it is important to recognize that a time perspective as long as 5–8 years may be needed to institutionalize WSA to ESD at a local school, which is a time span that is longer than usually suggested in the ESD literature (Müller et al., 2021).

In this chapter, we have not addressed the importance of having a well-functioning evaluation system providing important information about the quality of the work that is conducted and about the social life in a local school (Fullan, 2007). Therefore, we would like to conclude by pointing out that it is important to monitor the implementation process over time to be able to adjust the process and thereby hopefully ending with a WSA to ESD aiming to empower students' action competence for sustainability, as first intended. To fully monitor the implementation efforts and understand the underlying school improvement processes, it is recommended to gather data from the school leader, teacher, and student levels (Desimone, 2009). In addition, we would recommend monitoring how school organizational aspects are related to teaching practices and student learning outcomes to fully comprehend the WSA to ESD implementation process. Furthermore, data should be collected longitudinally during the implementation process (Boeve-de Pauw et al., 2022) and preferably used in a stimulated recall design in the continuing negotiating process with the stakeholders of the schools (Gericke & Torbjörnsson, 2022b). These basic ideas have been used in efforts to evaluate attempts to institutionalize WSA to ESD (see Boeve-de Pauw et al., 2022; Forssten Seiser et al., 2023; Gericke & Torbjörnsson 2022a, b; Olsson et al., 2022). Much more could be said about these issues, but that is a story for another chapter.

References

Berglund, T., & Gericke, N. (2018). Exploring the role of the economy in young adults' understanding of sustainable development. *Sustainability, 10*(8), 2738. https://doi.org/10.3390/su10082738

Berglund, T., & Gericke, N. (2022). Diversity in views as a resource for learning? Student perspectives on the interconnectedness of sustainable development dimensions. *Environmental Education Research, 28*(3), 354–381. https://doi.org/10.1080/13504622.2021.1980501

Blossing, U., & Ertesvåg, S. K. (2011). An individual learning belief and its impact on schools' improvement work–An individual versus a social learning perspective. *Education Inquiry, 2*(1), 153–171. https://doi.org/10.3402/edui.v2i1.21970

Blossing, U., Nyen, T., Söderström, Å., & Hagen Tønder, A. (2015). *Local drivers for improvement capacity*. Springer.

Boeve-de Pauw, J., Olsson, D., Berglund, T., & Gericke, N. (2022). Teachers' ESD self-efficacy and practices: A longitudinal study on the impact of teacher professional development. *Environmental Education Research, 28*(6), 867–885. https://doi.org/10.1080/13504622.2022.2042206

Cohen, D. K., & Mehta, J. D. (2017). Why reform sometimes succeeds: Understanding the conditions that produce reforms that last. *American Educational Research Journal, 54*(4), 644–690. https://doi.org/10.3102/0002831217700078

Desimone, L. (2009). Improving impact studies of teachers' professional development: Toward better conceptualizations and measures. *Educational Researcher, 38*(3), 181–199. https://doi.org/10.3102/0013189X08331140

Eames, C., Barker, M., Wilson-Hill, F., & Law, B. (2010). *Investigating the relationship between whole-school approaches to education for sustainability and student learning. A summary*. New Zealand Council for Educational Research.

Eilam, E., & Trop, T. (2010). ESD pedagogy: A guide for the perplexed. *The Journal of Environmental Education, 42*(1), 43–64. https://doi.org/10.1080/00958961003674665

Forssten Seiser, A., & Blossing, U. (2020). Actions and practice architectures for realising sustainable development by restructuring school organizations. *Forskning og forandring, 3*(2), 69–88. https://doi.org/10.23865/fof.v3.2457

Forssten Seiser, A., Mogren, A., Gericke, N., Berglund, T., & Olsson, D. (2023). Developing school leading guidelines facilitating a whole school approach to education for sustainable development. *Environmental Education Research, 29*(5), 783–805. https://doi.org/10.1080/13504622.2022.2151980

Fullan, M. (2007). *The new meaning of educational change* (4th ed.). Routledge, Teacher College Press.

Gericke, N. (2022). Implementation of education for sustainable development through a whole school approach. In G. Karaarslan Semiz (Ed.), *Education for sustainable development in primary and secondary schools: Pedagogical and practical approaches for teachers* (Sustainable Development Goals Series) (pp. 153–166). Springer. https://doi.org/10.1007/978-3-031-09112-4_11

Gericke, N., & Torbjörnsson, T. (2022a). Identifying capital for school improvement: Recommendations for a whole school approach to ESD implementation. *Environmental Education Research, 28*(6), 803–825. https://doi.org/10.1080/13504622.2022.2045256

Gericke, N., & Torbjörnsson, T. (2022b). Supporting local school reform toward education for sustainable development: The need for creating and continuously negotiating a shared vision and building trust. *The Journal of Environmental Education, 53*(4), 231–249. https://doi.org/10.1080/00958964.2022.2102565

Giddens, A. (1984). *The constitution of society: Outline of the theory of structuration*. Polity Press.
Hargreaves, A., & Fullan, M. (2012). *Professional capital: Transforming teaching in every school*. Teachers College Press and Ontario Principals' Council.
Harris, A. (2001). Building capacity for school improvement. *School Leadership and Management, 21*(3), 261–270. https://doi.org/10.1080/13632430120074419
Henderson, K., & Tilbury, D. (2004). *Whole school approaches to sustainability: An international review of sustainable school programs*. Report prepared by Macquarie University for the Australian Government Department of the Environment and Heritage.
Holst, J. (2023). Towards coherence on sustainability in education: A systematic review of whole institution approaches. *Sustainability Science, 18*(2), 1015–1030. https://doi.org/10.1007/s11625-022-01226-8
Hopkins, D. (2001). *School improvement for real*. Routledge Falmer.
Laurie, R., Nonoyama-Tarumi, Y., Mckeown, R., & Hopkins, C. (2016). Contributions of education for sustainable development (ESD) to quality education: A synthesis of research. *Journal of Education for Sustainable Development, 10*(2), 226–242. https://doi.org/10.1177/0973408216661442
Leo, U., & Wickenberg, P. (2013). Professional norms in school leadership: Change efforts in implementation of education for sustainable development. *Journal of Educational Change, 14*(4), 403–422. https://doi.org/10.1007/s10833-013-9207-8
Lotz-Sisitka, H., Wals, A. E. J., Kronlid, D., & McGarry, D. (2015). Transformative, transgressive social learning: Rethinking higher education pedagogy in times of systemic global dysfunction. *Current Opinion in Environmental Sustainability, 16*(17), 73–80. https://doi.org/10.1016/j.cosust.2015.07.018
Mathar, R. (2015). A whole school approach to sustainable development: Elements of education for sustainable development and students' competencies for sustainable development. In R. Jucker & R. Mathar (Eds.), *Schooling for sustainable development in Europe* (pp. 15–30). Springer.
McKeown, R., & Hopkins, C. (2007). Moving beyond the EE and ESD disciplinary debate in formal education. *Journal of Education for Sustainable Development, 1*(1), 17–26. https://doi.org/10.1177/097340820700100107
Midthassel, U. V., & Bru, E. (2001). Predictors and gains of teacher involvement in an improvement project on classroom management. Experiences from a Norwegian project in two compulsory schools. *Educational Psychology, 21*(3), 229–242. https://doi.org/10.1080/01443410120065441
Miles, M., Ekholm, M., & Vandenberghe, R. (1987). *Lasting school improvement: Exploring the process of institutionalization*. Acco.
Mogren, A. (2019). *Guiding principles of transformative education for sustainable development in local school organizations: Investigating whole school approaches through a school improvement lens* [Doctoral thesis, Karlstad University Studies].
Mogren, A., & Gericke, N. (2017a). ESD implementation at the school organization level, part 1—investigating the quality criteria guiding school leaders' work at recognized ESD schools. *Environmental Education Research, 23*(7), 972–992. https://doi.org/10.1080/13504622.2016.1226266
Mogren, A., & Gericke, N. (2017b). ESD implementation at the school organization level, part 2—investigating the transformative perspective in school leaders' quality strategies at ESD schools. *Environmental Education Research, 23*(7), 993–1014. https://doi.org/10.1080/13504622.2016.1226266
Mogren, A., & Gericke, N. (2019). School leaders' experiences of implementing education for sustainable development—Anchoring the transformative perspective. *Sustainability, 11*(12), 3343. https://doi.org/10.3390/su11123343
Mogren, A., Gericke, N., & Scherp, H.-Å. (2019). Whole school approaches to education for sustainable development: A model that links to school improvement. *Environmental Education Research, 25*(4), 508–531. https://doi.org/10.1080/13504622.2018.1455074
Müller, U., Hancock, D. R., Stricker, T., & Wang, C. (2021). Implementing ESD in schools: Perspectives of principals in Germany, Macau, and the USA. *Sustainability, 13*, 9823. https://doi.org/10.3390/su13179823
Olsson, D., Gericke, N., Sass, W., & Boeve-de Pauw, J. (2020). Self-perceived action competence for sustainability: The theoretical grounding and empirical validation of a novel research instrument. *Environmental Education Research, 26*(5), 742–760. https://doi.org/10.1080/13504622.2020.1736991
Olsson, D., Gericke, N., & Boeve-de Pauw, J. (2022). The effectiveness of education for sustainable development revisited - a longitudinal study on secondary students' action competence for sustainability. *Environmental Education Research, 28*(3), 405–429. https://doi.org/10.1080/13504622.2022.2033170
Rowe, F., & Stewart, D. (2009). Promoting connectedness through whole-school approaches: A qualitative study. *Health Education, 109*(5), 396–413. https://doi.org/10.1108/09654280910984816
Shallcross, T., & Robinson, J. (2008). Sustainability education, whole school approaches, and communities of action. In A. Reid et al. (Eds.), *Participation and learning* (pp. 299–320). Springer.
Somech, A., & Drach-Zahavy, A. (2007). Schools as team-based organizations: A structure-process-outcomes approach. *Group Dynamics: Theory, Research, and Practice, 11*(4), 305–320. https://psycnet.apa.org/doi/10.1037/1089-2699.11.4.305
Steffen, W., Broadgate, W., Deutsch, L., Gaffney, O., & Ludwig, C. (2015). The trajectory of the Anthropocene: The great acceleration. *The Anthropocene Review, 2*(1), 81–98. https://doi.org/10.1177/2053019614564785
Sterling, S. (2004). Higher education, sustainability, and the role of systemic learning. In P. B. Corcoran & A. E.

J. Wals (Eds.), *Higher education and the challenge of sustainability: Problematics, promise, and practice* (pp. 49–70). Springer.

Verhelst, D., Vanhoof, J., Boeve-de Pauw, J., & Van Petegem, P. (2020). Building a conceptual framework for an ESD-effective school organization. *The Journal of Environmental Education, 51*(6), 400–415. https://doi.org/10.1080/00958964.2020.1797615

Verhelst, D., Vanhoof, J., & Van Petegem, P. (2023). School effectiveness for education for sustainable development (ESD): What characterizes an ESD-effective school organization? *Educational Management Administration and Leadership, 51*(2), 502–525. https://doi.org/10.1177/1741143220985196

Wals, A. E. J., & Mathie, R. G. (2022). Whole school responses to climate urgency and related sustainability challenges. In M. A. Peters & R. Heraud (Eds.), *Encyclopedia of educational innovation*. Springer. https://doi.org/10.1007/978-981-13-2262-4_263-1

Wyn, J., Cahill, H., Holdsworth, R., Rowling, L., & Carson, S. (2000). MindMatters, a whole-school approach promoting mental health and wellbeing. *Australian and New Zealand Journal of Psychiatry, 34*(4), 594–601. https://doi.org/10.1080/j.1440-1614.2000.00748.x

Getting to a Whole School Approach: Lessons From School Effectiveness and School Improvement in ESD Research

5

Dries Verhelst, Anna Mogren, Jelle Boeve-de Pauw, and Peter Van Petegem

Key Message

The key message of this book chapter is that implementing a whole school approach (WSA) for education for sustainable development (ESD) is challenging for schools, and an enabling organizational context is essential for its success. The chapter suggests that incorporating insights from school effectiveness and improvement research can help bridge this gap and provide a better understanding of how to implement an effective WSA for ESD.

5.1 Introduction and Questions to Answer

In this chapter, we will explore how knowledge from the fields of school effectiveness and school improvement can support and challenge the study of whole school approaches (WSA). Understanding schools as organizations in implementing and sustaining a WSA to education for sustainable development (ESD) calls for a combined effectiveness and improvement perspective on the school, that is not inclined nor predisposed toward one or the other but maintains the strengths from both perspectives. In our attempt to shed light on this issue, we will visit horticulture metaphors while exploring what the seeds that get planted in schools need to grow into a healthy whole school approach, and ultimately support their students to build action competence for sustainable development.

Many scientific research and policy documents focus on the implementation of a whole school approach to effectively embed sustainability education into schools (e.g., Henderson & Tilbury, 2004; Mogren et al., 2019; UNESCO, 2020). Research has shown that collaboration, participation, and holistic school approaches, all important aspects of a whole school approach, lead to more effective learning, compared to hierarchical and individual learning environments. The Environment and School Initiatives (ENSI) network has a long history of investigating ESD practical interventions in terms of "what works" in praxis in various school contexts (Breiting et al., 2005; Mogensen & Schnack, 2010). In sustainability education, a WSA offers a theoretical perspective for a comprehensive approach (Hargreaves, 2008) guided by the following principles: structural support from the school organization, a holistic idea, internal structures and quality processes in a school's organization,

D. Verhelst (✉) · P. Van Petegem
Department of Training and Education Sciences, University of Antwerp, Antwerpen, Belgium
e-mail: Dries.Verhelst@uantwerpen.be

A. Mogren
Department of Pedagogical Studies, Karlstad University, Karlstad, Sweden

J. Boeve-de Pauw
Freudenthal Institute, Utrecht University, Utrecht, The Netherlands

A. E. J. Wals et al. (eds.), *Whole School Approaches to Sustainability*, Sustainable Development Goals Series, https://doi.org/10.1007/978-3-031-56172-6_5

student-centered planning and management, and proactive leadership (Mogren, 2019). Policy on the implementation of ESD also follows the perspective of a WSA as UNESCO highlights the importance of such an approach in their roadmap to ESD (UNESCO, 2020). Yet, it is not evident how to put this theoretical perspective into practice, making it difficult for a school organization to adopt a whole school approach to sustainability education. Like a gardener who wants to transform their backyard from a sleek lawn into a flourishing, sustainable, and thriving garden, the school organization adopting a WSA to sustainability education will need, in addition to the theoretical understanding, the right tools, hands-on practical knowledge, and room for trial and error. Inspiration for these tools can be found—or at least sought—in empirical educational effectiveness and improvement research. Focusing on educational effectiveness and improvement can aid in implementing a WSA to ESD as a specific way of school organizational development and improvement.

Much of the research interest in ESD has, until now, been oriented to how curricula and pedagogy are the basis for defining and shaping a whole school approach toward ESD (Boeve-de Pauw & Van Petegem, 2018; Hargreaves, 2008; Kuzmina et al., 2020), or in gardening terms: What plants and fruits do we want and how should we cultivate them. The aim of such studies is often to describe interconnections between different aspects of the WSA, teaching practices, and students' sustainability literacy. As they do not aim to nor permit investigation as to how such interconnections come to be or can be changed within the organizational reality, they remain distant from the practitioners trying to create a resourceful organizational context that fosters the implementation of a WSA. Despite the recognition by a number of ESD scholars (Mogaji & Newton, 2020; Mogren, 2019; Scott, 2013; Verhelst et al., 2020), the elements that shape such an organizational context remain fairly uncharted terrain for ESD research (Kuzmina et al., 2020; Verhelst et al., 2020). As the WSA essentially touches upon the school as an organization, attempting to implement a WSA without understanding how the school organization can enable this, would be the same as attempting to grow a garden without understanding the ecosystem it will grow in. In this chapter, we argue that the challenge of implementing a WSA is where we, as a research community, should direct our attention. By doing so, research can aid practitioners trying to understand their school organization and shape an enabling organizational context wherein a WSA to ESD can be implemented and anchored.

A first question that arises here is how this school organization should look and act in order to effectively implement a WSA to sustainability education. A promising, yet somewhat disregarded, perspective on the role of the school organization can be found within traditional school effectiveness research. Here, the school organization has been given an important role in facilitating educational effectiveness, thus recognizing its function as an enabler of effective education (Creemers & Kyriakides, 2010; Reynolds et al., 2014). The established field of school effectiveness has been going through a decades-long transformation, evolving into a synthesis of both school effectiveness (SE) and school improvement (SI) perspectives (Reynolds et al., 2014). This evolution combines the more formal, quantitative, organizational perspective of SE with the qualitative nature of SI and its focus on school culture and processes. Being a relatively young research area, ESD holds unique potential as it can benefit from the lessons learned in both fields of school effectiveness and school improvement in order to introduce ESD in schools in an effective manner that enables schools to embark on a journey of self-improvement. In fact, scholarly work on sustainability education and the role of the school organization within ESD applied aspects of both the SE and SI perspectives to their methodologies, thus making a first step toward a better understanding of how the school organization facilitates the implementation and outcomes of ESD (Mogren, 2019; Verhelst, 2021). Coming from an SI perspective, Mogren (2019) links ESD implementation to the school's general educational qualities and quality assurance process. The general quality criteria that are given specific importance to ESD-active schools are collaborative interaction, school improvement, student-centered education, cooperation with the local community, and proactive leadership (Mogren & Gericke, 2017). Coming from an SE perspective, Verhelst et al. (2020) developed a framework with a strong focus on tangible organizational characteristics of a school facilitating ESD effectiveness. Placing both perspectives into the equation fosters our understanding of how the organizational characteristics of a school enable an effective implementation of a whole school approach toward sustainability education.

Second, alongside the understanding of the organizational school context wherein the WSA has to be implemented, the everyday reality of education and schooling has to be taken into account (Mogren et al., 2019). Schools exist in a societal reality in which their organizational capacities can easily be overburdened. Simply adding sustainable development and sustainability education to schools' agenda is not only unfair but also its feasibility can be questioned if we want education to have an impact. Scott's (2009) question, "What can education do for sustainable development?" is more easily answered than put into practice. WSAs toward sustainability rely on the organizational capacities of the schools and the people within them. Although education plays a key role in anchoring sustainability in our societies, we should not unfairly assume that the entire responsibility for this lies with education and schools. Keeping this in mind, the second question Scott (2009) formulated might be more important: "What can sustainable development do for education?" As researchers, we ought to be conscious "that schools exist to educate young people, and they are not primarily agencies to drive sustainable development or any other social process" (Scott, 2009, p. 38). An understanding of the opportunities, pitfalls, and processes within a school organization is essential for an effective implementation of a whole school approach toward sustainability education.

5.2 The School Organization

A school transitioning toward a whole school approach for sustainable development requires the commitment of every actor at every level within that school: in the classroom with the students and teachers, the management, maintenance, etc. (Wals & Mathie, 2022). While essential for a successful and well-anchored WSA, the integrated coordination of all these actors is not self-evident. Simply assuming that a WSA will arise spontaneously if all the elements within the school decide on it would be the same as if our gardener decides that the neatly mown lawn has to spontaneously grow into a lush and biodiverse garden. If the gardener lets time pass, all sorts of things will indeed start to grow in the garden. However, after a certain period of time, they will notice that some plants wither, while others proliferate. They will have to consider how they can organize the different elements in their garden so that all species can thrive in an integrated and harmonic relationship.

A school implementing a WSA will have to look at different aspects within its organization and aim at effectively developing the necessary characteristics to enable a WSA to be anchored and implemented. Before thinking about how a school organization can facilitate the implementation of a WSA to sustainability, we need to agree about what we mean when we say "school organization." Whereas Wals and Mathie (2022) describe the flower model for a WSA which places vision, ethos, leadership, and coordination at the center, we hold that a wider view of the school as a professional organization with its own structures, resources, and aims offers a more fine-grained understanding of how the organization contributes to anchoring a WSA. With this, we do not intend to disconnect a WSA to ESD from the school organization, but rather argue that a WSA to ESD has its own specific orientation within the school organization. Also, aspects such as the school culture and ethos often remain hidden within the routines, structures, and ways of operating within the school and, therefore, might not be very tangible and thus hard to change for practitioners (Cameron & Quinn,

2011). In this organizational perspective, the school organization is a part of a whole school approach, but not the other way around. The school organization can be seen as the toolshed of the gardener, with all of the sowing seeds, seedlings, gardening tools, fertilizer, and other materials they will need to grow the garden. The gardener might even keep books on gardening and a sowing schedule in there. From a distance, one might find all the elements of a flourishing garden in that shed, but nonetheless, a lot of work and effort will still be needed in order to sow all the seeds and utilize all the tools and ideas. In a similar way, the school organization is the toolshed for a whole school approach: All elements of a WSA can be identified within the school organization, but in order for the school to be transformed toward a WSA, the organizational level of the school has to enable it.

Empirical research has shown that the school as an organization can enable effective sustainability education (Verhelst et al., 2022a). The framework of an ESD-effective school organization has been linked to student outcomes in ESD, namely the action competence of students in sustainable development. It offers a comprehensive overview of the different characteristics (i.e., tools in the toolshed) of a school for anchoring a WSA to sustainability education (Verhelst et al., 2022a). Before thinking about how to implement a WSA, we will visit each of the tools in our shed and have a look at the foundation on which the shed has been built (Fig. 5.1). The framework for which this toolshed is a metaphor is based on the conceptual literature study by Verhelst et al. (2020).

The framework distinguishes two levels within the school organization and consists of eight characteristics. On the first level, sustainable leadership and the resources of the school shape the field on which the rest of the organization can be built. Sustainable leadership is distinguished by a holistic, integrated view of the past, present, and future, both locally and globally, when setting out a strategic course for the school organization (Hargreaves & Fink, 2006). It will actively initiate and advance all characteristics of the organization so that continuous learning and development contribute to achieving the long-term holistic strategy of the school organization (Fullan, 2006; Hargreaves & Fink, 2006). Such leadership does not have to arise solely from the person of the school leader; multiple people within the school can take on sustainable leadership roles, regardless of their formal role within the organization (Verhelst et al., 2020). Nevertheless, support from the formal leadership within the school is often a requirement (Verhelst et al., 2021). Along with sustainable leadership, school resources make up the foundation of the organization. School resources refer to the available time, the people within the school organization, and the physical means such as infrastructure and finances. Sustainable leadership and the school's resources are in some way the foundation on which our toolshed stands. It is crucially important for the school as an organization that this foundation guarantees that the ground on which the shed is built "will not shift or be eroded with" in order to lock in the organizational characteristics of the school (Scott, 2013, p. 189).

In the toolshed of an ESD-effective school, six characteristics of a school organization can be found at the second level, all of which can be used as instruments in establishing a WSA. First, *pluralistic communication* is a reflective and critical dialogue among actors within the school organization. This style of communication leaves room for different opinions and perspectives and actively seeks to understand the perspective of others, even when this conflicts with one's own ideas. Next, an ESD-effective school promotes *supportive relations* among members of the school team, between the school and other schools, and with external partners. *Democratic decision-making* is a very powerful tool in anchoring a WSA within the school organization. By involving all relevant stakeholders in the decision-making process at the level of the school organization, the implementation and continuous efforts for ESD could face less resistance. Furthermore, a *shared vision* provides the school organization with a comprehensive sustainability education and offers the motivational basis within the school team to keep investing in it. Then, *adaptability* enables the school organization to

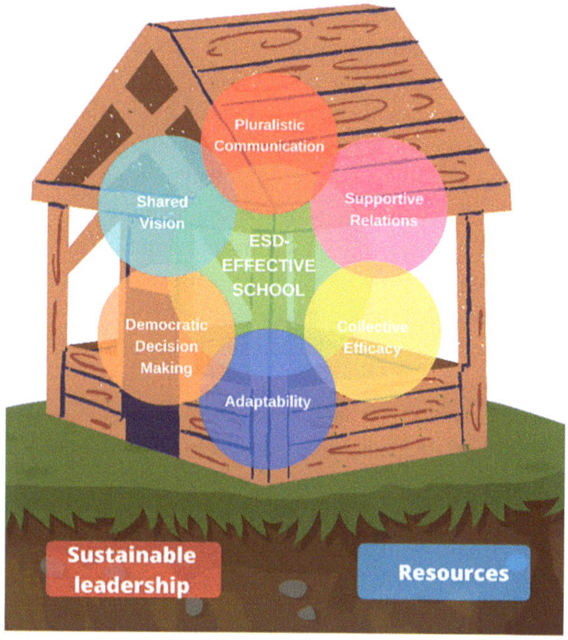

Fig. 5.1 The tool shed for an ESD-effective school

adequately adapt the organizational functioning to internal and external demands as necessary. The final tool in the shed is *collective efficacy,* referring to the confidence of the school team in their capacities to positively affect student learning related to sustainability.

We envision these eight characteristics as the foundation and tools for implementing a WSA for sustainability education within schools. The manifestation of these organizational characteristics offers an insight into the organizational culture of a school. An enabling organizational context, that could be created via the characteristics described above, is essential to implementing any school-wide approach, in addition to—or even as a condition for—the pedagogical aspects of sustainability education.

5.3 Use and Potential of School Effectiveness Studies in ESD

Implementing a WSA as a "continual reflexive process a school, with all its stakeholders, embarks on and commits to" (Wals & Mathie, 2022, p. 4), will require that all organizational characteristics (i.e., the tools in our toolshed) are directed at achieving that long-term strategic objective (Nikel & Lowe, 2010). "The degree of effectiveness is the extent to which stated (educational) aims are met" (Nikel & Lowe, 2010, p. 595). So an effective implementation of a WSA would mean that a school is able to "realize a holistic, systemic and collective effort toward sustainability as agreed upon by those affected" (Wals & Mathie, 2022, p. 2). When talking about a school striving for a strategic objective, the perspective of school effectiveness provides an array of opportunities to better understand the school's organizational functioning and outcomes. This would also be the case if that objective is implementing a WSA to sustainability. While offering ample opportunities for the research field, school effectiveness in sustainability education remains a divisive topic. Some argue that the nature of ESD, resolving around complex topics that involve different and potentially conflicting belief systems and values, makes a universal approach toward "good" education impossible (Van Poeck et al., 2018). If we interpret this narrowly, it implies that school effectiveness research, in which the achievement of predetermined educational goals or learning outcomes are used as a measure of the impact (effect) of educational efforts, is not possible within the domain of ESD since it would not be possible to put forward predetermined learning outcomes for ESD. Others propose conceptualizations of possible learning outcomes for ESD (e.g., Gericke et al., 2019; Olsson et al., 2020; Sass et al., 2021) which do allow for an exploration of the applicability of school effectiveness research in this domain, as we will argue below.

So, notwithstanding the above caution toward effectiveness research, the call for empirical evidence concerning the effects and outcomes of ESD exists and the argument that researching sustainability education ought to contribute to an understanding of what makes such an education effective is commendable (Boeve-de Pauw et al., 2015; Boeve-de Pauw & Van Petegem, 2018; Laurie et al., 2016; Verhelst et al., 2022a). There are, however, differences between the traditional school effectiveness perspective and the way it has been operationalized in research on school effectiveness in ESD. School effectiveness research studies which factors make schools "effective" (Nikel & Lowe, 2010), identifying those that influence the learning outcomes in relation to the objectives of the school and its education (Chapman et al., 2016). Traditionally, school effectiveness sought to identify the characteristics and factors that make schools effective in contributing to student learning (Chapman et al., 2016). For schools striving for largely similar goals, this would not be a notable issue since they could compare their organizations to one another to fine-tune them for the most effective configuration (Leithwood & Louis, 2006). Implementing a WSA toward sustainability requires an approach in which schools take their own specific context and goals into account. Since this is something that has to be shaped by everyone involved in the school, there will be little chance that a universal solution, appropriate for every school, exists. Yet, contemporary school effectiveness research is more conscious of a cultural understanding of schools' organizational processes, in contrast to a somewhat outdated focus on formal organizational elements (such as teacher–student ratio), with a shift from "input/output" to "input/process/output" (Reynolds et al., 2014). Consequently, an empirical and effectiveness-focused perspective can very well provide insight into the school's own processes and characteristics and give an indication to schools about how firm their ground is for the implementation of a WSA to sustainability education.

Identifying the tools (i.e., the different organizational characteristics, see Fig. 5.1) a school organization can invest in when implementing a WSA and having a clear understanding of the school's current characteristics, provides a good starting point to agree upon the baseline of a school's necessary preconditions for a WSA on ESD. Important here is that the focus of this effectiveness perspective is on the characteristics that the school organization intends to use in order to attain its goals. Like a gardener who checks that they have a good rake and enough potting soil, the school organization verifies that all the organizational characteristics required to implement a WSA are in place at the school. Recent studies have not only indicated that it is possible to measure the school characteristics of an effective school organization for sustainability education (Verhelst et al., 2022b) but they have also connected the characteristics of an ESD-effective school to action competence in ESD as a learning outcome for sustainability education (Verhelst et al., 2022a).

By making the organizational characteristics of a school implementing a WSA to sustainability education explicit, schools can be offered insight into their specific school characteristics, processes, and policies (Verhelst et al., 2022b). Analyzing the organization prior to implementing the change makes it possible to find out where the strengths of the school lie and which qualities they possess to transform their sustainability education into a school-wide approach. At the same time, the school can also gain insight into possible barriers for such a transformation (Verhulst & Lambrechts, 2015). Establishing such a baseline can provide a valid starting point for self-reflection and helps schools to set their own goals for organizational development. Knowledge about the starting position and the intended progress is especially essential since the lack of such knowledge could lead to a lack of consensus on ESD within the organization (Mogren, 2019). Mapping the starting position for implementation creates clarity and consensus regarding the expectations. If the gardener puts their seeds in the ground without checking what kind of substrate it is or which places get much or little sunlight, they cannot give the seeds every opportunity to grow. That said, identifying the requirements for effectiveness is a necessary but not sufficient condition for the school to develop and improve on this knowledge.

5.4 Use and Potential of School Improvement Studies in ESD

Even with all their tools in the toolshed, our gardener will still need to work with and understand the context of their garden. A lot of gardening needs to be done; by using the different tools in the toolshed and putting them to use toward the same goal, the gardener will reap the fruits of their efforts (Fig. 5.2). Likewise, the school has to put forth equivalent effort. The different organizational characteristics are present in the school and the school has to put these different parts together toward the bigger picture: the implementation of a WSA toward ESD. These processes have been the focus of research in school improvement research, which investigates how schools improve via the investigation of processes within schools (Chapman et al., 2016).

School improvement builds on a *totum pro parte* idea, which implies that the whole school organization is more than the sum of the different aspects and characteristics within the organization. School improvement research was developed as a response to one-dimensional, reactive judgments of schools (through student outcomes) that were present in traditional school effectiveness research (Reynolds et al., 2014). However, the development of school improvement research is more oriented to a coordination of effectiveness and improvement than on replacing one with the other (Reynolds et al., 2014). From its conception onwards, school improvement advocated an approach that improves supporting factors within the organization, such as school leader and teacher processes, in order to enhance stu-

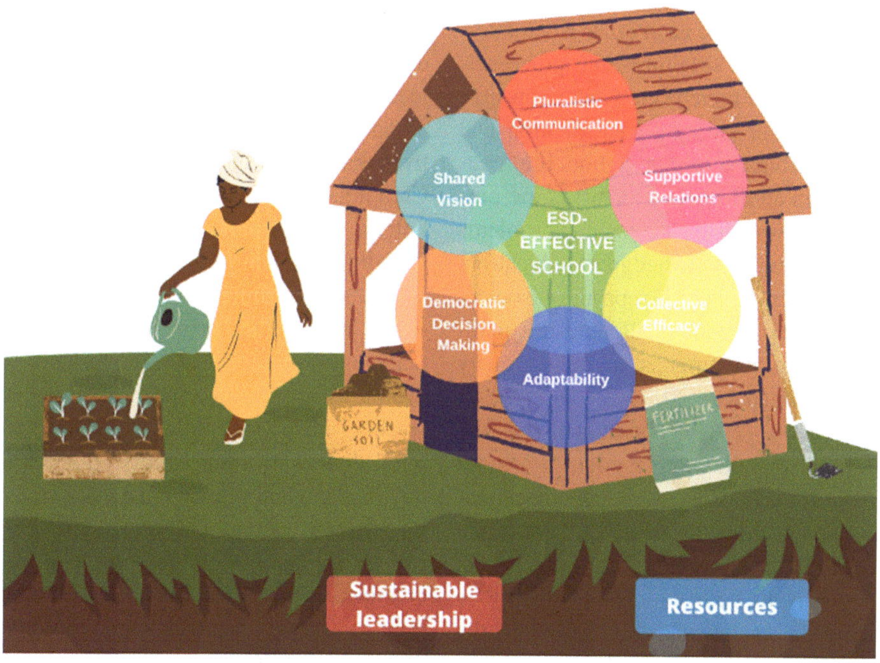

Fig. 5.2 Applying an improvement perspective on the different tools in the toolshed

dent learning while being reflective toward the contextual aspects (Creemers & Reezigt, 1997).

As argued by Von Bertalanffy (2003) and Scherp and Scherp (2007), keeping focused on the big picture of school improvement and inquiring about all contextual components that are supporting and facilitating education in schools' organizations, raised the quality in education in a more substantial way compared to what can be recognized in its individual parts. School improvement is generally put into practice through innovation and action (Fullan, 2006) rather than through the reactive action and knowledge obtained via school effectiveness research. Within ESD research, the school improvement perspective is often reflected in case studies of individual schools (Laurie et al., 2016) and international comparisons between schools and educational systems via qualitative methodologies (Iliško & Badyanova, 2014; Miškolci et al., 2016; Müller et al., 2021). School improvement research has allowed an in-depth description of good examples and role-model schools and aligns the school's own development within school improvement frameworks. The above-mentioned case studies and qualitative studies offer insight into how these existing parts of a school organization reinforce or oppose each other when interacting with contextual structures and processes, which goes beyond a mere sketch and allows for a holistic description of the transformation and improvement within the school. School improvement research in an ESD context serves the purpose of finding new solutions to societal problems and securing required knowledge in education. Education is often found to reproduce society rather than develop it. ESD challenges this paradigm and one of its aims is to continuously change education (Jucker, 2011).

Mogren et al. (2019) operationalized the Scherp school organization model, a school improvement tool, in order to aid the investigation of those educational aspects that drive transformation in ESD schools (Fig. 5.3). This model holds the ability to both understand quality in education by the established dimensions in the model and to also understand what drives increased quality in education by improvement processes. The analogy of the gardener's tools and labor provided a fitting way to understand the school effectiveness perspective, enabling us to tell which tools or which techniques contribute to more effective gardening. The Scherp school organization model, however, allows us to put on the school improvement lens to investigate the improvements in how skilled our gardener becomes over time and how well they learn to uses their tools. The quality of education as measured by the Scherp model is, in our metaphor, dependent on the gardener's skill at mastering a wide range of tools. They will increase the garden quality the better they can handle all these tools together as they become more skilled. Quality arises, for example, when they arrange plants and flowers growing together so they can share and divide sun and energy with each other in a way that makes them stronger together as an ecosystem than they would be as individual plants, similar to the holistic assumption in the Scherp model that indicates quality in education.

Looking at a school organization via the lens of the Scherp school organization model (Mogren et al., 2019), four dimensions can be distinguished: the holistic idea, routines and structures, professional knowledge creation, and the pedagogical practice. The holistic idea refers to the degree that schools have an articulated, holistic idea of their aims concerning student outcomes, along with the pedagogic methods and perspectives that should be applied to realize this vision. Creating, understanding, and implementing a holistic idea is part of the operationalization of this dimension in Scherp's model. The routines and structures dimension concerns the degree to which stability and security are maintained, and teaching is protected from disturbances through routines, scheduling, locations, and teachers' working units. The professional knowledge creation dimension refers to disturbances in everyday pedagogical practice that could indicate that the general educational (and particularly teaching) arrangements of the school do not match changes in the outer world or community, and thus should be adapted. Lastly, the pedagogical practice dimension indicates the learning and teaching situation in the school. Ideally, teachers

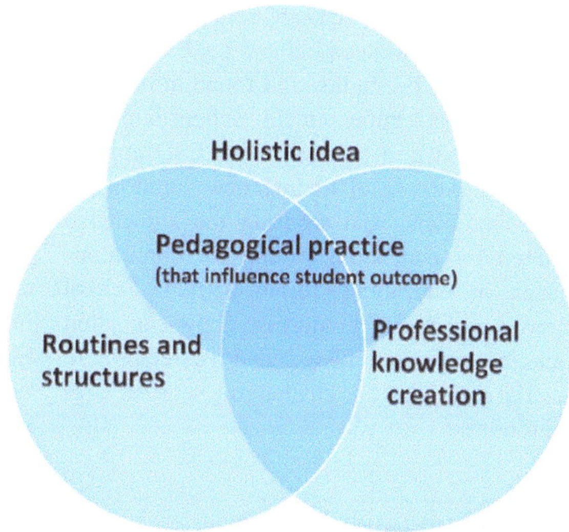

Fig. 5.3 The Scherp school organization model derived from school improvement research and introduced in an ESD context (Mogren et al., 2019)

and pupils create learning platforms together, handling situations in the learning interaction in a manner that promotes the students' learning of new knowledge. These four dimensions are indicators for quality within the school organization: The degree to which actors in the school (school leaders, teachers, and students) are motivated by a holistic idea, embracing the organizational routines and structures, professional knowledge creation, and the practical pedagogic work (Scherp, 2013). This is made visible in the model (Fig. 5.3) where the increased overlap of all the circles and an extended middle overlapping part indicates a raised quality in the school organization. The Scherp model reflects an understanding of education that Biesta (2013) and Jickling and Wals (2012) argue for—not only that education is a matter of receiving professional knowledge of known practices, but also that education should provide learners with a sense of responsibility and freedom to actively change the future of society. Thus, the Scherp model has been developed to clearly distinguish what progress in educational practice actually means in organizing schools. The dimension of professional knowledge creation stands for the degree of transformative organized education. A low connectedness to that dimension with the other dimensions of the model signals a hierarchal and structured school but one that lacks the drive for change and transformation. For instance, should our gardener not feel the urge to improve themselves when it comes to their professional gardening knowledge, and sticks to what they have always known, they may continue to use pesticides to get rid of pesky aphids. If, however, they do hold an improvement perspective toward gardening, the discrepancy between the desired results of their work (holistic idea) and their actual work (routines and structures) would lead to rethinking their professional knowledge and they would find that ladybugs are in fact also a great way to fight off aphids without hindering the pursuit of their holistic idea.

At the basis of the school improvement process lies a transformative education in which interventions and initiatives can sprout in order to foster organizational change toward educational quality. Identifying the discrepancy between any of the four dimensions of the Scherp model for school improvement acts as the starting point for initiating such improvements. The approach of transforming education by school improvement initiatives rather than merely measuring high quality via individual parts of an effective organization is thus thought to be progressive, systematic, and takes the whole school organization into account (Scheerens & Demeuse, 2005). The leading paper on WSA (Wals & Mathie, 2022) takes this a further step in its presentation of WSA, summarizing it as a tool for continuous collaborative processes of change to realize

ESD. This continuous change is often called the transformation of education and deals with processes that can be studied and found in research based on school improvement. School improvement research in a transformative ESD context serves the purpose of finding new solutions to societal problems and securing the knowledge required for education, aiming for continuous change in education (Jucker, 2011). School improvement research addresses how education could be organized in a reflective relation to material and time, thus challenging set structures and routines.

5.5 Combining School Effectiveness and School Improvement Toward a WSA

The contradictions and similarities between school effectiveness and school improvement are difficult to pin down. Both are, at the same time, closely related and yet opposed to each other. On the one hand, when school effectiveness research first emerged, researchers assumed, in a rather haughty way, that making their findings known was enough to make schools and teachers achieve effective change. Evidently, seeing teachers and schools merely as implementers of the knowledge arising from school effectiveness research, without providing the necessary ownership of this knowledge and support for its implementation, did not and does not work (Stoll et al., 2016). On the other hand, a continuous process of change within a transformative education demands a lot from the policy power and capacity of a school, policy, and support that may not be sufficiently present within the school organization. The expectation that all schools are able to shape such a trajectory sustainably is likely to be based on the same fallacy as that of the school effectiveness researchers. To illustrate it with the example of our gardener: Just telling them what the most efficient ways of gardening are or what tools to use will not enable them to apply them in the context of their own garden. At the same time, there can be no guarantees that they know completely on their own how to shape the process of effective gardening without receiving any kind of input or feedback. Improvement research would point out *how* the gardener can transform the small seedlings to floral splendor. *What* leads them to achieve the desired results is identified by effectiveness research. Striving toward a flourishing garden requires attention to both the what and the how of gardening. Overcoming the fallacies to which school effectiveness and school improvement fall victim on their own requires incorporating both approaches. The incorporation of school effectiveness and school improvement led the field of school effectiveness and improvement to evolve into an established research paradigm that generates theory via continuously developing methodologies with a strong focus on multiple measures of the outcomes and processes of "good" education (Chapman et al., 2016).

This contemporary school effectiveness and improvement perspective is starting to find its way to ESD research, despite earlier mentions of "toward effectiveness measurement in ESD." From a school effectiveness perspective, Verhelst et al. (2022a) answered the question of "what" makes a school ESD-effective in the form of tangible organizational characteristics of a school. The framework for an ESD-effective school, as described in Sect. 5.2 of this chapter, describes what makes a school ESD-effective based on conceptual foundations, qualitative empirical support, and quantitative statistical grounding. First, it provides a clear framework, useful for both research and practice, describing various organizational characteristics contributing to ESD within the school. Second, the development of a quantitative measurement instrument for this framework, the Education for Sustainable Development School Organization Questionnaire (ESD-SOQ) (Verhelst et al., 2022b), allows us to measure and investigate the characteristics in a reliable and valid way, thereby fostering our understanding. Moreover, making these characteristics measurable also allows for educational practitioners to become acquainted with what is under the hood of their school organization, offering valuable information on the tools they, as a school organization, have at their disposal.

Feedback and (self-)evaluation of the school's organizational functioning is essential to set the course for ESD implementation within schools (Mogren et al., 2019).

Mogren et al. (2019) answered the "how" question by studying the implementation of general educational qualities and quality assurance processes in schools, leading to the identification of four guiding principles of schools' organization that promote transformative ESD. The focus of the four guiding quality principles goes to the organization of the school, where "organization" is understood as a verb that can occur at different levels of the school (i.e., how is the school organized). In the framework of an ESD-effective school, in contrast, the school organization is seen as an *entity*, a structure, within the school. The first principle of school organization is collaborative interaction and school development (Mogren et al., 2019). This supports and actuates the idea that school development is an ongoing process of change and will raise the quality of the school if everyone takes responsibility for and supports the desired direction of education. The second principle, student-centered education, deals with the relationship between students and staff in education. Both the planning and the organizing of the school should consider the students' interests, conditions, and needs. Moreover, student-centered education enables students to practically connect their learning to a local or global context they find important, which contributes to higher quality. This connection to the local and global context is also apparent in the third principle: cooperation with the local community. Students must be encouraged and supported to cooperate with society by exchanging information and networking. Higher quality in education can provide insight into the workings of social stakeholders such as institutions of higher education, local decision-making organizations, enterprises, and cultural institutions. The last principle is proactive leadership and continuity, which describes a leadership style that is based on collective learning and implemented through scaffolding toward work on ESD. Problem-solving and conflict resolution are used to maintain continuity and to seek solutions among the employed teachers and other staff rather than seeking them from outside the school organization. These four quality principles make the organization and implementation of ESD within the school visible (Mogren & Gericke, 2017). They relate to how the school attunes its organization to the context and are thus context dependent by nature. It is precisely this context-dependence that illustrates the overlap between school effectiveness and school improvement. Answering the "what" provides insight into the context in which the "how" is answered.

As argued above, implementing a WSA to ESD requires a thorough understanding of the school organization and the school as an organization. The allegory of the gardener and their garden illustrates the need for answers to both the what and the how. They need to know what they start off with in terms of soil, climate, and tools in their toolshed in order to know how they can shape their gardening as an ongoing process of change, planning, and organizing the garden into a flourishing patch of nature, in connection to the local and global contexts, with a focus on learning and a step-by-step journey toward their dream garden.

Appropriate ways to implement a WSA to ESD can be unraveled via an effectiveness and improvement perspective by looking at what aspects of a WSA or ESD are already present in the school organization or in the schools' organizational functioning. A WSA to ESD would include components such as building capacity or institutional practices (Wals & Mathie, 2022), components that may or may not be present in the school as an organization. Effectiveness-oriented research could shed light on what characteristics of the school as an organization contribute to these components. As a hypothetical example, pluralistic communication as a characteristic of a school organization might make it easier to establish institutional practices as staff members can openly discuss these practices among themselves. Conversely, knowing that this characteristic is not yet strongly developed within the school might influence how they take on the challenge of implementing a WSA to ESD. While general knowledge on organizational characteristics is helpful for understanding schools as organizations, it remains important to acknowledge the unique position and context of each school. Characteristics or practices of one school are not always transferable to other schools, although there is evidence that many aspects are, in fact, generalizable (Jarl et al., 2017). Each school is unique, although school organizations can obtain testable insights into what might be important general structures and processes for specific educational purposes by studying other schools (Hargreaves, 2008).

Next to the guidelines for implementation, school effectiveness and improvement would allow schools to adopt actions and practices which foster desirable outcomes for students. From a school improvement perspective, this can be done by using certain actions or functions in the school organization argued to contribute to high-quality education to improve the outcome for the students (Mogren, 2019). By linking the organizational characteristics of a school to student outcomes of ESD, Verhelst et al. (2022a) answered the question of what makes a school organization ESD-effective. Combining both the SI and SE approaches results in a valuable holistic approach to what schools and education can do in the pursuit of educational quality. A gardener striving for a flourishing garden will always be attentive to both perspectives. Without looking to the intended results and the tools that they intend to use, they cannot make statements about the quality of their garden, and without knowing how they want to strive for a flourishing garden and acknowledging what they value, they cannot define quality in their gardening.

5.6 How We See School Effectiveness and School Improvement Anchored in ESD Research

Scott (2009) asked, "What can education do for sustainable development?" and "What can sustainable development do for education?" In his later work, Scott (2013, 2015) has studied ESD further in the way the whole school body aligns to ESD. His theoretical contribution suggests that a "firm ground" is needed to find answers to his questions. Scott (2013) means that the school organization plays a pivotal role in cultivating the firm ground which helps schools to be open to change by an established vision, to allocate resources appropriately, and, where the organization is part of a social learning community, to provide a systematic view of the world. The idea that education has an active role in changing societies is easier to put forward in relation to theories than to operationalize. Research has shown that schools are often overwhelmed by the amount of change and new agendas thrust upon them.

Answering Scott's (2009) first question on education's contributions to a sustainable society requires us as researchers to move away from the idea that school effectiveness and school improvement as research paradigms are sufficient to implement a transformative form of education. As Stoll et al. (2016) indicate:

> Researchers who want their effectiveness findings to be taken seriously have to find ways in which to help practitioners to engage with them in such a way that they can make the kind of meaning that enables them to use these findings to enhance their practice (Stoll et al., 2016, p. 363).

Research on answering what education can do for sustainable development has to take a step back and indicate how research can aid education in implementing education for sustainable development. Our proposed combination of school effectiveness and school improvement tools could incorporate temporary ESD goals, highly relevant to a local context (as understood and operationalized in practice by school improvement tools) within a highly effective environment for learners that are taught in collaborative, participatory, and flexible environments (identified by frameworks of ESD effectiveness). Yet, this perspective with a focus on effectiveness also has its challenges within the area of ESD, particularly concerning the pluralistic and democratic foundations of ESD (Öhman & Östman, 2019). These foundations are to some conflicting with the "normative" tradition of effectiveness research (Van Poeck et al., 2018). However, as we laid out in the section on school effectiveness, we do not strive toward the narrow "normative" perspective on school effectiveness. By seeing effectiveness as the degree to which a school is able to meet their aims, an effectiveness perspective can facilitate the pluralistic and democratic principles of ESE. In this reasoning, the school itself decides where it wants to go as an organization and the characteristics of the school organization can be improved in order to reach that goal.

Bringing together research and practice into an effective synergy is not a unique challenge within ESD. School effectiveness and improvement research has already taken several steps to bridge this gap. These steps also show the compatibility of the retrospective school effectiveness thinking and the change-oriented improvement perspective. Stoll et al. (2016) formulated suggestions about how to incorporate research into school and classroom practices, such as manageable units of meaning in accessible formats. Our metaphor of the flourishing garden shows that some tools related to effective ESD and some processes to align with the local context are a positive way forward. However, quality in education and the way our garden is developing can be improved by means of an increased understanding of the possible tools and skills needed in gardening. In line with Stoll et al. (2016) idea of presenting possible ways forward in practice, we claim that neither effective nor improvement characteristics of ESD play against each other; instead, they strengthen each other as ESD implementation is operationalized. Another suggestion by Stoll et al. (2016) on how to translate research into practice is to explore the reactions and responses to research findings in ways that connect with experiences in the local context. The school improvement model of Scherp opens up room for such discussions through the professional school development dimension where teachers, school leaders, and students plan for changes in a collaborative manner is also suggested by Stoll et al. (2016). Also, self-analysis and the work of prioritizing first or next steps in development is one of the suggestions to use research in practical education by Stoll et al. (2016). The literature shows that there is more than one way to reach well-functioning ESD in schools (Mogren, 2019; Verhelst et al., 2021). Some schools start with the inner structures to establish support in the school organization (more related to school effectiveness) while other schools reach out to society to make education highly relevant for students' learning (more in line with school improvement). Mogren et al. (2019) show that the combination of the two in a school they studied is leading to a WSA of high ESD quality. The inner structures of ESD enabled student initiatives to engage with society, leading to the reciprocal transformation of both society and education.

The second question put forward by Scott (2009), "What can sustainable development do for education?" is given an answer by our approach to study education developing in practice based on the combined theory of school effectiveness and improvement. New ways of organizing education are needed to make learners active participants in their schools. We claim that our combined model drives a transformation of education toward collaboration, participation, and action, by using tools well-rooted in education (effectiveness and improvement) to embrace ESD in its operationalized form. In ESD, temporal, static structures in the school organization constantly need evaluation to make sure they support the improvement process toward ESD. Schools are also at different places in the implementation process and face different conditions. Different schools will need different drivers, but the knowledge they gain in the field will help them discover more general patterns of operationalized ESD as WSA. In that sense, sustainable development can offer education guidance on transforming itself toward new educational practices and perspectives, linking together the first and second questions of Scott (2009).

To implement a whole school approach for sustainability education, we do not only need the school organization to be open to change toward ESD, as Scott (2013) describes the "firm ground" of ESD, but also need it to be well functioning and able to practice in the way exemplified by the two ESD models of effectiveness and improvement in this chapter. Accordingly, we propose that the firm ground of ESD also needs to be fertile. Looking at a WSA to ESD via school effectiveness and school improvement allows for the identification of what is likely to enhance ESD within the school and how this is likely to be done, thus providing a firm and fertile ground (Mogren, 2019; Verhelst, 2021).

References

Biesta, G. (2013). Receiving the gift of teaching: From 'Learning from' to 'Being taught by'. *Studies in Philosophy and Education, 32*(5), 449–461.

Boeve-de Pauw, J., & Van Petegem, P. (2018). Eco-school evaluation beyond labels: The impact of environmental policy, didactics and nature at school on student outcomes. *Environmental Education Research, 24*(9), 1250–1267. https://doi.org/10.1080/13504622.2017.1307327

Boeve-de Pauw, J., Gericke, N., Olsson, D., & Berglund, T. (2015). The effectiveness of education for sustainable development. *Sustainability, 7*(11), 15693–15717. https://doi.org/10.3390/su71115693

Breiting, S., Mayer, M., & Mogensen, F. (2005). *Quality criteria for ESD-Schools: Guidelines to enhance the quality of education for sustainable development.* Austrian Federal Ministry of Education, Science and Culture.

Cameron, K. S., & Quinn, R. E. (2011). *Diagnosing and changing organizational culture: Based on the competing values framework.* Wiley.

Chapman, C., Reynolds, D., Muijs, D., Sammons, P., Stringfield, S., & Teddlie, C. (2016). Educational effectiveness and improvement research and practice: The emergence of the discipline. In C. Chapman, D. Muijs, D. Reynolds, P. Sammons, & C. Teddlie (Eds.), *The Routledge international handbook of educational effectiveness and improvement: Research, policy, and practice* (pp. 1–24). Routledge.

Creemers, B., & Kyriakides, L. (2010). Using the dynamic model to develop an evidence-based and theory-driven approach to school improvement. *Irish Educational Studies, 29*(1), 5–23. https://doi.org/10.1080/03323310903522669

Creemers, B., & Reezigt, G. (1997). School effectiveness and school improvement: Sustaining links. *School Effectiveness and School Improvement, 8*(4), 396–429. https://doi.org/10.1080/0924345970080402

Fullan, M. (2006). The future of educational change: System thinkers in action. *Journal of Educational Change, 7*(3), 113–122. https://doi.org/10.1007/s10833-006-9003-9

Gericke, N., Boeve-de Pauw, J., Berglund, T., & Olsson, D. (2019). The sustainability consciousness questionnaire: The theoretical development and empirical validation of an evaluation instrument for stakeholders working with sustainable development. *Sustainable Development, 27*(1), 35–49.

Hargreaves, L. G. (2008). The whole-school approach to education for sustainable development: From pilot projects to systemic change. *Policy & Practice - A Development Education Review, 6.*

Hargreaves, A., & Fink, D. (2006). *Sustainable leadership.* Jossey-Bass.

Henderson, K., & Tilbury, D. (2004). *Whole-school approaches to sustainability: An international review of sustainable school programs.* ARIES.

Iliško, D., & Badyanova, Y. (2014). A case study of ESD implementation: Signs of sustainable leadership. *Discourse and Communication for Sustainable Education, 5*(1), 38–48. https://doi.org/10.2478/dcse-2014-0004

Jarl, M., Andersson, K., & Blossing, U. (2017). Success or failure? Presenting a case selection strategy for studies of school improvement. *Education Inquiry, 8*(1), 17–32. https://doi.org/10.1080/20004508.2016.1275177

Jickling, B., & Wals, A. E. J. (2012). Debating education for sustainable development 20 years after Rio: A conversation between Bob Jickling and Arjen Wals. *Journal of Education for Sustainable Development, 6*(1), 49–57. https://doi.org/10.1177/097340821100600111

Jucker, R. (2011). ESD between systemic change and Bureaucratic Obfuscation: Some reflections on environmental education and education for sustainable development in Switzerland. *Journal of Education for Sustainable Development, 5*(1), 39–60. https://doi.org/10.1177/097340821000500109

Kuzmina, K., Trimingham, R., & Bhamra, T. (2020). Organisational strategies for implementing education for sustainable development in the UK primary schools: A service innovation perspective. *Sustainability, 12*(22), 9549. https://www.mdpi.com/2071-1050/12/22/9549

Laurie, R., Nonoyama-Tarumi, Y., McKeown, R., & Hopkins, C. (2016). Contributions of Education for Sustainable Development (ESD) to quality education: A synthesis of research. *Journal of Education for Sustainable Development, 10*(2), 226–242. https://doi.org/10.1177/0973408216661442

Leithwood, K., & Louis, K. S. (2006). *Organizational learning in schools.* Taylor & Francis.

Miškolci, J., Armstrong, D., & Spandagou, I. (2016). Teachers' perceptions of the relationship between inclusive education and distributed leadership in two primary schools in Slovakia and New South Wales (Australia). *Journal of Teacher Education for Sustainability, 18*(2), 53–65. http://search.ebscohost.com/login.aspx?direct=true&db=eric&AN=EJ1131209&site=ehost-live

Mogaji, I. M., & Newton, P. (2020). School leadership for sustainable development: A scoping review.

Journal of Sustainable Development, 13(5). https://doi.org/10.5539/jsd.v13n5p15

Mogensen, F., & Schnack, K. (2010). The action competence approach and the 'new' discourses of education for sustainable development, competence and quality criteria. *Environmental Education Research, 16*(1), 59–74. https://doi.org/10.1080/13504620903504032

Mogren, A. (2019). *Guiding principles of transformative education for sustainable development in local school organisations: Investigating whole school approaches through a school improvement lens* [Doctoral thesis, Karlstad University]. Karlstad.

Mogren, A., & Gericke, N. (2017). ESD Implementation at the school organisation level, Part 1—Investigating the quality criteria guiding school leaders' work at recognized esd schools. *Environmental Education Research, 23*(7), 972–992. https://doi.org/10.1080/13504622.2016.1226265

Mogren, A., Gericke, N., & Scherp, H.-Å. (2019). Whole school approaches to education for sustainable development: A model that links to school improvement. *Environmental Education Research, 25*(4), 508–531. https://doi.org/10.1080/13504622.2018.1455074

Müller, U., Hancock, D. R., Stricker, T., & Wang, C. (2021). Implementing ESD in schools: Perspectives of principals in Germany, Macau, and the USA. *Sustainability, 13*(17), 9823. https://www.mdpi.com/2071-1050/13/17/9823

Nikel, J., & Lowe, J. (2010). Talking of fabric: A multi-dimensional model of quality in education. *Compare: A Journal of Comparative and International Education, 40*(5), 589–605. https://doi.org/10.1080/03057920902909477

Öhman, J., & Östman, L. (2019). Different teaching traditions in environmental and sustainability education. In K. V. Poeck, L. Östman, & Ö. Johan (Eds.), *Sustainable development teaching: Ethical and political challenges* (pp. 70–82). Routledge.

Olsson, D., Gericke, N., Sass, W., & Boeve-de Pauw, J. (2020). Self-perceived action competence for sustainability: The theoretical grounding and empirical validation of a novel research instrument. *Environmental Education Research, 26*(5), 742–760. https://doi.org/10.1080/13504622.2020.1736991

Reynolds, D., Sammons, P., De Fraine, B., Van Damme, J., Townsend, T., Teddlie, C., & Stringfield, S. (2014). Educational effectiveness research (EER): A state-of-the-art review. *School Effectiveness and School Improvement, 25*(2), 197–230. https://doi.org/10.1080/09243453.2014.885450

Sass, W., Boeve-de Pauw, J., De Maeyer, S., & Van Petegem, P. (2021). Development and validation of an instrument for measuring action competence in sustainable development within early adolescents: The action competence in sustainable development questionnaire (ACiSD-Q). *Environmental Education Research, 27*(9), 1284–1304. https://doi.org/10.1080/13504622.2021.1888887

Scheerens, J., & Demeuse, M. (2005). The theoretical basis of the effective school improvement model (ESI). *School Effectiveness and School Improvement, 16*(4), 373–385. https://doi.org/10.1080/09243450500234567

Scherp, H.-Å. (2013). *Lärandebaserad skolutveckling. Lärglädjens förutsättningar, förverkligande och resultat* [Learning based school development. The joy of learning, conditions, implementation and results]. Studentlitteratur.

Scherp, H.-Å., & Scherp, G.-B. (2007). *Lärande och skolutveckling : Ledarskap för demokrati och meningsskapande* (91-7063-105-0 (ISBN) 14038099 (ISSN)). Karlstad University Studies, Issue. http://urn.kb.se/resolve?urn=urn:nbn:se:kau:diva-700

Scott, W. (2009). Judging the effectiveness of a sustainable school. *Journal of Education for Sustainable Development, 3*(1), 33–39. https://doi.org/10.1177/097340820900300110

Scott, W. (2013). Developing the sustainable school: thinking the issues through. *The Curriculum Journal, 24*(2), 181–205. https://doi.org/10.1080/09585176.2013.781375

Scott, W. A. H. (2015). Exploring a transformative orientation to sustainability in universities: A question of loose and tight framings. *Environmental Education Research, 21*(6), 943–953. https://doi.org/10.1080/13504622.2014.954238

Stoll, L., Earl, L., Anderson, S., and Schildkamp, K. (2016). Educational effectiveness and improvement research, and teachers and teaching. In C. C. D. Muijs, D. Reynolds, P. Sammons, & C. Teddlie (Eds.), *The Routledge International Handbook of Educational Effectiveness and Improvement: Research, policy, and practice* (pp. 348–364). Routledge.

UNESCO. (2020). *Education for sustainable development: A roadmap*. https://www.gcedclearinghouse.org/sites/default/files/resources/200782eng.pdf

Van Poeck, K., König, A., & Wals, A. E. J. (2018). Environmental and sustainability education in the Benelux countries: Research, policy and practices at the intersection of education and societal transformation. *Environmental Education Research, 24*, 1–16. https://doi.org/10.1080/13504622.2018.1477121

Verhelst, D. (2021). *Sustainable schools for sustainable education: Characteristics of an ESD-effective school* [Doctoral thesis, University of Antwerp]. Antwerp.

Verhelst, D., Vanhoof, J., Boeve-de Pauw, J., & Van Petegem, P. (2020). Building a conceptual framework for an ESD-effective school organization. *The Journal of Environmental Education, 51*(6), 400–415. https://doi.org/10.1080/00958964.2020.1797615

Verhelst, D., Vanhoof, J., & Van Petegem, P. (2021). School effectiveness for education for sustainable development (ESD): What characterizes an ESD-effective school organization? *Educational Management Administration & Leadership*. https://doi.org/10.1177/1741143220985196

Verhelst, D., Vanhoof, J., de Maeyer, S., Sass, W., & Van Petegem, P. (2022a). Enabling effective education for sustainable development: Investigating the connection between the school organization and students' action competence. *The Journal of Environmental Education, 53*, 1–15. https://doi.org/10.1080/00958964.2022.2072797

Verhelst, D., Vanhoof, J., & Van Petegem, P. (2022b). Development and validation of the education for sustainable development school organisation questionnaire. *Environmental Education Research, 28*(2), 241–259. https://doi.org/10.1080/13504622.2021.2007219

Verhulst, E., & Lambrechts, W. (2015). Fostering the incorporation of sustainable development in higher education. Lessons learned from a change management perspective. *Journal of Cleaner Production, 106*, 189–204. https://doi.org/10.1016/j.jclepro.2014.09.049

Von Bertalanffy, L. (2003). General system theory. In D. Ammond (Ed.), *The science of synthesis exploring the social implications of general system theory* (pp. 103–142). University Press of Colorado.

Wals, A. E. J., & Mathie, R. G. (2022). Whole school responses to climate urgency and related sustainability challenges. In M. A. Peters & R. Heraud (Eds.), *Encyclopedia of educational innovation* (pp. 1–8). Springer Singapore. https://doi.org/10.1007/978-981-13-2262-4_263-1

6 Dimensions and Conditions of the Development of a Sense of Empowerment in a Whole School Approach

Émilie Morin, Geneviève Therriault, and Barbara Bader

Key message
First, this work can be useful to link theories in climate change education with youth voices. Second, it can help to find ways to reinforce the sense of empowerment of youth when it comes to climate change. Finally, it can be useful to better plan a whole school approach based on what youth demands to feel empowered.

6.1 Youth Voices About Their Sense of Empowerment

Given the urgent need to address climate change, it is necessary to empower youth. Since youth are particularly affected by climate change and its short-, medium-, and long-term repercussions, they should be recognized as full-fledged actors in climate change education (UNESCO, 2020; Ballet et al., 2013). This means, for example, that they must be given the freedom to create, change, and influence the course of events in accordance with their priorities, goals, and values. The education system must offer youth this kind of freedom of agency and other types of freedoms related to their well-being and opportunities (Sen, 2010). These different freedoms can reinforce their sense of empowerment and lead youth to act if that is their goal.

The whole school approach (WSA) may be seen as the starting point for a systemic and holistic redesign of the school as an institution (Hargis et al., 2021; Mogren, 2019; Wals & Mathie, 2022). UNESCO (2017b) has proposed "whole school approaches to climate action" that include aspects of school governance, teaching and learning, resource management, and school-based and extracurricular activities in partnership with the community. Various schools and extracurricular actors should thus be mobilized in such actions and support should be given to concrete youth initiatives. In our opinion, it is crucial to base this type of approach on what young people tell us. Therefore, we would like to propose here some relationships between what strengthens youth's sense of empowerment from their perspective (Morin, 2021) and emerging theories about the WSA. Based on the comments gathered from youth and on the capability theoretical approach, we will point out some very interesting links between the WSA and the empowerment of students, as well as ways to implement the WSA in certain schools in Québec.

É. Morin (✉) · G. Therriault
Université du Québec à Rimouski, Rimouski, QC, Canada
e-mail: emilie_morin02@uqar.ca; Genevieve_Therriault@uqar.ca

B. Bader
Université Laval, Québec, QC, Canada
e-mail: barbara.bader@fse.ulaval.ca

A. E. J. Wals et al. (eds.), *Whole School Approaches to Sustainability*, Sustainable Development Goals Series, https://doi.org/10.1007/978-3-031-56172-6_6

6.2 Epistemologies and Approaches in Climate Change Education for Empowerment

It is usually acknowledged in the environmental education research community that some schools, whether in the province of Québec (Canada) or in other parts of the world, do not achieve their climate change education objectives, particularly those aimed to encourage active eco-citizenship, build competencies such as systemic analysis and critical thinking, and give hope (Bader et al., 2013; Blaze Corcoran & Osano, 2009; Elshof, 2010; Field et al., 2019; Hayward, 2021; Jickling & Wals, 2018; Kwauk, 2020; Pruneau et al., 2015; UNESCO, 2017a; Zeyer & Kelsey, 2013). Despite the socio-constructivist approach to teaching and learning in Québec's education department curriculum (Ministère de l'Éducation, du Loisir et du Sport, 2007a), a more positivist/empirical-realistic epistemological and pedagogical perspective and a transmissive science and environmental education still seem to emerge all too often in the classroom (Baillet & Therriault, 2021; Charland et al., 2009). However, this epistemology severely limits the development of competencies for critically analyzing environmental science and the work of scientists or experts, including the competency for understanding that scientific knowledge is socially constructed in a specific context. This dominant epistemology in today's education systems creates a clash between the culture of youth (that is, their personal experience, life history, and initial conceptions) and the culture of Western science as usually represented by schools (Zeyer & Kelsey, 2013).

Therefore, even with the best of pedagogical intentions to develop young people's knowledge about climate change, a teaching method that is based to a large extent on a so-called positivist epistemology and a transmissive conception of teaching can ultimately lead some youth to develop what Zeyer and Kelsey (2013) have named "environmental depression." In their analysis, presenting youth with an accumulation of facts relating to environmental issues that are all too often alarmist tends to have the opposite effect of that desired. In fact, it contributes to reduce the empowerment of the learner by making them feel overwhelmed by complex, large-scale social and environmental problems that cannot be resolved easily and are way beyond their capabilities. The issue of climate change is of an exceedingly complex and interdisciplinary nature that should be discussed in the classroom (Bader et al., 2013; Schreiner et al., 2005). Adopting a socio-constructivist, critical, and transformative epistemological stance can thus help to further empower youth (Bader et al., 2014; Jickling & Wals, 2013, 2018; Morin et al., 2019; Zeyer & Kelsey, 2013).

We would have much to gain in secondary school by adopting a different approach to climate change education (Glithero, 2015; Sadler et al., 2007). Preferably, we should do more than simply teach facts, but should develop a more critical and transformative science and environmental education aimed to empower youth (Bader et al., 2017; Elshof, 2010). Various recommendations have already been made in this regard (Field et al., 2019; Kwauk, 2020; Sauvé, 2018). Kwauk (2020) insists that schools lack the support they need to properly integrate climate change education both at organizational and pedagogical levels.

Over the past few years, UNESCO (2016; 2017a, b; 2019; 2020) has suggested practical steps and directions for climate education that we think could foster the development of youth's sense of empowerment. For example, the new roadmap for education for sustainable development (ESD) (2020) proposes, as a priority action area (area 4), to "recognize young people as key actors in addressing sustainability challenges and the associated decision-making processes" (p3). It has also suggested that the role of educational institutions be redesigned by focusing more on the community, that steps be taken to promote the cooperation and solidarity of all actors around causes defended by youth, and that the latter be asked to fully participate in all the stages involved in implementing environmental projects. However, many aspects essential to foster a sense of empowerment seem to be lacking in schools, particularly in the learning objectives related to sustainable development goal (SDG) to fight against climate change (SDG 13) (UNESCO, 2017a). One such aspect encompasses the emotional dimensions of learning that are tied to youth well-being. These dimensions seem to be fundamental to the ongoing discourse about changes in individual behavior, which is nonetheless accompanied by a proposed reorganization of societal structures.

Therefore, we believe that a whole school approach to climate change is relevant (Hargis et al., 2021; UNESCO, 2017b; Wals & Mathie, 2022). This type of approach was proposed during sessions of professional development in ESD that mobilized various actors from Québec's education sector (Potvin & Bader, 2020). The project enabled us to reaffirm the appropriateness of key ESD competencies and to identify the types of school projects usually conducted in this area in Québec while proposing ways to enrich them in a collaborative manner.

This global approach has several different facets. At the heart of this school culture of action and engagement are an ethical stance and values that respect nature, life, solidarity, and social justice. Preference must thus be given to educational actions based on these values. The school curriculum must make considerable room for projects that are meaningful to youth because they are closely connected to their lives, their territory and community, and their concerns in a dialogical, critical, reflexive, and creative manner. This would reinforce the appropriateness of our approach, which involves listening to and considering the needs and expectations of youth so that schools can take them into account more fully in order to increase young people's sense of empowerment. In regard to teaching practices and important learnings, we have advocated for many years for interdisciplinary pedagogical practices, systemic analyses, and a critical and creative stance. In particular, we would like to renew the concept of environmental sciences and the relationship between the scientific knowledge of youth and teachers (Bader et al., 2014). Our goal would be to encourage youth to consider themselves as being competent and entitled to enter into a dialogue with the actors concerned, while enabling them to act individually and collectively as citizens or professionals in order to change the way we live and consume. As for the management of schools, it must respect nature, reduce its energy and carbon footprint, be open to differences, and know how to reconcile them in an interesting manner. Lastly, the WSA must be based on professional development and ongoing dialogue with stakeholders.

6.3 Clarification of Concepts Pertaining to Empowerment in Climate Change Education

Many environmental education researchers (Blanchet-Cohen & Brunson, 2014; Dimick, 2012; Mackey, 2012; Payne, 2015; Schreiner et al., 2005) defend the goal of social participation (where power is shared equitably among the various actors) rather than that of change in individual behavior (where youth are expected to reproduce a social construct predefined by adults). Striving for social participation implies that youth are truly "empowered" (Hayward, 2021). However, there does not appear to be a consensus among the aforementioned authors on what is meant by the term "empowerment". Some of these researchers refer to a *sense of self-efficacy* or *personal efficacy* (Corner et al., 2015; Goldman et al., 2017; Mead et al., 2012; Tsevreni, 2011), and others to *agency* (Blanchet-Cohen, 2008; Doyle, 2015; Glithero, 2015; Hayward, 2021; Lotz-Sisitka, 2016; Oliveira et al., 2015; Vongalis-Macrow, 2013) or *empowerment* (Birdsall, 2010; Blanchet-Cohen & Brunson, 2014; Dimick, 2012; Hayden et al., 2011; Schreiner et al., 2005). Nonetheless, it is necessary to theorize in detail and identify the limits of concepts identified as being relevant by researchers in this field if we want our educational practices in climate change education and, more broadly, in our secondary schools to be based on those concepts.

One approach in particular that has driven the research presented here seems especially appropriate for imagining an educational system that is characterized to a greater extent by social and environmental justice and is likely to give youth more freedom as well as the ability to act in ways they value. Known as the *capability approach*, it was developed by Sen (1985a, b) and Nussbaum (2000) and could easily be part of a transformative, sociocritical, and socioconstuctivist epistemological posture that also targets social participation in EE. It brings together three concepts that are particularly prevalent in the scientific literature in the area of climate education for empowerment, namely, *empowerment*, *agency*, and a *sense of self-efficacy*.

6.3.1 The Capability Approach

Sen (1985a), an economist and philosopher, conceptualized the capability approach in the 1980s to illustrate the relevance of considering the real opportunities and freedoms that a person have at their disposal in order to accomplish what they value instead of focusing essentially on individual access to resources (Comim et al., 2011). This approach, which goes beyond education and is in keeping with the perspective of improving human conditions, makes it possible to take an interest not only in justice in resource distribution but also in the other types of freedoms. In developing this approach, Sen (1985a) sought to defend the idea that social structures should always be designed with the goal of developing individual capabilities.

Although the capability approach provides theoretical tools that we cannot discuss in detail here, it should be noted that we divide these capabilities into two main categories (Morin, 2021): (1) capabilities that refer to opportunity and well-being freedoms and (2) capabilities that refer to process and achievement freedoms, with the latter referring directly to what is meant by the term *agency*. The capability approach is based on the idea that there are individuals who are more or less vulnerable or more or less free, who may or may not have to contend with injustice. Depending on their situation, we would like to empower them either by increasing their agency, that is, their ability to act, or by giving them opportunities and freedoms to enjoy fair and equitable well-being (Alkire, 2008; Gangas, 2016; Ibrahim & Alkire, 2007).

Comim et al. (2011) illustrate how appropriate it is to use the capability approach with youth given that they are not consulted or considered very often in the organization of social structures. In accordance with this approach, youth must be seen as full-fledged social actors, able to develop all of the capabilities they need to actively participate in society when they want to. They must be consulted on ways to boost their capabilities and reduce the injustices they face, particularly because they are the generation that will be most affected by the impacts of climate change. These are the main theoretical principles that have guided our reflection in this chapter, which is intended to underscore the relevance of certain aspects of the WSA, given the concerns expressed by youth and that affect their sense of empowerment.

6.3.2 Concepts Related to the Sense of Empowerment

As shown in Fig. 6.1, agency derives from process freedom and achievement freedom. Agency is what allows a person to act based on what he or she value. Achieving it depends directly on a person's sense of self-efficacy (Bandura, 2003). Without this sense, agency is diminished or even nonexistent because it constitutes the belief of an individual in his ability to act.

Empowerment is broader. It is both a process aimed at increasing agency and access to resources and goods, as well as a result of that increase. It is also a process aimed at boosting individual capabilities, be they directly related to an action or a state, in other words, the opportunity to act. Empowerment thus leads to the development of well-being and opportunity freedoms, together with process and achievement freedoms.

As shown in Fig. 6.1, the sense of empowerment is understood as being the way a person feels in regard to the various freedoms he or she has, their well-being and opportunity freedoms,

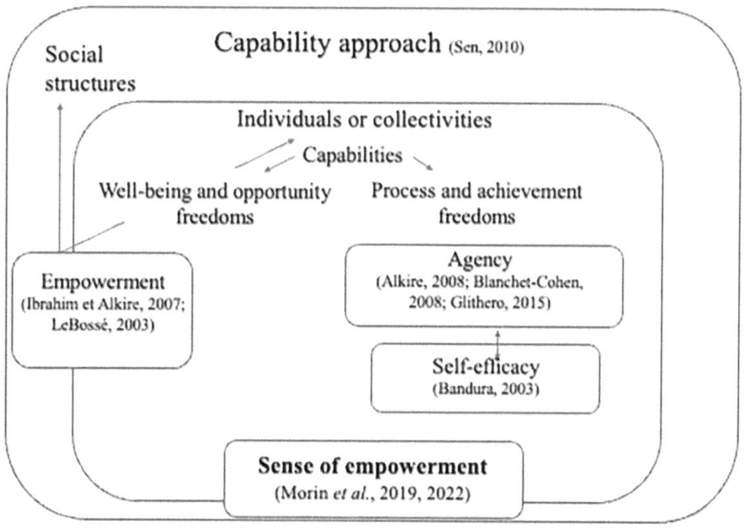

Fig. 6.1 The capability approach and the emergence of a sense of empowerment

and their process and achievement freedoms. This concept, whose principal dimensions were to be defined by the research presented here (Morin, 2021), is particularly well suited to explaining why a person wishes to act or not. It is in contrast to agency, which inevitably involves an action or a sense of self-efficacy that refers to a feeling toward a specific action. In our opinion, the sense of empowerment is particularly relevant within the context of the problem of climate change since this context makes people (especially youth) feel vulnerable and requires us to consider not only the process and achievement freedoms of individuals but also their opportunity and well-being freedoms.

6.4 Methodology

The methodological approach recommended for this research is in keeping with a qualitative/interpretive paradigm (Savoie-Zajc, 2018) and a socioconstructivist epistemological posture (Creswell & Poth, 2018). This posture involves reconciling the conceptualization of the sense of empowerment with the way in which youth, as social actors, interpret their own reality. In this sense, it aims to overturn the epistemic injustice faced by youth when they voice their concerns with a view to making recommendations for the education system.

The target population of this research consisted of young Quebecers at the end of secondary school. The secondary schools where the youth were interviewed were chosen on the basis of criteria similar to those used by Karpudewan et al. (2015) and Dawson (2015), namely (1) the public or private status of schools; (2) the environment, whether urban, rural, or semi-rural; (3) the socioeconomic background; (4) the number of students in the school; and (5) the presence or the absence of a particular program focusing on environmental education in the school. The objective was to obtain the greatest variability possible in the participants' sense of empowerment, until theoretical saturation. We specify to the teachers who help us that the students we choose should not necessarily be engaged but should be interested in discussing about climate change and their sense of empowerment. A total of 29 French-speaking youth, comprising 18 girls and 11 boys, aged 15–17, were met. Overall, we conduct ten ($N = 10$) semi-structured interviews in five secondary schools in spring 2019. The first part of the group interview protocol was designed to remind the youth of times in their lives when they felt very empowered or, on the contrary, not at all empowered. A discussion was then held to highlight what a "sense of empowerment" meant for them. In the second part of the interview, more specific questions were asked about the youth's sense of empowerment in the face of climate change, along with follow-up questions. For example, we asked: "Do you feel free to act on climate change? Do you feel able to act on climate change?" The interviews were during lunch, in a classroom and last for approximately 1 h. The analysis conducted was essentially thematic (Paillé & Mucchielli, 2016), and *NVivo 12* software was used in view of the size of the corpus of data. The themes and subthemes obtained were the dimensions of the sense of empowerment. During the final step of the analysis, inspired by Creswell and Poth (2018), we took a step back to make sense of these themes and subthemes. Connections between broader units of abstraction and the scientific literature were listed. All the items in this list were then grouped by *family resemblance* (Wittgenstein, 1961). To make the results more accessible, these groupings were synthesized. They made it possible to identify three main findings for empowering youth in the face of climate change.

6.5 Dimensions of the Sense of Empowerment and Links with the WSA

As mentioned earlier, various links can be identified between the dimensions of the sense of empowerment that emerged from what was said by the youth we met with (Morin, 2021) and the WSA (Wals & Mathie, 2022). In this section, we present the findings that we believe are more in keeping conceptually and empirically speaking with the different components of the WSA defined by Wals and Mathie (2022).

6.5.1 Vision, Ethos, Leadership, and Coordination

To bring about changes in discourses and practices in school, it is essential that the actors in each school community share the same vision of environmental education at both pedagogical and administrative levels. As specified by Mogren (2019) and Wals and Mathie (2022), the school culture needs to align with this shared vision.

The young people interviewed in the course of our research mentioned various needs regarding school leadership. First of all, they talked about the actions that should be taken by schools to ensure that their students have a minimum of facilities and material resources for boosting their sense of empowerment. Among other things, they said that they would like schools to be better adapted to environmental action and be equipped at the very least with reusable dishware, composting systems, bike racks, and reusable bottles, instead of beverage vending machines, etc. Some became angry when they talked about their school's material resources, including the fact that there were no recycling bins or that recycling bins were present but their contents were simply thrown into the garbage. This situation was distressing for the youth we met with and did not help them build a sense of empowerment.

The youth we interviewed also called for environmental action opportunities, meaningful projects, resources, support for their initiatives, and consistent, environmentally friendly actions on the part of school authorities. They also said that they wanted schools to help them make enlightened choices, develop their self-confidence, and foster their sense of responsibility. As well, some called for greater recognition for youth engagement. As shown by the following statement, the youth saw this as a way to develop their sense of empowerment:

> Student engagement isn't promoted enough in our school. And especially, no one praises those who do get involved. Just having someone give you a little pat on the back and say "You did it! Congratulations!" or "Thank you" encourages you to continue (translation, youth, age 14).

Such recognition should also be provided to schools that make efforts and get involved in an environmentally friendly manner.

6.5.2 Curriculum

The education systems in various countries have implemented school curriculums and promote educational approaches that may or may not address the challenges related to the climate emergency. For example, in Québec, one of the three aims of the Québec Education curriculum is organized around helping students to "become empowered" (Ministère de l'Éducation, du Loisir et du Sport, 2007a, p9). This idea is based essentially on students' know-how and knowledge. Although very relevant, it does not seem to be put into practice to any great extent in many of the schools attended by the youth we interviewed.

In fact, knowledge is presented as a fundamental basis for empowerment in the curriculum concerned. Knowledge about climate change is made available, in particular, to 16 of age youth teachers in the following general education curriculums: *Science and Technology* (MELS, 2007b) and *Environmental Science and Technology* (Ministère de l'Éducation, du Loisir et du Sport, 2007c). It is also available for the 17 of age youth in the *Contemporary World* curriculum (Ministère de l'Éducation, du Loisir et du Sport, 2007d). As part of general education in *Science and Technology*, teachers may choose to discuss the issue of climate change, but they are not obliged to do so. As a result, when young people have finished their compulsory schooling at age 17, they may have had little exposure to the climate change issue in the different curriculums because none of them are devoted entirely to environmental education.

That being said, most of the youth we interviewed said that they had explored the question of climate change in school, while only a very small minority said that they had not discussed it at all. However, this did not prevent many of them from criticizing the way that climate change is dealt with in school. As with a positivist epistemological approach, it seems that the knowledge that is taught is rarely put into context or linked to the youth's preoccupations, territory, and community (Morin et al., 2021; Morin, 2021). Therefore, nearly half of the youth interviewed said that *knowledge* was the only thing they heard about in school when it came to climate change. They also deplored the fact that they were mainly required to write about the subject rather than to *take action*. A sociocritical, transformative, and socio-constructivist approach where youth play an active role in choosing themes or actions they value (Bader et al., 2017) would seem to be more in keeping with the WSA.

Generally speaking, when the youth were asked what they wanted in order to feel more capable of taking action in regard to climate change, they asked for a course on climate change at school. The debate over the need for a curriculum on the environment made its way into the news in 2019 when youth in Québec, who had demonstrated in the wake of the climate strikes organized by young Greta Thunberg, made this their main request to the government in place.

In addition to the need for more relevant knowledge and the creating of a specific course on the environment, the discussions with the youth revealed that these young people did not have enough opportunities to talk about their feelings on climate change. However, many of them said they felt upset, powerless, anxious, depressed, distressed, pessimistic, disappointed, overwhelmed, mad, indignant, lacking in freedom, fearful, under pressure, etc. In contrast, feeling free, being hopeful, looking forward to a brighter future, and believing in a better response on the part of the community are feelings that can strengthen a person's sense of empowerment.

We believe it is essential to be able to create a forum for discussion within curriculums, where youth can say how they feel. Philosophy for children, youth arts, and youth literature are all ways that can foster discussion. However, very full curriculums, frequent evaluations, and rigid timetables—particularly in secondary school—make it very difficult for teachers, even the most motivated ones, to put all that is needed in place. It seems desirable to have education curriculums that give more freedom to students and teaching staff, provided the latter receive appropriate support and training in climate change education.

6.5.3 Pedagogy and Learning

According to Wals and Mathie (2022), the type of pedagogy used coupled with conceptions of teaching and learning constitute another dimension of the WSA. As mentioned earlier, a transformative approach to climate change education can have very beneficial effects in a school. In our opinion, this type of approach seems to be well aligned with what the youth we met with said, especially because we feel they are being subjected to a form of epistemic injustice (Demers et al., 2018). Confidence, recognition, commitment, and their successes should be considered fundamental in this transformative socio-critical education on environmental issues.

According to some of the youth we met with, the fact that they feel encouraged to acquire knowledge about climate change without, however, being able to take action and be recognized for their actions, undermines their freedom. As one girl explained, learning about climate change is usually viewed favorably, but when this learning leads to action, some adults in positions of authority consider their action futile.

> I feel free, yes and no, to act to address climate change. We're encouraged to come up with ideas, do things, and develop our own view of climate change. But there's always someone, like the school principal or a member of the political party in power, who tells you that you have good ideas and encourages you, but then says to themselves: "look at that girl making all those efforts for nothing" (translation, youth, age 16).

However, some of the youth we interviewed had had a very different experience. They drew attention to examples of environmental initiatives at the school they attended. Most of these youth felt that they had a certain amount of power at their school and were perfectly capable of working with the school administration or teachers to develop environmental initiatives. According to one girl, the fact that teaching staff had confidence in the students and gave them responsibilities as well as the freedom to act in their school setting was enough to convince them that they had the capacities needed to act:

> When someone tells you to do what you want because they have confidence in you, you can't ask for more power to act than that! I've come to understand that I'm in charge. This makes me resourceful, self-sufficient, and accountable for what I do. And since I'm responsible for others as well, that gives me a certain autonomy. Of course, you're proud of yourself, but at the same time you gain self-confidence because you tell yourself that if someone gives you power, it's because you can handle it (translation, youth, age 14).

The youth we met with said that they were satisfied, happy, or proud with regard to their actions, primarily those that had had visible results. Others said that they felt good about or enjoyed being involved. However, according to many of these youth, the successes of others were not discussed very much if at all in school, even when they were very relevant. More attention should be focused on their successes because this would help youth to develop a sense of empowerment and to be granted more importance.

6.5.4 Institutional Practices and Bureaucracy

Several of the youth we interviewed complained about the complex structures and bureaucracy needed to make changes that might be readily applicable. Some of them admitted that they found it very hard to deal with this bureaucracy: "What's so annoying is that it takes so long to do even simple things. That makes me mad." (translation, youth, age 14).

The youth we met with mentioned many very worthy and relevant initiatives that could be implemented in schools. Some of them said that they had talked to school administrators about these initiatives and had received strong support. Others expressed indignation about the way in which the adults gave to much importance to the supposedly complexity of changes the youth propose.

Other youth were aware that their lack of choices or opportunities could be due to a single individual, group, government, or organization. These youth said that they also felt helpless in such situations:

> I felt very empowered when the school administration lifted the ban on taking part in protests, because prior to that I had felt really powerless in dealing with the administrators. You belong to the administration during school hours, so you can't decide to skip classes or you'll suffer the consequences . . .You really feel stripped of your power to act because your school prevents you from acting (translation, youth, age 16).

Many of the youth in our sample had been prevented from taking part in the Friday climate protests and they talked about this regularly during the interviews. Most of them said that it had had a direct impact on their sense of empowerment. As recommended by Mogren (2019), proactive leadership by the school organization is important to create a structure that facilitates transformative education.

6.5.5 Valuing Community and Intergenerational Connections

According to the youth we interviewed, establishing emancipatory relations with the various actors in the community surrounding a school seems to be essential for developing their sense of empowerment. Another element that emerged very clearly from the statements analyzed was a strong desire for solidarity.

Many of the youth we met with stressed first and foremost the importance of working together in order to address environmental issues. This helped them to feel less isolated:

> If there's no cooperation, you don't have any power (translation, youth, age 16).
> You need mutual aid to feel empowered (translation, youth, age 14).
> When you feel empowered, you feel like you're almost of the same caliber as others, you know what you have to do and you can share it with others (translation, youth, age 14).

Those youth had a fairly strong sense of empowerment. They already seemed aware of the surrounding community and the opportunities for assistance offered by it. One girl in particular seemed to be able to reflect upon the basis of the social solidarity she felt and promoted in all spheres of her life: "You mustn't forget, in any case, that you're never alone in life. You're the one who makes the decisions . . . but you're never alone. You're never alone, even when you make personal decisions" (translation, youth, age 14).

Many of the youth we interviewed said that they would like to feel listened to, understood, and supported to a greater extent. They also recognized that such values are essential for creating a fairer society and that individuals would feel more able to act if they felt important to others and were listened to when they made choices.

Our analysis identified two other values that can give youth a sense of empowerment in the face of climate change, namely, empathy and respect. According to two young people, empathy can help us to better understand others and the situations in which they live so that we can help them more effectively. It can give rise to a multitude of learnings and address the most vulnerable needs.

Many of the statements we analyzed talked about "other generations" and, particularly, older generations. The youth we interviewed found it hard when older generations do not take responsibility for climate change. Some of them deplored the fact that those generations pass the burden to younger people and thus fail to shoulder their responsibilities: "We should be able to see the older generations as setting the example, but no. They flood us with information in the hope that we'll take care of things, but why can't they take the initiative?" (translation, youth, age 15). According to one youth, the best way to increase a person's sense of empowerment is to encourage them to stop thinking that it's up to others to act: "To make me feel more empowered, I'd like people to stop saying that future generations will do something." (translation, youth, age 14).

Regardless of the form it takes, shared responsibility is important for putting an end to injustices. Achieving it requires thinking outside the box, getting rid of often deep-seated habits, and being open minded. In this sense, connections with the community around a school or outer society (Mogren, 2019) are essential for strengthening solidarity and showcasing models where older generations forcefully engage with younger ones.

6.5.6 Capacity Building and Continued Professional Development of All School Staff

As mentioned by Wals and Mathie (2022), teaching staff must build various capacities if they want to take part in a WSA. They will have to work with various partners in the community, be resilient in the face of change. and have a good capacity to adapt to unforeseen events. The staff will also have to defend the relevance of this type of approach before different audiences. When we interviewed the youth taking part in our study, they told us which capacities they deemed essential for increasing their sense of empowerment. These capacities also seem to be in keeping with what can be expected of teaching staff. For example, they said that they believe it is necessary to be able to convince, influence, and educate others, as well as to argue a point of view. When asked "If anything was possible, what would you like to have at your disposal for addressing climate change in the way you want?", they mentioned "influence" on several occasions as a capacity that should be developed.

The ability to *make efforts* also seems to be important for developing a sense of empowerment. Efforts seem to be contagious and should be encouraged. However, they should have positive results if they are to be meaningful and generate hope. In this sense, determination, perseverance, and efforts are necessary, but not sufficient.

Courage is another sub-theme related to the aspect of capacity. For example, courage can be needed to do what is right. Therefore, it is important "to step out of your comfort zone" (translation, youth, age 14), "to not be afraid of what people will think" (translation, youth, age 14), "to have mental fortitude" (translation, youth, age 16), and "to even defy rules and standards at times."

All of these capacities, be they the ability to convince, influence, educate, or persuade, or remain determined, persevere, make efforts, or display courage, are largely a matter of impressions or feelings. According to what the young people we interviewed said, these capacities depend a great deal on how they are perceived by others and on the actions that others take. Seeing others make efforts in the wake of your own or knowing that they share your determination can completely change the way you view your capacities.

6.6 Three Main Findings for Empowering Youth in the Face of Climate Change

The first finding that emerged from our results is the need to consider in climate change education not only what actions can be taken by youth and thus their agency in regard to the problem but also their well-being and the opportunities made available to them (Sen, 2010).

Taking well-being and opportunities into account also recalls the importance of considering the emotional dimensions of climate change education when it comes to empowerment, in order to prevent and mitigate feelings of despair, fear, urgency, or fatalism that are still overly present among some youth (Zeyer & Kelsey, 2013). In that regard, Kelsey and Armstrong (2012) put forward the hypothesis that asking youth to prematurely address an issue like climate change, which is an environmental problem well beyond their control, could contribute to reducing their empowerment. We support this idea and show that some of the youth we met with had a very limited sense of empowerment owing to the scope of the climate change problem and their lack of control over it. This is one of the climate education for empowerment challenges mentioned by Schreiner et al. (2005) that schools must strive to address. Therefore, adopting a different approach to dealing with climate change in schools is no longer an option but a necessity, especially when we think that certain current education practices exacerbate feelings of helplessness, discouragement, and even fear (Elshof, 2010).

The second finding that emerged is the link between the importance attached to prescribing changes in individual behavior and the decrease in young people's sense of empowerment in regard to climate change. The young people we met with, like those met with by Kenis and Mathijs (2012), widely criticized an educational model where youth are expected to reproduce a social construct predefined by adults (Jickling & Wals, 2008). Since youth consider that they are among the individuals most affected by climate change (they will experience the effects to a greater extent) (Corner et al., 2015; Schreiner et al., 2005), their sense of empowerment will be reduced if they feel they are not being listened to or are not considered credible in decision making on this question, but have to comply with directives that they believe are lacking in scope.

Our results also show that the development of a sense of empowerment would be fostered by a more participatory citizen-action social model (Glithero, 2015; Hayward, 2021; Sadler et al., 2007), where there is an equitable sharing of power. Therefore, the youth we interviewed identified hierarchical social systems based on authoritarian political and social structures and major, but socially accepted, power relations (Jickling & Wals, 2008) as being at the root of a reduced sense of empowerment. These youth felt that they were regularly subjected to power dynamics that they did not consider legitimate, whether in regard to educational institutions (Demers et al., 2018), governments (Fielding & Head, 2012), multinationals (Kenis & Mathijs, 2012), or even adults in general. Ensuring that youth are directly and truly involved in making decisions that concern them in their school seems to be totally consistent with the WSA and should be considered a priority.

Similarly, it is interesting to note that some of the youth we met with criticized the social norm that values, for example, overconsumption and individualism, as well as capitalism. These youth, who recognized that it is important to do what is fair, to break free from social norms or their comfort zone, and to overcome the lack of support from others (Blanchet-Cohen, 2008), were also perfectly aware of the choices and responsibilities we have as individuals and members of society. They recognized the need to change habits, call the system in place into question, consider the most vulnerable individuals, and provide them with the opportunities and all the capabilities (Sen, 2010) needed to ensure greater social and environmental justice.

Since the vast majority of the youth we interviewed recognized that schools are accountable in the fight against climate change, education seems to be a promising way to call the system into question, imagine a more viable future, and take concrete action for the community. For some of the youth, it is impossible to not be affected by the injustices associated with the climate crisis. Therefore, critical reflection and systemic analysis (UNESCO, 2017a) encouraged by schools might facilitate the necessary social and ecological transition. Some of the youth we interviewed in one of the schools seemed to have completed this critical reflection and systemic analysis phase and were now looking forward to a more ecologically viable future. The fact that they experienced environmental successes on a daily basis and were part of a school community where they felt free and where adults had confidence in them encouraged them to have a strong sense of empowerment when it came to dealing with climate change.

The third and last finding that emerged from the results is the important role of solidarity in the development of a sense of empowerment. As in Hayward (2021) and Glithero (2015), some of the youth we met with called for climate change education that builds on the fairer and more equitable social participation of youth (at the level of consultation and decision making as well as action). Such social participation should be encouraged and developed in school and conducted for the well-being of the community around the school (Hargis et al., 2021; Hayward, 2021; UNESCO, 2020). In so doing, values such as solidarity or, as mentioned in particular by Sauvé (2018), benevolence should be valued in school. To that end, youth must be recognized as valuable, credible, and legitimate interlocutors.

Therefore, it seems logical that youth's sense of empowerment will increase when group activities, particularly those that bring all generations together, are conducted and encouraged. It will then feel like responsibility is being shared and that the different actors in the school community are concerned as well. According to some of the youth we met with, schools are responsible for promoting this type of activity.

6.7 Conclusion

It is morally untenable that formal education as usually provided in formal school settings helps to increase the anxiety and discouragement of youth in the face of climate change (Elshof, 2010). To develop a real sense of empowerment in schools when it comes to climate change, it must be acknowledged that youth are free to act or not based on what they believe should be done. Various freedoms (or capabilities) must be granted to youth as part of a WSA so that they can feel empowered in the face of climate change.

Several initiatives are already underway in schools in Québec, thanks to the unrelenting efforts of teachers, school administrators, and other school employees. These initiatives must be recognized and promoted in order to inspire other school communities to follow suit. Political authorities must provide these communities with adequate support in order to reinforce this aspect of their school-based and extracurricular initiatives.

In conclusion, it should be noted that we have a tremendous need for education if we are to deal with climate change. However, the world of education (particularly, teaching staff and school administrators) also requires freedom, resources, and recognition if it is to achieve that goal. As Glithero (2015) said, teaching staff also need appropriate preservice teacher training and professional development in environmental education as well as an organizational structure that supports this type of education. In that sense, the WSA should serve as a source of inspiration.

Acknowledgments This research was funded by the Social Sciences and Humanities Research Council of Canada (SSHRC). We also want to thank Alison McGain who translated this chapter from French to English.

References

Alkire, S. (2008). *Concepts and measures of agency*. Oxford. Retrieved September 12, 2022, from https://ophi.org.uk/working-paper-number-09/

Bader, B., Jeziorski, A., & Therriault, G. (2013). Conception des sciences et d'un agir responsable des élèves face aux changements climatiques. *Les Dossiers des Sciences de l'Éducation, 29*, 15–32. https://doi.org/10.4000/dse.81

Bader, B., Morin, É., Therriault, G., & Arseneau, I. (2014). Rapports aux savoirs scientifiques et formes d'engagement écocitoyen d'élèves de quatrième secondaire face aux changements climatiques. *Revue francophone du développement durable, 4*, 171–190.

Bader, B., Therriault, G., & Morin, É. (2017). Engagement écocitoyen, engagement scolaire et rapport aux savoirs. In L. Sauvé, I. Orellana, C. Villemagne, & B. Bader (Eds.), *Éducation, Environnement, Écocitoyennenté. Repères contemporains* (pp. 81–100). Presses de l'Université du Québec. Retrieved September 12, 2022, from https://www.puq.ca/catalogue/livres/education-environnement-ecocitoyennete-3109.html

Baillet, D., & Therriault, G. (2021). Entre ce qu'ils pensent, ce qu'ils disent et ce qu'ils font, quelles articulations? Analyse des croyances épistémologiques, des conceptions pédagogiques et des pratiques d'enseignants québécois du secondaire en sciences naturelles et en sciences humaines et sociales. *Phronesis, 10*(2-3), 129–152. https://doi.org/10.7202/1081789ar

Ballet, J., Bazin, D., & Pelenc, J. (2013). *Justice environnementale et approche par les capabilités*. ESIA. Retrieved September 12, 2022, from https://halshs.archives-ouvertes.fr/halshs-01071203

Bandura, A. (2003). *Auto-efficacité. Le sentiment d'efficacité personnelle* (J. Lecompte, Trans.). De Boeck Université.

Birdsall, S. (2010). Empowering students to act: Learning about, through and from the nature of action. *Australian Journal of Environmental Education, 26*, 65–84. https://doi.org/10.1017/S0814062600000835

Blanchet-Cohen, N. (2008). Taking a stance: Child agency across the dimensions of early adolescents' environmental involvement. *Environmental Education Research, 14*(3), 257–272. https://doi.org/10.1080/13504620802156496

Blanchet-Cohen, N., & Brunson, L. (2014). Creating settings for youth empowerment and leadership: An ecological perspective. *Child and Youth Services, 35*(3), 216–236. https://doi.org/10.1080/0145935X.2014.938735

Blaze Corcoran, P., & Osano, P. M. (2009). *Young people, education, and sustainable development. Exploring principles, perspectives, and praxis*. Wageningen Academic Publishers.

Charland, P., Potvin, P., & Riopel, M. (2009). L'éducation relative à l'environnement en enseignement des sciences et de la technologie: une contribution pour mieux Vivre ensemble sur Terre. *Éducation et*

Francophonie, 37(2), 63–78. Retrieved September 12, 2022, from https://www.erudit.org/fr/revues/ef/2009-v37-n2-ef3580/038816ar/

Comim, F., Ballet, J., Biggeri, M., & Iervese, V. (2011). Introduction—Theoretical foundations and the book's roadmap. In M. Biggeri, J. Ballet, & F. Comim (Eds.), *Children and the capability approach* (pp. 3–21). Palgrave Macmillan. https://doi.org/10.1057/9780230308374

Corner, A., Roberts, O., Chiari, S., Völler, S., Mayrhuber, E. S., Mandl, S., & Monson, K. (2015). How do young people engage with climate change? The role of knowledge, values, message framing, and trusted communicators. *Wiley Interdisciplinary Reviews: Climate Change, 6*(5), 523–534. https://doi.org/10.1002/wcc.353

Creswell, J. W., & Poth, C. N. (2018). *Qualitative inquiry and research design. Choosing among five approaches* (4th ed.). Sage.

Dawson, V. (2015). Western Australian high school students' understandings about the socioscientific issue of climate change. *International Journal of Science Education, 37*(7), 1024–1043. https://doi.org/10.1080/09500693.2015.1015181

Demers, S., Clermont, A., Lefrançois, D., Éthier, M.-A., & Potvin, M. (2018). Quel rapport au droit les codes de conduite d'écoles secondaires impliquent-ils pour les jeunes? *Éducation et Sociétés, 2*(42), 101–118. https://doi.org/10.3917/es.042.0101

Dimick, A. S. (2012). Student empowerment in an environmental science classroom: Toward a framework for social justice science education. *Science Education, 96*(6), 990–1012. https://doi.org/10.1002/sce.21035

Doyle, W. (2015). The many faces of agency. *Cultural Studies of Science Education, 10*(2), 275–279. https://doi.org/10.1007/s11422-014-9610-3

Elshof, L. (2010). Transcending the age of stupid: Learning to imagine ourselves differently. *Canadian Journal of Science, Mathematics and Technology Education, 10*(3), 232–243. https://doi.org/10.1080/14926156.2010.504483

Field, E., Schwartzberg, P., & Berger, P. (2019). *Canada, climate change and education: Opportunities for public and formal education (Formal report for learning for a sustainable future).* York University Printing Services. Retrieved September 12, 2022, from http://lsf-lst.ca/media/National_Report/National_Climate_Change_Education_FINAL.pdf

Fielding, K. S., & Head, B. W. (2012). Determinants of young Australians' environmental actions: The role of responsibility attributions, locus of control, knowledge and attitudes. *Environmental Education Research, 18*(2), 171–186. https://doi.org/10.1080/13504622.2011.592936

Gangas, S. (2016). From agency to capabilities: Sen and sociological theory. *Current Sociology, 64*(1), 22–40. https://doi.org/10.1177/0011392115602521

Glithero, E. (2015). *Exploring the development of student agency from the perspective of young canadian eco-civic leaders* [Doctoral dissertation, Ottawa University]. Retrieved September 12, 2022, from https://ruor.uottawa.ca/bitstream/10393/32335/1/Glithero_Elizabeth_2015_thesis.pdf

Goldman, D., Pe'er, S., & Yavetz, B. (2017). Environmental literacy of youth movement members—is environmentalism a component of their social activism? *Environmental Education Research, 23*(4), 486–514. https://doi.org/10.1080/13504622.2015.1108390

Hargis, K., McKensie, M., & LeVert-Chiasson, I. (2021). A whole institution approach to climate change education. In R. Iyengar & C. Kwauk (Eds.), *Curriculum and learning for climate action* (pp. 43–66). https://doi.org/10.1163/9789004471818_004

Hayden, M., Houwer, R., Frankfort, M., Rueter, J., Black, T., & Mortfield, P. (2011). Pedagogies of empowerment in the face of climate change uncertainty. *Journal for Activist in Science and Technology Education, 3*(1), 118–130. Retrieved September 12, 2022, from https://jps.library.utoronto.ca/index.php/jaste/article/view/21202/17274

Hayward, B. (2021). *Children, citizenship and environment. SchoolStrike Edition.* Earthscan/Routledge. https://doi.org/10.4324/9781003000396

Ibrahim, S., & Alkire, S. (2007). *Agency and empowerment: A proposal for internationally comparable indicators.* Oxford. Retrieved September 12, 2022, from https://ophi.org.uk/wp-content/uploads/OPHI-wp04.pdf

Jickling, B., & Wals, A. E. J. (2008). Globalization and environmental education: Looking beyond sustainable development. *Journal of Curriculum Studies, 40*(1), 1–21. https://doi.org/10.1080/00220270701684667

Jickling, B., & Wals, A. E. J. (2013). Probing normative research in environmental education. In R. B. Stevenson, M. Brody, J. Dillon, & A. E. J. Wals (Eds.), *International handbook of research on environmental education* (pp. 74–86). Routledge. https://doi.org/10.4324/9780203813331

Jickling, B., & Wals, A. E. J. (2018). Globalization and environmental education: Looking beyond sustainable development. In A. Reid (Ed.), *Curriculum and environmental education: Perspectives, priorities and challenges* (pp. 221–242). Routledge, Taylor and Francis Group.

Karpudewan, M., Roth, W.-M., & Abdullah, M. N. S. (2015). Enhancing primary school students' knowledge about global warming and environmental attitude using climate change activities. *International Journal of Science Education, 37*(1), 31–54. https://doi.org/10.1080/09500693.2014.958600

Kelsey, E., & Armstrong, C. (2012). Finding hope in a world of environmental catastrophe. In A. E. J. Wals & P. Blaze Corcoran (Eds.), *Learning for sustainability in times of accelerating changes* (pp. 187–200). Wageningen Academic Publishers. https://doi.org/10.3920/978-90-8686-757-8

Kenis, A., & Mathijs, E. (2012). Beyond individual behaviour change: The role of power, knowledge and strategy in tackling climate change. *Environmental*

Education Research, 18(1), 45–65. https://doi.org/10.1080/13504622.2011.576315

Kwauk, C. (2020). *Roadblocks to quality education in a time of climate change*. Center for Universal Education. Retrieved September 12, 2022, from https://www.brookings.edu/wp-content/uploads/2020/02/Roadblocks-to-quality-education-in-a-time-of-climate-change-FINAL.pdf

Lotz-Sisitka, H. (2016). Reviewing strategies in/for ESD policy engagement: Agency reclaimed. *The Journal of Environmental Education, 47*(2), 91–103. https://doi.org/10.1080/00958964.2015.1113915

Mackey, G. (2012). To know, to decide, to act: The young child's right to participate in action for the environment. *Environmental Education Research, 18*(4), 473–484. https://doi.org/10.1080/13504622.2011.634494

Mead, E., Roser-Renouf, C., Rimal, R. N., Flora, J. A., Maibach, E. W., & Leiserowitz, A. (2012). Information seeking about global climate change among adolescents: The role of risk perceptions, efficacy beliefs, and parental influences. *Atlantic Journal of Communication, 20*(1), 31–52. https://doi.org/10.1080/15456870.2012.637027

Ministère de l'Éducation, du Loisir et du Sport. (2007a). *Programme de formation de l'école québécoise. Enseignement secondaire, 2e cycle. Parcours de formation générale. Parcours de formation générale appliquée*. Retrieved September 12, 2022, from http://www.education.gouv.qc.ca/fileadmin/site_web/documents/education/jeunes/pfeq/PFEQ_presentation-deuxieme-cycle-s econdaire.pdf

Ministère de l'Éducation, du Loisir et du Sport. (2007b). *Programme de formation de l'école québécoise. Enseignement secondaire, 2e cycle*. Parcours de formation en Science et technologie. Retrieved September 12, 2022, from http://www.education.gouv.qc.ca/fileadmin/site_web/documents/education/jeunes/pfeq/PFEQ_science-technologie-deuxieme-cycle-secondaire.pdf

Ministère de l'Éducation, du Loisir et du Sport. (2007c). *Programme de formation de l'école québécoise. Enseignement secondaire, 2e cycle*. Parcours de formation en Science et technologie/Science et technologie de l'environnement. Retrieved September 12, 2022, from http://www.education.gouv.qc.ca/fileadmin/site_web/documents/education/jeunes/pfeq/PFEQ_science-technologie-environnement.pdf

Ministère de l'Éducation, du Loisir et du Sport. (2007d). *Programme de formation de l'école québécoise. Enseignement secondaire, 2e cycle. Monde contemporain. 4 unités*. Retrieved September 12, 2022, from http://www.education.gouv.qc.ca/fileadmin/site_web/documents/education/jeunes/pfeq/PFEQ_monde-contemporain-4-unites.pdf

Mogren, A. (2019). *Guiding principles of transformative education for sustainable development in local school organisations. Investigating whole school approaches through a school improvement lens* [Doctoral dissertation, Karlstad University].

Morin, É. (2021). *L'étude du sentiment de pouvoir agir de jeunes du Québec face aux changements climatiques: Dimensions et conditions favorables à son développement à l'école secondaire* [Doctoral dissertation, Université du Québec à Rimouski].

Morin, É., Therriault, G., & Bader, B. (2019). Le développement du pouvoir agir, l'agentivité et le sentiment d'efficacité personnelle des jeunes face aux problématiques environnementales sociales et environnementales: apports conceptuels pour un agir ensemble. *Éducation et socialisation, 51*. Retrieved September 12, 2022, from http://journals.openedition.org/edso/5821

Morin, É., Therriault, G., Bader, B., & Dumont, D. (2021). Éducation à l'environnement et au développement durable du Saint-Laurent: une démarche éducative pour susciter les apprentissages et développer le pouvoir agir de jeunes Québécois du secondaire. *Canadian Journal of Environmental Education*. Retrieved September 12, 2022, from https://cjee.lakeheadu.ca/article/view/1694/0

Nussbaum, M. (2000). *Women and human development: The capabilities approach*. Cambridge University Press. https://doi.org/10.1017/CBO9780511841286

Oliveira, A. W., Rogers, P., Quigley, C. F., Samburskiy, D., Barss, K., & Rivera, S. (2015). Environmental agency in read-alouds. *Cultural Studies of Science Education, 10*(2), 247–274. https://doi.org/10.1007/s11422-013-9531-6

Paillé, P., & Mucchielli, A. (2016). L'examen et l'analyse phénoménologiques des données d'entretien. In P. Paillé & A. Mucchielli (Eds.), *L'analyse qualitative en sciences humaines et sociales* (pp. 143–159). Armand Colin Éditeur.

Payne, P. G. (2015). Critical curriculum theory and slow ecopedagogical activism. *Australian Journal of Environmental Education, 31*(2), 165–193. https://doi.org/10.1017/aee.2015.32

Potvin, A., & Bader, B. (Eds.). (2020). *Guide d'action. Accompagner une démarche intégrée de développement durable en milieu scolaire*. Retrieved September 12, 2022, from https://ieds.ulaval.ca/linstitut/demarche-eds/outils-et-documents-sur-les-odd/

Pruneau, D., Kerry, J., Langis, J., & Léger, M. (2015). Améliorer les programmes canadiens de sciences et technologies au primaire par l'ajout de compétences du 21e siècle. *Revue canadienne de l'éducation, 38*(3), 1–23. Retrieved September 21, 2022, from https://journals.sfu.ca/cje/index.php/cje-rce/article/view/1932/1775

Sadler, T. D., Barab, S. A., & Scott, B. (2007). What do students gain by engaging in socioscientific inquiry? *Research in Science Education, 37*(4), 371–391. https://doi.org/10.1007/s11165-006-9030-9

Sauvé, L. (Eds.). (2018). *Stratégie québécoise d'éducation en matière d'environnement et d'écocitoyenneté*. Retrieved September 12, 2022, from https://www.coalition-education-environnement-ecocitoyennete.org/wp-content/uploads/2019/07/Strategie-Edition-complete.pdf

Savoie-Zajc, L. (2018). La recherche qualitative/interprétative. In T. Karsenti & L. Savoie-Zajc (Eds.), *La recherche en éducation. Étapes et approches* (4th ed., pp. 191–218). ERPI.

Schreiner, C., Henriksen, E. K., & Kirkeby Hansen, P. J. (2005). Climate education: Empowering today's youth to meet tomorrow's challenges. *Studies in Science Education, 41*(1), 3–49. https://doi.org/10.1080/03057260508560213

Sen, A. K. (1985a). *Commodities and capabilities*. North Holland.

Sen, A. K. (1985b). Well-being, agency and freedom: The Dewey lectures 1984. *The Journal of Philosophy, 82*(4), 169–221. Retrieved September 12, 2022, from https://www.jstor.org/stable/2026184?seq=1

Sen, A. K. (2010). L'idée de justice (P. Chemla, trad.). Paris : Flammarion.

Tsevreni, I. (2011). Towards an environmental education without scientific knowledge: An attempt to create an action model based on children's experiences, emotions and perceptions about their environment. *Environmental Education Research, 17*(1), 53–67. https://doi.org/10.1080/13504621003637029

UNESCO. (2017a). *L'éducation en vue des objectifs de développement durable. Objectifs d'apprentissage*. UNESCO. Retrieved September 12, 2022, from https://unesdoc.unesco.org/ark:/48223/pf0000247507

UNESCO. (2017b). *Faire face aux changements climatiques. Guide sur l'action-climat destiné aux établissements scolaires*. UNESCO. Retrieved September 12, 2022, from https://unesdoc.unesco.org/ark:/48223/pf0000247820

UNESCO. (2019). *Country progress on climate change education, training and public awareness. An analysis of country submissions under the United Nations framework convention on climate change*. UNESCO. Retrieved September 12, 2022, from https://unesdoc.unesco.org/ark:/48223/pf0000372164

UNESCO. (2020). *L'éducation au développement durable: feuille de route*. UNESCO. Retrieved September 12, 2022, from https://unesdoc.unesco.org/ark:/48223/pf0000374891

UNESCO, and United Nations Framework Convention for Climate Change. (2016). *Action for climate empowerment. Guidelines for accelerating solutions through education, training and public awareness*. UNESCO. Retrieved September 12, 2022, from https://unfccc.int/files/cooperation_and_support/education_and_outreach/application/pdf/action_for_climate_empowerment_guidelines.pdf

Vongalis-Macrow, A. (2013). How the concept of agency aids in teaching about sustainability. *Educational Research and Reviews, 8*(18), 1642–1649. Retrieved September 12, 2022, from https://academicjournals.org/journal/ERR/article-full-text-pdf/8EFC9FF5578

Wals, A. E. J., & Mathie, R. G. (2022). Whole school responses to climate urgency and related sustainability challenges. In M. A. Peters & R. Heraud (Eds.), *Encyclopedia of educational innovation*. Springer. https://doi.org/10.1007/978-981-13-2262-4_263-1

Wittgenstein, L. (1961). *Les investigations philosophiques*. Gallimard.

Zeyer, A., & Kelsey, E. (2013). Environmental education in a cultural context. In R. B. Stevenson, M. Brody, J. Dillon, & A. E. J. Wals (Eds.), *International handbook of research on environmental education* (pp. 206–212). Routledge. https://doi.org/10.4324/9780203813331

7 Accelerating Change-Making: Reflections on Embedding Regenerative Practices in School Climate Action

Ellen Field, Michèle Andrews, John Hannah, Eleonor Kerr, Drew Stephens, and Alison Elliott

Key Message
Schools have the opportunity, and arguably a moral imperative, to create an aspirational vision for a healthy, thriving world. This chapter shares commonalities, challenges, and emerging opportunities of three schools' journeys as they work to accelerate climate action through whole school approaches with efforts that aim beyond sustainability toward restoring and regenerating the world around us.

7.1 Introduction

In this chapter, administrators and teachers from three independent schools in Canada and at different stages of whole school sustainability describe and critically reflect on their journeys. The schools joined the pilot cohort of the Climate Action Accelerator Program (CAAP) that was launched in January 2022. As part of the program, each school has made a commitment to create and implement bold, high-impact, whole school climate action plans. In doing so, they are striving to become leaders in the K-12 sector and within their communities, aiming to be at the forefront of educational responses focused on a just and regenerated world. These three schools are all at different stages of their whole school approach journeys of integrating sustainability and climate action. These schools each started their journey prior to joining CAAP. The CAAP allows them to form a network with other schools pursuing the same goals and requires that schools develop a committed governance team that aligns with a whole school framework.

7.1.1 Why Whole School Approaches to Climate Action Are Needed

Young people primarily learn about climate change through formal schools (Field et al., 2019); however, most often climate change education occurs within science subjects and often predominantly focuses on cognitive knowledge, specifically centering scientific literacy in terms of the physical mechanisms of climate change and the validity of climate science (Henderson, 2019; Wynes & Nicholas, 2019). While a cogni-

E. Field (✉)
Lakehead University, Orillia, Canada
e-mail: efield@lakeheadu.ca

M. Andrews
DoorNumberOne.org, Maitland, Canada
e-mail: michele@doornumberone.org

J. Hannah · E. Kerr
Hillfield Strathallan College, Hamilton, Canada

D. Stephens
Southridge School, Surrey, Canada

A. Elliott
Trinity College School, Port Hope, Canada

A. E. J. Wals et al. (eds.), *Whole School Approaches to Sustainability*, Sustainable Development Goals Series, https://doi.org/10.1007/978-3-031-56172-6_7

tive understanding of climate change mechanisms and impacts is imperative, it does not attend to the whole person and often lacks an action-oriented engagement with socio-emotional or justice-focused dimensions of climate change education (Grewal et al., 2022; Hargis & McKenzie, 2020; McKenzie, 2021; UNESCO, 2019; Wynes & Nicholas, 2019). A national evaluation of climate change education policy in Canada (Bieler et al., 2018) showed shallow engagement with climate change, an overwhelming focus on energy efficiency upgrades in schools, and a lack of holistic responses to climate change. Bieler et al.'s research identifies gaps between physical operations and learning. From our experience, this gap occurs when school facility departments do not communicate directly with teaching staff, causing missed learning opportunities. We believe that students would be more likely to see their learning as responsive to the unfolding climate crisis if they experience these changes not only in physical operations but across school governance and in classroom teaching, signifying that their schools are taking climate action seriously.

The delayed leadership of formal education systems has been echoed in the streets by youth climate activists in countries around the world (Han & Anh, 2020), specifically by youth groups advocating for increased climate change education (Climate Education Reform British Columbia, 2021; Teach for the Future, 2021), including the first-ever student-written education bill, *Climate Education Bill*, in the UK (2022). Currently, 39% of countries have a national-level law, policy, or plan specifically focused on climate change education (Benavot & McKenzie, 2022). In Canada, delayed policy leadership is also evident in the limited number of explicit and mandatory climate change curriculum expectations in regional curriculum documents (Field et al., 2023). Other recent research has benchmarked rates of youth climate anxiety internationally, reporting that 60% of youth respondents across countries are very to extremely worried about climate change (Hickman et al., 2021). In this same study, climate anxiety and distress were correlated with perceived inadequate government responses and associated feelings of betrayal (Hickman et al., 2021). A recent Canadian study found that Canadian youth, similar to the Hickman et al. (2021) study, reported feeling afraid (67%), anxious (63%), and powerless (58%) in relation to climate change, which are all normal emotional responses to the climate crisis and for most young people are not pathological (Clayton et al., 2017). In addition, this Canadian study explored young people's perspectives on the education system and found that 65% of young people surveyed believe that the education system in Canada should be doing "more" or "a lot more" to educate young people about climate change. Text responses to an open-ended question on how the education system could better support young people's learning indicated that young people would like an increase in climate change content, a focus on teaching solutions and taking action, and mental health supports provided within the education system.

Given that young people primarily learn about climate change through school, spend a large majority of their time at school, and are also experiencing high rates of negative climate emotions, we propose that schools are essential spaces for not only learning about the climate crisis but also for students to engage in active learning, innovating, experimenting, and taking action that addresses local climate issues. The climate crisis, along with the many intersecting social crises of our time, is emerging and evolving and require that all generations learn about and engage with them synchronously rather than in isolation. There is an amplifying opportunity for schools to be sites of climate learning and action not just for the members of the school but for the surrounding community (Facer, 2012). Research has also demonstrated that the ripple effect of climate change education within schools extends beyond students into families and communities (Lawson et al., 2019). K-12 schools have the potential to be centers for learning and action that help achieve the necessary tipping points in a whole-society approach, inclusive of technical, instrumental, and adaptive transformation, through adopting whole school approaches, community-focused climate actions, and integrating green skills for the climate action framework (Kwuak & Casey, 2021). Young people involved in the transformation of their schools into institutions leading in climate action will undoubtedly experience some offset to negative climate emotions; moreover, we believe that young people witnessing adults, such as school leaders and teachers, taking climate action in pursuit of securing more stable climate futures is one of the most powerful antidotes to negative climate emotions and youths' accompanying feelings of betrayal (Hickman et al., 2021).

7.1.2 Climate Action Accelerator Program: A Whole School Approach

The Climate Action Accelerator Program, launched in January 2022, provides an iterative approach and process for schools to improve their climate action through schools by deepening students' understanding of the climate crisis, developing a whole school climate action plan, establishing benchmarks from which future change can be measured, and participating in a community of practice with other school teams. Participating schools make a 3 year commitment to collaborate with other schools to create and begin implementing a whole school climate action plan. After 1 year, the CAAP enrolled 21 schools from five provinces across Canada. These schools are all independent and are members of the Canadian Accredited Independent Schools (CAIS) association. They are all not-for-profit organizations, registered charities, fee-based, and are governed by an independent board of governors. Many, though not all, independent schools are very established with significant endowments and alumni/ae support. Participating schools create a multidisciplinary team with at least one senior administrator, one faculty member, one representative from the facilities team, and two students. School teams have in many cases grown to be a dozen or more people, including many students. The peer support, sharing of ideas, building capacity, and the level of accountability brought by making this commitment together are seen as helpful to making significant progress.

The CAAP begins with the concept of regeneration—the idea that we must not strive merely to maintain or sustain the current state or strive to do "less bad," but whole school climate action today can, and should, be about healing, repairing damage, restoring, and creating a future of thriving and flourishing for all. The inspiration for this approach with schools came from the work of Jason F. McLennan, who created the Living Building Challenge, a program first published in 2006 out of the not-for-profit he founded, the International Living Future Institute (ILFI). The Living Building Challenge invites all those working in the built environment to design new buildings and renovate existing ones, with the intention of making them regenerative, using the metaphor of the flower: flowers draw their energy from the sun, nutrients and water from the ground, and their waste provides food for future flowers and supports the health of the ecosystem in which they are growing. Examples of Living Building Challenge projects, as well as many related initiatives for products and communities, can now be found all over the world (International Living Future Institute, 2022). The CAAP also references the work of Bill Reed, a planning consultant, author, and leader in the field of regenerative development for over two decades (2007; 2023), the work of Damon Gameau, an actor, director, and activist from Australia who founded Regen Studios as a catalyst for building the regeneration movement (2023), and the work of Paul Hawken, activist and author, founder of Project Drawdown (2017) and Regeneration (2021). The concept of regeneration, along with themes of urgency, literacy, equity, and hope that underpin the program, has proven to be immediately inspiring for participants. School participants within the CAAP are asked to consider all their actions with the question, *"What if every action you took, every decision you made, contributed to a thriving life for all people and all the rest of nature?"*

Donella Meadows (1999), in her work on systems change, identified a ranked set of leverage points for intervention. She identified the greatest opportunity for leverage as the "paradigm out of which the system—its goals, structure, rules, delays, parameters—arises" and with "the power to transcend paradigms" (p. 3). The CAAP attempts to help school leaders, formal and informal, to adopt this new paradigm of regeneration to drive faster, deeper, and more effective systems change toward whole school climate action. The CAAP theory of change is that transforming schools can inspire school community members, including parents, grandparents, alumni/ae, neighbors, and anyone else connected into becoming change-makers who can influence climate action in business, society, and government today. In this way, the CAAP definition of "whole school" is potentially more aspirational than other frameworks, with its emphasis on regeneration and its striving to influence multiple generations within the communities connected to the schools to expedite the transition to a zero-carbon, just, regenerated world.

The theme of hope is introduced alongside the idea of regeneration at the outset of the program, and is carried throughout the CAAP using the work of Elin Kelsey and her book *Hope Matters: Why Changing the Way We Think Is Critical to Solving the Environmental Crisis* (2019). Using Kelsey's model for evidence-based hope, CAAP participants are invited to seek out evidence of where there is already healing, restoration, and regeneration happening in their own communities and around the world. Adults and students alike find this to be a very helpful, motivating practice that provides a greater sense of confidence as they make the mental shift toward regenerative practices.

A strategic framework for whole school climate action was developed for use in the CAAP represented by a common pattern found in nature: the hexagon (Fig. 7.1). From honeycombs to corals and from turtle shells to dragonfly eyes, the hexagon is nature's elegant solution to maximizing volume and strength while minimizing material use.

At the core of a plan, a school leading in climate action must have a compelling aspirational vision of their school as an agent of change in transition to a just and regenerated world. The vision has the power to galvanize the school community and provide inspiration for their climate plan. The vision should be reflected in the school governance and leadership, with clear responsibilities and accountabilities, supported by performance metrics shared throughout the school community that track progress toward defined short- and longer-term goals.

In addition to the vision, there are three dimensions and six strategies that comprise the CAAP framework and that help guide schools in operationalizing their transition to whole school climate action. The three dimensions are briefly outlined below.

1. Organizational Culture

The culture of a school is defined by the shared values, social norms, and practices within the organization. To achieve their compelling vision, schools must ensure that the paradigm of regeneration is embedded in their culture and visible throughout the school in the physical space, the curricular and co-curricular programming, school operations, and governance.

2. Collective Learning

A whole school approach, led by cross-disciplinary teams, is optimal for collective learning. Everyone in the school community needs to engage in climate change education, learning from and with each other.

3. Physical Space

The physical space includes school operations, all resources used on campus, the facilities, and the natural environment stewarded by the school. This physical dimension implicitly teaches and communicates the school's values to users through interactions with the space and its features. It is the context for the learning experience and, as such, should reflect the compelling vision for sustainability throughout.

The framework includes six strategies, two in each dimension, for schools to create and implement their plans, as displayed in the outer ring of hexagons in the figure above: create a healthy synergetic system, communicate strategically,

Fig. 7.1 Climate action accelerator program's strategic framework

collaborate in cross-disciplinary teams, use a student-centred approach, share leadership, and design space as a teacher (DoorNumberOne.org, 2022).

The CAAP framework is similar to other whole school frameworks and was specifically influenced by Barr et al.'s model published by The Center for Green Schools at the US Green Building Council (2014). Barr et al. (2014) created a guiding framework to advance whole school sustainability efforts, which has three components, Organizational Culture, Physical Place, and Educational Program, and, within each, three principles to help schools develop their approach to whole school sustainability. More recently, Wals and Mathie (2022) orient a whole school approach as a flower. They guide schools toward taking a systems approach in understanding the opportunities and interrelatedness of working with all the stakeholders in a school community to transform education in the context of emerging global challenges. Building capacity for systems thinking is a common core principle found in all three frameworks, along with many others in practice, in driving successful whole school approaches. Moreover, Holst's (2022) conceptual framework of whole school approaches, which was developed from synthesized literature, identifies a set of core principles (coherence, continuous learning, participation, responsibility, long-term commitment) and seven areas of action, including interactive and participative governance, cross-disciplinary orientation of curriculum and learning, the process of (re-) designing operations and campus management, community-orientation, fostering competencies through capacity building, and clear and consistent communication. Overall, there is much alignment and commonalities across the whole school approach frameworks derived from literature reviews and the CAAP's programmatic design.

7.2 Three Schools' Journeys Toward Whole School Climate Action

This chapter is a critical reflection from three schools that have worked for years or even decades on sustainability projects and have now completed their first year in the CAAP. Within the cases, it is clear that each school is at a different stage of whole school sustainability (early, intermediate, and advanced), and each school is prioritizing sustainability and climate actions that are context-specific. The schools volunteered to participate in the reflective process and in the writing of this chapter. The process was coordinated by Dr. Ellen Field, Assistant Professor in the Faculty of Education at Lakehead University, and Michèle Andrews, Co-founder and Executive Director of DoorNumberOne.org, the organization running the CAAP. The team met several times to discuss the specific school contexts, successes, and challenges. All conversations were recorded, and transcripts of the discussions inform the Discussion section of the chapter. All participants spoke authentically about their school's journeys in order to co-learn with each other and with the aim to share their process with schools beyond the CAAP through this publication.

7.2.1 Southridge School

Southridge School, founded in 1995, is situated on 17 acres in South Surrey, British Columbia, on the unceded native Coast Salish territories of the Kawantlen, Katzie, Tsawwassen, and Semiahmoo First Nations. It serves 680 students in kindergarten through Grade 12/University Entrance. Southridge's whole school approach has been catalyzed by the head of school, who has made the school's sustainability initiatives a clear priority.

Under the leadership of the assistant head of school, a team of 16, including faculty, staff, and students, works with the school community across ten areas of impact, including teaching and learning, building and grounds, and school culture. The school's approach to developing its Environmental Stewardship Plan (ESP) was research-based, using the results of environmental audits completed in 2014 and 2019. With the goal of integrating student-initiated environmental projects into a whole school strategic framework that encompasses all facets of operations, a fundamental cornerstone of the ESP is the belief that coordinated efforts within a common understanding of shared purpose are more impactful than well-intentioned initiatives operating independently from one another. As such, the perspective Southridge will explore in this chapter is a systems approach (as defined by Wals & Mathie, 2022) to the development and implementation of a whole school environmental/climate action plan.

Stemming from its first environmental audit completed in 2014, Southridge positioned its climate action efforts from two distinct and separate perspectives: school operations and student initiatives. Over the next 5 years, numerous initiatives were drawn from the audit and implemented with the intention to become a more environmentally responsible school community. Indeed, the efforts of the school from both perspectives received support from the board of governors, and financial resources were made available to move the school forward. In 2018, and in alignment with the school's preparation for its 25th anniversary, Southridge looked to the future and began to write the next chapter of its story. With a long-term outlook in their sights, the board of governors and head of school designed a 10 year strategic plan to focus on equipping each student for their path ahead, developing a creative and supportive environment for learning and work, and connecting its diverse community. With a distinct and sincere interest in how the school would prepare its students to become globally aware and engaged citizens, the head of school began to consider the extent to which environmental stewardship would play a role in equipping the school's students for the future.

In alignment with the school's mission to educate students who make a difference in the world, the head of school used the 2014 environmental report as a basis to form the content of its next environmental audit. The 2019 audit highlighted many areas of progress the community had made over the 5 years, particularly in the areas of food services and information technology, and it also identified points for continued focus, growth, and improvement (in particular, co-curricular programming, housekeeping services, and waste management).

The points for continued growth and improvement from the 2019 audit report were used in alignment with the 2 year "getting ready" phase of the school's 10 year strategic plan. From an environmental stewardship point of view, the "getting ready" phase meant creating a mini strategic plan to unite distinct and separate climate action efforts—school operations and student initiatives—in a new and purposeful whole school approach. The beauty of the design process was in how it brought together voices from all areas of the school to engage in thoughtful discussion based on evidence contained in the audit report. Over the course of the next 6 months (from November 2019 to April 2020), a team of 16 people wrote Southridge's first strategically aligned ESP; and in September 2021, they began to implement the first series of 15 key initiatives.

Even though the output was a comprehensive and engaging ESP, one of the challenges the team faced in writing it was finding common ground on issues that were, up to that point in time, clearly the jurisdiction of only certain people or departments. Welcoming distinct voices into iterative discussions to generate a whole school perspective was a hurdle that created tensions and tested the team's ability to truly collaborate. It took practice to listen openly to the opinions of nonexperts who did not have practical experience in certain areas. For example, teachers offering their opinions on sustainable purchasing practices the school *should* be following created an oppositional dynamic that was difficult to overcome. With time and after many honest conversations and authentic group interactions, it became evident that everyone was working toward a common goal. The trust that blossomed from these kinds of interactions helped ease the transition from a disparate approach of well-intentioned initiatives operating independently from one another (and with questionable influence on the school's environmental culture) to a more impactful whole school systems approach of coordinated efforts around a common understanding of shared purpose.

The implementation of the school's ESP went very well, largely due to the model of shared leadership it inspired. One of the most significant benefits of taking on a whole school approach to environmental stewardship and climate action is that people recognize that the only way to help move the school forward is to contribute to the execution of the plan. People in designated leadership roles intentionally stepped back from advancing each key initiative on their own, and they provided the space and opportunity for nondesignated leaders to decide how best to approach the initiatives. Team members had more ownership over the plan and felt a greater sense of accomplishment and excitement about the changes that took place over the course of the year, which translated into enthusiasm for continuing the work in the following year. It also garnered appreciation for the work people do in all departments across the school, which helped build an even stronger sense of shared leadership and shared goals.

Southridge administration took some time after the CAAP Year One to align its Environmental Sustainability Plan with the Climate Action Accelerator Strategic Framework. This analysis showed that a systems approach to environmental stewardship is strongly represented in the school's work on the building and grounds, as well as in other areas of school operations such as sustainable purchasing. It also showed that key initiatives in the Southridge plan are equally represented across Organizational Culture, Collective Learning, and Physical Space.

However, one important area of focus for the future will be providing professional development and education opportunities for board members. An excellent generative discussion at a spring 2022 board meeting revealed a desire to make better-informed environmental and climate action decisions in the future. The current annual board education plan for 2022–2023 has two meetings designated as "environmental." The first meeting will focus on what it means to shift an organization from using a green initiative-based program to using a regenerative paradigm model. The second meeting will be designed to help position the school's key environmental performance indicators against industry standards, as defined by progressive Canadian independent schools. In the end, the board will have a greater appreciation for what leading schools across the country are doing and how Southridge can position itself as a contributing member of the growing number of schools that are recognizing their role in creating a regenerative future.

A second area of focus will be on revisiting the vision for the work. In 2020, the committee responsible for creating the ESP spent considerable time thinking about its vision to help the school have a better understanding of possible future directions. Taking into account that the initiatives the committee planned to undertake at the time were relatively modest, the original vision read, "Given its emphasis on making a difference in the world, Southridge intentionally and proactively contributes to an environmentally sustainable future." Southridge is reconsidering this vision statement in light of its work in the CAAP. Providing some education for the board of governors in 2022–2023 will help facilitate that process. Since developing a compelling vision will have a financial impact on the school, it is important for the board to possess sufficient background knowledge about climate action and environmental regeneration to contribute in a well-informed way to the development of an appropriately positioned vision that is more aspirational and compelling than what is currently in place.

7.2.2 Hillfield Strathallan College

Hillfield Strathallan College (HSC) is a Canadian independent school with 1280 learners from prekindergarten to Grade 12. The campus is nestled on 50 acres in the traditional territories of the Erie, Neutral, Huron-Wendat, Haudenosaunee, and Mississaugas. This land is covered by the Dish With One Spoon Wampum Belt Covenant, which was an agreement between the Haudenosaunee, Anishinabek, and other Indigenous nations to share and care for the resources around the Great Lakes. HSC is committed to honoring this covenant by working alongside Indigenous neighbors and building a sustainable relationship with the land. In this chapter, the work of reimagining and redeveloping the campus grounds as a Learning Landscape will be explored as a way to facilitate building a culture of sustainability and integrating it into curricular and co-curricular programs.

7.2.2.1 A Not-So-Novel Idea...

There was a time, not too long ago, when Canadian schools did not have gymnasiums (New World Encyclopedia, 2022). As surprising as it may seem, school assemblies, theatrical performances, and physical education classes did not have a specialized instructional space a century ago. Fast-forward one hundred years, and gyms have become an essential school feature based on their many educational benefits. However, there's another potential piece of school infrastructure that has been overlooked, one that can more significantly impact student learning and engagement. Over the last two decades, research has shown that looking outside the school to the school grounds as a learning space provides a highly effective opportunity to increase student learning and wellness (numerous articles within Children and Nature Network Research Library). In 2009, HSC set out to create outdoor classroom spaces and biodiversity enhancements that would fully integrate into all grades, understanding that this could become the twenty-first century's educational innovation. While the built infrastructure covers approximately half of the campus, the remaining land consists of lawns, playing fields dotted with ornamental trees, and an apple orchard remnant of the farm that operated on the land a century ago. Reintroducing native biodiversity led to the emergence of a new vision of the campus as a Learning Landscape. Rather than a passive backdrop of open space, the grounds are now being treated as a new piece of school infrastructure for twenty-first-century learners. Its features support and enhance student learning and social development, academic achievement, creativity, focus, and mood, along with staff job satisfaction and wellness, climate change resilience and adaptation, and more.

7.2.2.2 Meet the Learning Landscape

The typical unbuilt campus area, along with the property surrounding many suburban North American schools, has lawns, some ornamental plantings, and sports fields lined with invasive nonnative trees. The Learning Landscape reenvisions these areas to leverage new opportunities for growth and development in students across all ages and subject areas. It began with reintroducing naturalized areas, which act as co-teacher, provide an ecological function to support local biodiversity, and even help mitigate some impacts of climate change. The landscape includes an arboretum and living laboratory, a conservation corridor, nature sanctuaries, wellness spaces, and outdoor classrooms. This work also affords using these spaces for year-round outdoor activity and physical education classes.

7.2.2.3 Setting the Course

In 2009, HSC embarked on a strategic plan with the goal of becoming a leader in sustainability and environmental responsibility. The school leadership opened the door to student and staff initiatives related to biodiversity or sustainability. Initiatives were supported and encouraged, whether they were tied to curriculum, infrastructure change, or day-to-day facility decisions. A global body of research on the impact of outdoor learning and activity in nature has developed in the last 15–20 years; it provides the rationale for looking outdoors at the unbuilt parts of campus as a Learning Landscape (see Children and Nature Network research library (n.d.) for numerous research articles on the benefits of outdoor learning: https://research.childrenandnature.org/).

7.2.2.4 Wayfinding Led by Students

Staff and students across HSC have played an important role in developing new landscape elements. Students are enthusiastic about the tasks, and, more importantly, their involvement builds skills and agency that give them hope for their future in a rapidly changing world. Younger students thrive in the Learning Landscape, and the revised junior school science program is based on a gardening curriculum. Grade 6 students have been proposing campus biodiversity enhancement projects for 5 years. Grade 9 students investigate what actions are taken on the campus that could support particular species at risk. Senior biology students stratify, germinate, and plant specialized plant species native to Ontario, and they are developing a campus native tree inventory to then calculate greenhouse gas sequestration potentials. There are boundless opportunities for creative, imaginative, experiential, and hands-on learning, as well as opportunities to collaborate with community partners. Students have a deep bond and well-developed sense of place and purpose, and they know that their school is solving problems locally and modeling solutions to global challenges.

Cross-curricular integration, inquiry, and project-based learning work well in natural spaces. The HSC campus showcases biodiversity and leverages green infrastructure for cost savings and climate change resilience and mitigation. Still, above all else, it enhances student achievement and both student and teacher health and well-being. Ongoing naturalization projects at HSC directly support the 2021–2030 United Nations Decade of Ecosystem Restoration initiative and the broader Sustainable Development Goals. These projects help represent and build an understanding of the local presettlement habitats on the site while shifting focus to situating experiential learning on campus and reducing greenhouse emissions from busing that would have been used for field trips. Just as gymnasiums went from "new concept" to "essential school facility" through the twentieth century, HSC is looking forward to a future where a Learning Landscape will be a crucial part of every school, and all students and faculty will have the opportunity to reap its many benefits. A cultural shift is happening that is purposely and deliberately supported by leadership and driven by the initiative of staff and students.

HSC joined the CAAP in the first cohort of schools to lead system-wide thinking and to become change agents to model what is possible. There is a commitment to continue to leverage their growing capacity to make better decisions, stimulate a more comprehensive whole school approach to strategies, fully integrate student voices, and collaborate and share best practices that they aspire to co-design and implement in a community of like-minded leading independent schools across Canada. The CAAP is a platform for sharing ideas, strategies, and best practices, and, equally important, it is a space where school leaders can articulate and discuss what is really meant by bold action that inspires hope for a better future. Peer schools in the CAAP will take back a stronger and more compelling voice to individual schools and continue to build the momentum of a whole school approach within a collective of transparent and shared CAIS (Canadian Accredited Independent School) approaches. Since joining CAAP, there has been heightened appreciation at HSC that every degree of warming matters, every year matters, every choice matters, every action matters, and hope matters.

7.2.2.5 Compelling Vision

The HSC Campus Master Plan and Strategic Plan 2022 offer a response to the climate emergency, in particular on the theme of purposeful and sustainable learning environments. This response ensures that the entire community is aligned with the overall goals of significantly reducing the school's footprint, role-modeling regenerative behaviors, and working toward a carbon-negative future. It is recognized that no part of this problem will be solved in isolation, and everyone in the community will have to innovate, support, and sacrifice to achieve these goals.

7.2.2.6 Organizational Culture

HSC is a large and physically dispersed set of four schools, all operating within one school, and with a very large catchment area. HSC is working to ensure that the values of sustainability and climate action are woven throughout the school. Implementing concepts from the CAAP, such as an overall climate action plan, sustainability survey, ongoing professional development, communications, and student leadership roles, reinforces the school's commitment to environmental sustainability and climate action.

7.2.2.7 Collective Learning

HSC is bridging curriculum areas and looking for continuums that will thread through curricular and co-curricular learning and strengthen climate actions. HSC is a platinum EcoSchool and has established student committees to lead student actions. The sometimes-hidden talents of students, faculty, staff, and community members are being tapped. Programming is emerging and evolving to be more data-driven, action-oriented, and project-based and, as a result, is having more impact on more students. Preparing students for an uncertain future includes developing skills and mindsets that are aligned with a progressive school ground element like the Learning Landscape, where they have agency to actively support biodiversity and climate change adaptation, mitigation, and resilience and to do so on the other side of their classroom wall. Planting oak savannah species, managing runoff into waterways and wetlands, and investigating new solutions for increasing food security are examples that enrich the student experience.

7.2.2.8 Physical Space

In 2007, HSC began to focus on sustainability with initiatives to measure and sort waste, replace light bulbs, consume less, and investigate energy reduction behaviors that could be implemented across campus. Now, HSC is turning its attention to actively model regenerative systems and infrastructure, including gardens, food forests, representative natural habitats, solar arrays, and electrification, and ensure the entire 50 acres are used as outdoor classrooms and an intentional campus-wide Learning Landscape that serves desired learning outcomes. A baseline greenhouse gas inventory was established from 2019 data that will drive many of the operational decisions moving forward. During the last major capital campaign and construction of a new senior school, climate action-related elements were introduced. A Green School Steering Committee was established to determine priorities and renew conversations around energy conservation, waste reduction, green infrastructure, and other ideas that environmentally offered "more good" rather than just "less bad." Several significant stormwater bioswales have been added to intercept parking lot run-off, and building features were planned that support sustainable innovations in the future.

Moving forward, HSC will continue to strive to be a leader in climate action and environmental sustainability through education, construction, and ongoing operations. The large catchment area and corresponding bus fleet remain a challenge to resolve. There is a commitment that each new building on campus will be designed and operated to a very high environmental standard. Similar to the learning landscapes outside, the built environment will be used as a learning tool for environmental stewardship and climate change mitigation. This will serve current and future students across every grade and curriculum area by enhancing and enriching educational experiences, supporting biodiversity and human wellness, and improving the impacts of climate change on campus and within the community. HSC intends to make a meaningful impact on the built environment on campus to support the goal of carbon neutrality by 2030.

7.2.3 Trinity College School

Trinity College School (TCS), founded in 1865, is situated on a one-hundred-acre campus in Port Hope, Ontario, on the traditional territories of the Wendake-Niowentsïo, Mississauga, Anishinabe Waki, and Haudenosaunee. The student population of approximately 600 students from grades five through 12/Advanced Placement includes over three hundred boarding students and approximately three hundred day students. TCS has been working on environmental sustainability initiatives since the early 1970s and has taken a whole school approach since the mid-2000s. It has certified with EcoSchools Canada since 2012, achieving its platinum level each year since 2017, and has received several awards, including the Canada Green Building Council's "Greenest School in Canada" award in 2018 and a Canada Clean 50 Top Project award in 2019. The following will provide an overview of the TCS approach to greening school culture through incentivized student challenges, sustainability awards, the MOGO (more good) framework, a portfolio of special curriculum initiatives occurring beyond the academic classroom, and retrofitting aged infrastructure to fight climate change and turn awareness to action.

In the early 2000s, with global headlines broadcasting ever-growing concerns about environmental degradation and leading independent schools in Canada and the United States beginning to invest in school greening programs, TCS established a formalized environmental coordinator position. Over the next 20 years, the role and its reach evolved, as did the approach to achieving meaningful change. The leadership provided through the position has helped shift the lens from project-based one-offs to business-style sustainability planning to a whole school model situated in the overall TCS strategic plan, with responsive strategic directions and both board and budgetary support. TCS transitioned from "yes, we do Earth Day and recycle" to working in all elements of "green campus, green curriculum, green culture" to asking, "how can our school systems regenerate our living systems such that we have healthy people and a healthy planet in the future?"

Initially, project-based, student-focused initiatives such as the junior school outdoor classroom program and senior school green cup challenge were implemented to transform environmental activities from fringe to normalized. Both initiatives strive to engage and empower all students instead of just a select club, and they directly support the school's mission to develop habits of the heart and mind for a life of purpose and service. Both have also flourished over the better part of two decades through vision, perseverance, hope, and reflective adaptation.

It became apparent in the late 2000s that a much broader approach was needed in order to ensure TCS was progressing on the path to being a green school (Chapman, 2012). A facilities-focused environmental audit was completed in 2012 to establish a baseline of utility usage and other footprint components such as waste, transportation, and purchasing. TCS's first sustainability plan was created in response to the audit with 140 goals set across ten target areas under a new, multi-stakeholder-created sustainability vision and mission. Throughout the mid-2010s, a multi-stakeholder sustainability committee was established with youth and adults from the school. The committee identified and implemented many initiatives identified as "low-hanging fruit," accomplishing 75% of the goals in the plan. See a list of some of those initiatives in the Table 7.1.

TCS came to a time of reckoning, however, after winning two significant green school awards. The celebration came together with questioning, "What did these accolades really mean? Were we really walking the talk? If there were more impactful and innovative ways that we could improve, how could we be the greenest school?" The school has come a long way in achieving the goal of normalizing green and sustainable practices in the school culture and reducing the campus footprint, but work is needed to continue to make green visible, desirable, and doable within the school culture.

Table 7.1 TCS early sustainability initiatives

Collective learning	Physical space
Although these examples are listed in two of the three CAAP strategic dimensions for whole-school climate action, the school's unifying Vision and Mission for sustainability, as well as their systemic approach, leads to impact across the school's Organizational Culture as well.	
Establishing sustainability awards and other incentives Increasing the number of senior students in green leadership roles to 15% of the student body Running internal sustainability professional development with all staff and in both operational and educational departments Promoting external professional development Engaging youth in external, multi-school workshops Educating parent groups and board Connecting senior school students to community environmental projects and groups through the service learning program Establishing senior school co-curricular programs that focus on trail maintenance, bike maintenance, and gardening Raising EcoSchools certification level to the highest achievable level	Retrofitting lighting systems Improving building automation systems Replacing windows and insulation Installing solar photovoltaics and solar hot water Creating a "green revolving fund" system for using project savings to fund future projects Expanding tree-planting, gardening, and habitat restoration programs Condensing bussing Establishing no-idle zones Establishing organics collection and better recycling programs Farming Sourcing more food within a 20 km radius

A multi-stakeholder barriers analysis was conducted to determine what might be holding TCS back from a greener campus and greener curriculum. The analysis determined that blocks to walking the green school talk continued to exist, and there would need to be new strategies to address these identified areas. Many of the barriers were typical: time, budget, competition between aims, and linear, siloed approaches. Some of the barriers to a greener culture were unexpected. The insights gained, from students in particular, indicated that appropriate "green" behavior was being limited by certain mindsets (privilege, entitlement, apathy, guilt paralysis, and rogue elements), unsustainable lifestyles, incorrect or limited knowledge, and, most surprisingly, peer pressure. Furthermore, the very concepts of "green" and "sustainable" were still unclear to many and/or linked to actions that made people feel restricted in their lives.

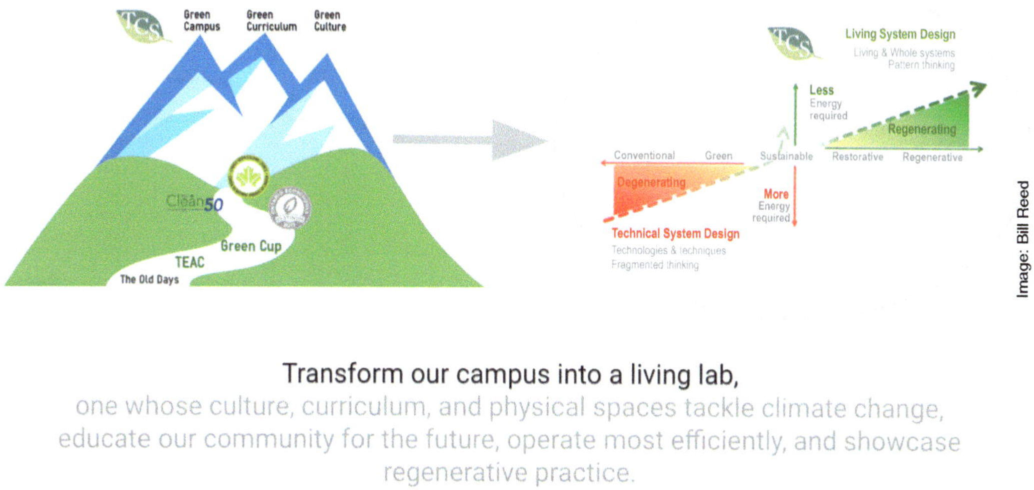

Fig. 7.2 TCS path forward as a three-peaked mountain. Note: Image on the right is of Reed's (2007) spectrum of conventional to regenerative design

TCS reframed the path forward in the image of a three-peaked mountain (Fig. 7.2). The present was described as being at basecamp on sustainability mountain, with the school having already climbed for some time and now looking toward climbing further over the various boulders and crevasses on the way to a green campus, green curriculum, and green culture. The MOGO (do more good, not less bad) campaigns were born as a motivating principle, and systems-thinking workshops were piloted for students and teachers. A new sustainability audit was initiated to showcase the last decade's improvements and provide direction for a large-scale sustainable infrastructure fundraising campaign. Recent infrastructure projects aimed at reducing the school's greenhouse gas footprint included adding solar hot water preheating to one of the boiler systems, doubling the solar photovoltaic array such that it will produce 10–12% of campus electricity, and installing electric vehicle charging stations.

TCS has mapped its many initiatives along a spectrum based on Bill Reed's description of conventional practices to sustainability through to restoration and regeneration (2007) (Fig. 7.3). In doing so, their community more clearly understands the opportunities to stretch their activities further.

As TCS emerged from the COVID-dictated programming lull, joining the CAAP provided a meaningful opportunity to share learning and mentor schools in earlier phases of the whole school approach planning. Most significantly, the discourse on and sharing of regenerative practice research has led to a paradigm shift: the top of sustainability mountain is in fact beyond green or sustainable; it is regeneration. This thinking is now informing the development of a new climate action and sustainability plan which will add regenerative elements to the mitigation and adaptation tactics. A new tagline, "healthy planet healthy people," has been launched to guide student, staff, and community focus. This tagline will help unify programming and integrate diversity, equity, inclusion, belonging, and justice into the climate action and sustainability plan.

7.3 Discussion

This section outlines common processes or approaches across the schools that we noted were important for moving the schools along their cli-

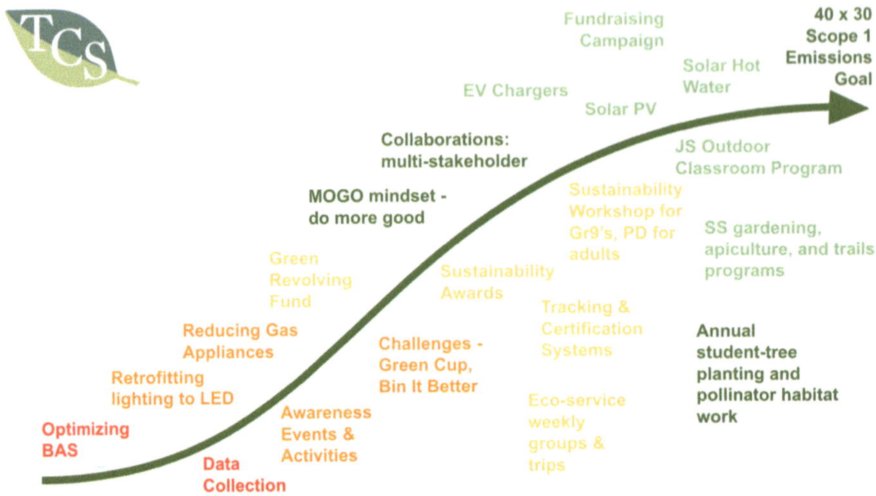

Fig. 7.3 TCS activities mapped according to Bill Reed's conventional to regenerative scale (2007). Note: Color coding aligns with Reed's (2007) color coding in Fig. 7.2

mate action and whole school journeys. The section is followed by a discussion of shared challenges and emerging opportunities.

7.3.1 Commonalities

7.3.1.1 Regenerative Practice as Paradigm Shift

Through our conversations, it became increasingly clear that all three schools are approaching a climate action-focused whole school approach as a paradigm shift or "whole system redesign" (Wals & Mathie, 2022) throughout the organizational culture, collective learning, and physical spaces of their schools. The teachers and administrators spoke of their aims and hopes for their work in the CAAP to bring them closer to transforming their schools toward a regenerative model of education and school operations. Hillfield Strathallan College and Trinity College School both have long histories of environmental sustainability. Their participation in the CAAP, along with their new paradigm of regeneration and commitment to be leaders, has them currently exploring how to integrate their climate action more centrally in their strategic plans. Southridge School, while in an earlier stage of its climate action journey, is approaching its strategic plan in an integrated and relational way by considering multiple global influences: the climate crisis, Indigenous Truth and Reconciliation, and pluralism, which includes equity, diversity, inclusion, and justice considerations. All three schools discussed the importance of holistic strategic plans that link student well-being, equity, diversity, and inclusion, and the climate crisis in order for the school to be responsive and act in regenerative ways to current societal challenges.

The conversation also discussed the challenge of working with multi-year strategic plans that do not traditionally allow for flexibility within the timeframe to adapt to emerging crises. These strategic plans try to balance the integrated and relational components while also being specific enough to identify objectives and strategies. Questions such as, "At what levels do these global influencers affect the school and should be reflected in the strategic plan?" and, "How do we ensure objectives are specific but will also still be relevant in a few years' time?" were discussed. Through discussions with the schools and some of the illustrative examples provided, these schools are addressing the need to respond to emerging and urgent climate challenges not simply as something to be implemented, a box to be ticked, or a siloed program, but rather as a catalyst of fundamental change that impacts all aspects of their educational program and operations. This approach requires the involvement of all actors that affect or are affected by a school's activities, including competencies such as systems thinking, inter- and transdisciplinary thinking, and anticipatory thinking, alongside attention to socio-emotional aspects and dealing with anxiety and fear (Wals & Mathie, 2022). This type of thinking represents the amplifying opportunity when schools take climate action seriously.

7.3.1.2 Leadership and Organizational Culture

For these three schools, and the rest of the schools in the CAAP, the challenge will be to ensure that the paradigm shift of regeneration becomes an unshakeable part of the school culture. To be successful, more and more of the administration, staff, students, and parents will need to shift their own mindsets. In our discussions with the three schools, a committed principal/head of school was considered essential for schools to take on a whole school approach to climate action. We discussed the many competing priorities and complexities that school administrators must balance. A commitment to climate action from the principal is critical for continually prioritizing its importance and seeing how climate change education and action intersect with other priorities such as equity, diversity, and inclusion, fundraising campaigns, or wellness initiatives. The consistent commitment of a principal will ensure that climate change is embedded into strategic plans; included in professional learning plans, accountabilities, hiring, and orientation processes; integrated into the physical space of the school; and communicated internally and externally.

The schools also discussed the importance of an inclusive and representative process in the development of a strategic plan and the act of sharing leadership in its implementation. One teacher reflected on her school's journey to becoming more systemic and reflective: "Twenty years ago, the adults viewed *the school system* as the students, building, and grounds and that somehow the adults were outside of the system." She noted, "Now, it is not the case; we look at the staff, board, and parents as part of our school system." These three schools seem to be taking a systemic approach where they are considering key aspects of schooling (curriculum, pedagogy and learning, professional development, school-community relationships, school practices, ethos, vision, and leadership) simultaneously (Wals & Mathie, 2022).

The schools approaching climate action as a regenerative paradigm shift through a whole school approach aligns with previous schools' successes on whole school sustainability. Some of the critical factors for school success within whole school approaches to sustainability, as evidenced through case studies, evaluation, and research, include taking a future-oriented perspective on the school's culture and practice, embracing complexity over direct problem-solving, and distributed leadership (Tilbury & Gavin, 2022). The schools participating in the CAAP seem to be demonstrating these factors in their progress through the program.

7.3.2 Challenges

For all three schools, integrating climate change into the curriculum remains a challenge to varying degrees. For Southridge, the province of British Columbia (BC) has a reformed competency-based curriculum that moves away from content knowledge descriptors. BC teachers have indicated that there are more opportunities to integrate climate change education and action. The other two schools in the province of Ontario identified the rigidity of the curriculum as limiting for integrating transdisciplinary climate action inquiry projects within instruction classes, as well as the limits of climate change expectations within the curriculum. This is not unique in Canada or to secondary schools attempting to implement whole school approaches in other countries (Bosevska & Kriewaldt, 2020). A national study on climate change education practices in Canada showed that there is limited class time spent on climate change content with between 33% (closed-sample) and 59% (open-sample) of teachers reporting teaching any climate change in a national survey. For the teachers who do integrate climate change content, most students experience 1–10 h of instruction per year or semester (Field et al., 2019).

All three schools also identified knowledge gaps among school actors as barriers. Southridge is actively facilitating environmental professional development at focused school board meetings to ensure school decision-makers are informed of the urgency of the climate crisis. Knowledge gaps among parents, staff, and teachers were also identified as a barrier by all three schools. These gaps align with findings from a national climate change education study, which found that only 32% of closed-sample teachers feel that they have the knowledge and skills to teach about climate change, indicating the need for professional development, classroom resources, and curriculum policy (Field et al., 2019).

Going forward, schools will face many other challenges in achieving regenerative and responsive systems change. Reflecting on the most effective leverage points for systems interventions, Meadows (1999) identifies rules, incentives, information access, and feedback loops as necessary for a system to go through a paradigm shift. Applying a regenerative lens to existing policies and reflecting on the reward and recognition systems for faculty, staff, and students are two ways to help drive systems change. Another area of particular interest for the CAAP leaders is in the area of feedback loops and information. The CAAP leaders are looking to create a dashboard of metrics across the three dimensions of culture, learning, and physical space for schools to identify goals and track and communicate their progress toward their short- and longer-term climate action plan goals.

In a review of the schools' journeys through the CAAP, the schools' commitments to the CAAP and their whole school approach will most likely result in the schools having higher-quality school improvement processes and a more coherent organization structure with greater potential to support pedagogical and teaching practices around climate action than other schools based on comprehensive research conducted on performances of whole school approaches and school improvement (Morgen et al., 2019).

7.4 Conclusion

These three schools are exceptional examples of motivated schools that are voluntarily taking climate action seriously. Unfortunately, climate change education and action often rely on the competence, dedication, commitment, and enthusiasm of devoted teachers and administrators (Eames, 2017; Nicholls, 2016; Whitehouse, 2017), and there is a lack of regulation or policy accountability. All three schools recognized and discussed at length that educational change and culture shifts are slow processes, yet they see that students and the world need climate action now. They strive to be sites of regenerative practice in a time of existential vulnerability, uncertainty, complexity, and ambiguity. The CAAP is still in the early stages, and the schools enrolled are working to achieve both an understanding and an application of regenerative whole school climate action. The aspirational vision to catalyze climate action not only in schools but through them in the community and society will certainly take time to unfold. In the meantime, these schools are important sites for educators, administrators, students, parents, and communities to learn from and be inspired by as we all forge ahead to build a viable and healthy future.

References

Andrews, M., & Delhaes, D. (2022). *Strategic framework for high-impact,whole school climate action*. DoorNumberOne.org. https://doornumberone.org/whole-school-caap-resources/

Barr, S., Cross, J., & Dunbar, B. (2014). *The Whole-school sustainability framework: Guiding principles for integrating sustainability into all aspects of a school organization*. Institute for the Built Environment at Colorado State University. The Center for Green Schools, U.S. Green Building Council. https://centerforgreenschools.org/sites/default/files/resource-files/Whole-School_Sustainability_Framework.pdf

Benavot, A., & McKenzie, M. (2022). Indicators interactive data platform - primary and secondary education. *Monitoring and Evaluating Climate Communication and Education Project*. https://mecce.ca/data--platform/idp/?_sft_idp_cce_approach=1-1-primary-secondary-education-2

Bieler, A., Haluza-Delay, R., Dale, A., & McKenzie, M. (2018). A national overview of climate change education policy: Policy coherence between subnational climate and education policies in Canada (K-12). *Journal of Education for Sustainable Development, 11*(2), 63–85.

Bosevska, J., & Kriewaldt, J. (2020). Fostering a whole-school approach to sustainability: Learning from one school's journey towards sustainable education. *International Research in Geographical and Environmental Education, 29*(1), 55–73. https://doi.org/10.1080/10382046.2019.1661127

Chapman, P. (2012). *Greening America's Schools: A roadmap for fostering sustainable schools for the 21st century*. National Association for Independent Schools.

Children and Nature Network. (n.d.). *Research Library* [contains leading research on the benefits of outdoor learning]. https://research.childrenandnature.org/

Clayton, S., Manning, C., Krygsman, K., & Speiser, M. (2017). *Mental health and our changing climate: Impacts, implications, and guidance*. American Psychological Association and ecoAmerica.

Climate Education Bill, Bill 197, UK Parliament (2022, January). https://publications.parliament.uk/pa/bills/cbill/58-02/0197/210197.pdf

Climate Education Reform British Columbia. (2021). *Reform to transform campaign*. https://www.climate-educationreformbc.ca/

Eames, C. (2017). Climate change education in New Zealand. *Curriculum Perspectives, 37*(1), 99–102. https://doi.org/10.1007/s41297-017-0017-7

Facer, K. (2012). Taking the 21st century seriously: Young people, education and socio-technical futures. *Oxford Review of Education, 38*(1), 97–113.

Field, E., Schwartzberg, P., & Berger, P. (2019). *Canada, climate change and education: Opportunities for public and formal education* [Technical report]. https://www.researchgate.net/publication/337111645_Canada_Climate_Change_and_Education_Opportunities_for_Public_and_Formal_Education

Field, E., Spiropolous, G., Nguyen, A., & Grewal, R. (2023). Climate change education within Canada's regional curricula: A systematic review of gaps and opportunities. *Canadian Journal of Educational Administration and Policy*. https://journalhosting.ucalgary.ca/index.php/cjeap/article/view/74980

Gameau, D. (2023). *Regenerating Australia*. Regenerators Studio. https://theregenerators.org/regenerating-australia/

Grewal, R., Field, E., & Berger, P. (2022). Bring climate injustices to the forefront: Learning from the youth climate justice movement. In E. Walsh (Ed.), *Equity and social justice in climate change education: Exploring social and ethical dimensions of environmental education* (pp. 39–70). Routledge. https://www.routledge.com/Justice-and-Equity-in-Climate-Change-Education-Exploring-Social-and-Ethical/Walsh/p/book/9780367344702

Gym – New World Encyclopedia. (2022). Newworldencyclopedia.org. https://www.newworldencyclopedia.org/entry/Gym

Han, H., & Ahn, S. W. (2020). Youth mobilization to stop global climate change: Narratives and impact. *Sustainability, 12*(10), 4127. https://www.mdpi.com/2071-1050/12/10/4127

Hargis, K., & McKenzie, M. (2020). *Responding to climate change education: A primer for K-12 education*. The Sustainability and Education Policy Network. https://sepn.ca/wp-content/uploads/2021/01/SEPN-CCEd-Primer-January-11-2021.pdf

Hawken, P. (Ed.). (2017). *Drawdown: The most comprehensive plan ever proposed to reverse global warming*. https://drawdown.org/

Hawken, P. (2021). *Regeneration: Ending the climate crisis in one generation*. https://regeneration.org/solutions

Henderson, J. (2019). Learning to teach climate change as if power matters. *Environmental Education Research, 25*. https://doi.org/10.1080/13504622.2019.1660309

Hickman, C., Marks, E., Pihkala, P., Clayton, S., Lewandowski, R. E., Mayall, E. E., et al. (2021). Climate anxiety in children and young people and their beliefs about government responses to climate change: A global survey. *The Lancet Planetary Health, 5*(12), e863–e873. https://www.sciencedirect.com/science/article/pii/S2542519621002783

Holst, J. (2022). Towards coherence on sustainability in education: A systematic review of Whole Institution approaches. *Sustainability Science*, 1–16. https://link.springer.com/article/10.1007/s11625-022-01226-8

International Living Future Institute. (2022). *Living building challenge*. https://living-future.org/lbc/

Kelsey, E. (2019). *Hope matters: Why changing the way we think is critical to solving the environmental crisis*. Greystone Books.

Kwuak, C., & Casy, O. (2021). *A new green learning agenda: Approaches to quality education for climate action*. Centre for Universal Education at Brookings. https://www.brookings.edu/wp-content/uploads/2021/01/Brookings-Green-Learning-FINAL.pdf

Lawson, D. F., Stevenson, K. T., Peterson, M. N., Carrier, S. J., Strnad, R. L., & Seekamp, E. (2019). Children can foster climate change concern among their parents. *Nature Climate Change, 9*, 458–462. https://doi.org/10.1038/s41558-019-0463-3

McKenzie, M. (2021). Climate change education and communication in global review: Tracking progress through national submissions to the UNFCCC Secretariat. *Environmental Education Research, 27*, 1–20. https://doi.org/10.1080/13504622.2021.1903838

Meadows, D. (1999). *Leverage points: Places to intervene in a system*. The Sustainability Institute. https://donellameadows.org/wp-content/userfiles/Leverage_Points.pdf

Mogren, A., Gericke, N., & Scherp, H. Å. (2019). Whole school approaches to education for sustainable development: A model that links to school improvement. *Environmental Education Research, 25*(4), 508–531. https://doi.org/10.1080/13504622.2018.1455074

Nicholls, J. A. (2016). *Understanding how Queensland teachers' views on climate change and climate change education shape their reported practices* [PhD Dissertation]. Retrieved from: https://researchonline.jcu.edu.au/48975/

Reed, B. (2007). Shifting from 'sustainability' to regeneration. *Building Research and Information, 35*(6), 674–680. https://doi.org/10.1080/09613210701475753

Teach The Future. (2021). *Prepare students for tomorrow, teach the future today*. https://www.teachthefuture.org/

Tilbury, D., & Galvin, C. (2022). *Input paper: A whole school approach to learning for environmental sustainability*. European Commission. https://education.ec.europa.eu/document/input-paper-a-whole-school-approach-to-learning-for-environmental-sustainability

UNESCO. (2019). *Country progress on climate change education, training, and public awareness: An analysis of country submissions under the United Nations framework convention on climate change*. https://unesdoc.unesco.org/ark:/48223/pf0000372164

Wals, A. E., & Mathie, R. G. (2022). Whole school responses to climate urgency and related sustainability challenges: A perspective from Northern Europe. In M. A. Peters & R. Heraud (Eds.), *Encyclopedia of educational innovation*. Springer Nature. https://link.springer.com/referenceworkentry/10.1007/978-981-13-2262-4_263-1

Whitehouse, H. (2017). Point and counterpoint: Climate change education. *Curriculum Perspectives, 37*(1), 63–65. https://doi.org/10.1007/s41297-017-0011-0

Wynes, S., & Nicholas, K. A. (2019). Climate science curricula in Canadian secondary schools focus on human warming, not scientific consensus, impacts or solutions. *PLoS One, 14*(7). https://doi.org/10.1371/journal.pone.0218305

How can a Whole School Approach to Sustainability be Inclusive to All Learners?

8

Katarina Rončević and Marco Rieckmann

Key Message
This chapter contributes to raising awareness of potential noninclusive blind spots in the implementation of WSAs and offers suggestions that can, at best, serve as inspiration for future research and school development practices. It invites to deal with the question of to which extent WSAs are limited from the outset by discrepancies between education systems, policies, and governance, and at the same time from school practices. It reinforces inclusive pedagogy in ESD to make WSAs accessible for all learners.

8.1 Introduction

One overall task of Education for Sustainable Development (ESD) might be to prepare the young generation—regardless of the socioeconomic and cultural background, (dis)ability, race, or sexual orientation—for current and future sustainability-related challenges. However, to ensure the achievement of transformative education for sustainable development and thus to encourage participation by all learners, an inclusive approach is needed. Precisely because of the high complexity of sustainability issues and the greater proximity to sustainability issues among some learners than others, ESD in practice is often accompanied by exclusion mechanisms (Ideland & Malmberg, 2014; Jordt Jørgensen et al., 2020; O'Donoghue & Roncevic, 2020).

Against this backdrop, there has been increased discussion of ESD and inclusive education in recent years, but the two concepts have been discussed and promoted separately and from different angles. It is thus unsurprising that there has been little research into the development of a common perspective that combines ESD and inclusive education with practical experience of *inclusion-oriented* ESD and that there is, therefore, a research gap (Rončević & Rieckmann, 2024; Vierbuchen & Rieckmann, 2020; Böhme, 2019). Instead of the sometimes also commonly used term "inclusive ESD," we use the term "inclusion-oriented ESD." This is to emphasize that the aim is to use principles of inclusion and inclusive education, such as differentiation of goals, in ESD and thus to make it more accessible to all learners.

Embedding ESD for all learners effectively within formal education goes way beyond the classroom. A whole school approach (WSA) can thus help mainstream inclusive education for all (Hue & Karim, 2022) and thus can conclusively mainstream inclusion-oriented ESD within the formal school system and beyond (all areas of school development, as well as [preservice] teacher training and community-based activities). However, although a WSA is aimed at all

K. Rončević (✉) · M. Rieckmann
Department of Education, University of Vechta, Vechta, Germany

A. E. J. Wals et al. (eds.), *Whole School Approaches to Sustainability*, Sustainable Development Goals Series, https://doi.org/10.1007/978-3-031-56172-6_8

learners per se, it does not necessarily reach all learners. It is therefore a matter of revealing exclusion mechanisms and further developing ESD and thus also the WSA with pedagogical approaches from inclusive education.

Against the background that a WSA to implementing ESD from the perspective of inclusive education is a desideratum—both in research and in practice—we conducted expert interviews to gain insights into different perspectives and experiences with regard to success factors, and to discuss the WSA through the lens of inclusion-oriented ESD. To this end, 11 experts (six from Germany and five from outside of Germany) were interviewed. This chapter will highlight requirements for WSAs and ways of implementing ESD for all learners, regardless of special needs, and focus on the development of competences to address the challenges of the twenty-first century.

8.2 Inclusion-Oriented ESD

Approaches to ESD that explicitly and effectively seek to be truly inclusive of all learners are not widely used in the context of formal education. Rather, ESD, on the one hand, and inclusive education, on the other, have been discussed and put into practice separately from each other. While inclusive education emphasizes the goal of participation by all learners and promotes diversity, ESD focuses on strengthening competences through participation in—and contribution to—sustainable development (Böhme, 2019). However, ESD and inclusive education have extensive and normative overlaps in their approaches and should, therefore, not be considered separately from each other: for both concepts, the ideas of empowerment, human rights, human development, and social justice are central (Vierbuchen & Rieckmann, 2020; Böhme, 2019; Böhme & Führing, 2014). Inclusion-oriented ESD is characterized by inclusive practices (based on inclusive pedagogy) within heterogeneous learning environments.

Learning environments, and teaching and learning methods in ESD, need to meet all learners' individual needs. As reflected in the "living schools" concept, ESD can act as a catalyst to reconnect students with nature; when students interact with other living things, it helps them realize that they can develop agency and have a positive impact on the world through their compassion and actions. This environmental lens also provides an opportunity for the seamless infusion of human health and well-being, indigenous perspectives, social justice, diversity, and inclusion into our daily learning culture (Cuthbert, 2020).

To make ESD more assessable for all learners, it is important to take a closer look at the concept of inclusive education itself. It is a field that has led to contradictory discussions. In empirical research, the concept of inclusion is often understood as the teaching of children with special educational needs and those without in mixed groups (Grosche, 2015). However, this understanding falls far too short (Grosche, 2015; Ainscow, 2007; Göransson & Nilholm, 2014; Hinz, 2002; Wocken, 2014); it often leads to the denouncing of the broader understanding of inclusion to avoid any risk of stigmatization (Wocken, 2010), which may result from the differentiation of learners with disabilities from those without. On the other hand, some argue that a differentiated diagnostic perspective may provide a decisive basis for any (inclusive) pedagogical initiatives in teaching and support (Neumann & Lütje-Klose, 2020; Ricken & Schuck, 2011). This broader understanding of inclusion reinforces a fundamentally individualizing perspective within process-oriented and support-focused diagnostics (support needs of highly vulnerable students and those with special risk factors) in an inclusive school (Lindmeier & Lütje-Klose, 2018). An inclusive school that takes this perspective provides process-oriented support for learners with so-called special needs, for example, differential cognitive competences, deafness and difficulty with hearing, issues with vision and visual perception, and gifted learners, by making targeted adaptations to teaching methods and learning objectives. Groups of learners with special learning needs are in themselves highly heterogeneous: cognitive competences, for instance, can refer to a particular dependence "on understandable and simplified language, clear structures and small amounts of exercises" (Engagement Global and bezev, 2017, p. 14). Adaptations also need to be made for gifted learners and stimuli provided to address their individual abilities and strengthen their competences. Similarly, deaf and hard-of-hearing learners also represent a very wide group, ranging from those whose first language is sign language (e.g., deaf learners with deaf parents) to those with mild to moderate hearing loss; teaching approaches will vary immensely for different types of learners within groups characterized by specific learning needs. The challenge or potential for conflict that arises is that the promotion of special education within inclusive learning environments can lead to the stigmatization of learners with special needs. The challenge here is to balance inclusive education and special needs education in such a way that support for learners with special needs is provided when needed—in an inclusive learning environment. Consequently, inclusion-oriented ESD can never fit within a one-size-fits-all pedagogical approach. Instead, the focus is on individualization and contextualization of learning.

However, certain criteria may help with the planning and adapting of inclusion-oriented ESD. It is helpful to address some basic questions in the first instance (Vierbuchen & Rieckmann, 2020): how heterogeneous is the group regarding special learning needs? What structural or learning support is required? How should the learning environment and teaching materials be designed to enable all learners to participate in the learning process?

We refer to the broader understanding of inclusion, namely, the principle of diversity that encompasses all individual abilities, cultural and socioeconomic backgrounds, sexual orientation, religion, gender, language, and ethnicity. In this way, ESD aims to avoid the reproduction of stereotypes, especially on the part of the so-called Global North with regard to the Global South, for example, in schoolbooks and learning materials—to ensure that interaction is fair, nondiscriminatory, and cross-cultural. If we understand inclusion as a process that espouses the principle of diversity with the aim of providing an excellent education for all learners, (inclusive) education will be focused much more strongly on the individualization of content and teaching methods. This of course also includes people with special learning needs. Here, inclusion is process-oriented and offers diagnostics focusing on support within inclusive schools; it requires adaptations to be made to ensure learners with so-called disabilities can access and participate in education and to ensure that all learners receive the best educational support for their individual (learning) needs. Activities, materials, methods, and approaches need to be adapted to take individual learning needs into account (bezev, 2017), for instance by providing tactile materials and additional audio material for blind learners; by adding subtitles to video material and by using sign language for deaf or hard-of-hearing learners; and by shortening task lengths and differentiating teaching methods and subject matter for learners who require assistants (due to cognitive difficulties).

Implementing ESD for all learners is also characterized by the need for a variety of teaching and learning methods, such as self-discovery learning, action- and solution-oriented learning; the use of different (digital) media; and so on. This approach can attract criticism. ESD mostly promotes more open learning and teaching methods, which in inclusive learning environments can be challenging for certain types of learners (Vierbuchen & Rieckmann, 2020). Learners who require a more defined or structured approach, or more detailed guidance to enable them to engage openly and independently with sustainability issues could quickly become overwhelmed by a more open and self-guided approach to learning. The guiding principle could, therefore, be to be as inclusive as possible and to provide adaptations and assistance where necessary. O'Donoghue and Rončević (2020) highlight the principles of inclusive pedagogy as developed by Kullmann et al. (2014) on the basis of the work of Klafki (2007), such as acceptance of the individuality of all learners, integration of the teaching of individualized curricula, adaptation of content, adaptive teaching, co-teaching, cooperation between teachers, and cooperation with students to create communities. To make these principles more effective in the classroom and a reality for all, they should be shared by all involved with schools (teachers, school administrators, parents, students, and others) and communicated to the locales in which the schools are situated. This also applies to ESD, which can achieve its goals and approaches only if it goes beyond specific learning environments and reaches out to individuals who are not directly engaged with schools.

It quickly becomes clear that ESD needs to go beyond the classroom to ensure that inclusion is not limited to classroom settings. Values such as appreciation of diversity and a fair and sustainable environment need to be reflected upon and experienced by all stakeholders throughout the school. Consequently, inclusion-oriented ESD takes broader areas and a broader range of stakeholders into consideration, going beyond teachers and learners and reaching out to municipalities and communities where schools are located, parents, and/or local NGOs, so that inclusion-oriented ESD does not end either at the door of the classroom or at the school gates. This points to the importance of a whole school approach to inclusion-oriented ESD.

8.3 Whole School Approach to Inclusion-Oriented ESD

A whole school approach (WSA) can be understood as an approach to ESD that "calls for sustainable development to be integrated throughout the formal [...] curriculum in a holistic manner, rather than being taught on a standalone basis" (Hargreaves, 2008, p. 1). Specifically, the WSA can "advocate [...] active and participatory learning, a hallmark of ESD, and call [...] for the entire school, including students, educators and administrators, to be actively engaged in working towards a sustainable school with ESD fully integrated into the curriculum as the driving factor" (ibid.). WSAs can promote the creation of "structured environments and modifications, [which] will be most effective if implemented consistently across the school in both classroom and general school environments" (Roberts & Webster, 2020, p. 5).

The WSA does not provide a given structure to be followed or a blueprint for a sustainable school in itself, but it offers direction to help individual schools evaluate their practices and formulate guidelines and challenges to help develop new practices and activities aimed at achieving greater sustainability (Mathar, 2014). Although WSAs "must be closely related to schools' own quality assurance culture in education" (Mogren et al., 2019, p. 3, see also Hargreaves & Shirley, 2009; Scott, 2009, 2013), several action-based processes for implementing WSAs (UNESCO, 2014) have also been described (Gough, 2005; Behinderung und Entwicklungszusammenarbeit (bezev), 2019; Greenpeace, 2021). Some studies summarize the qualities required to deliver WSAs: according to Gough (2005), they require coherence, policies, transparency, practice, and continuing professional development. Mogren and Gericke (2017) conclude that the key aspects include collaborative interaction, school improvement, student-centered education, cooperation with the local community, and proactive leadership.

Mogren et al. (2019) introduce Sherp's model of school improvement and identify four aspects of school organization (holistic concept, routines and structures, professional knowledge creation, and practical pedagogy). Regarding professional knowledge creation, Mogren et al. (2019, p. 5) point out that this "is important for developing new understanding of learning and teaching, so it is strongly linked to critical reflection in education and understanding different values that underpin different views of reality." Furthermore, they argue that disruptions in everyday life indicate that, in general, education and teaching practices are not appropriate to current circumstances in the world or the community and should be adapted. In particular, consideration of the community seems to be of great significance for the impact and efficacy of WSAs. Wals and Benavot (2017, p. 7) argues that "to ensure pro-environmental outcomes, schools must be embedded in their communities, seeking to influence not only the views and actions of learners who walk their halls, but also the decisions made by policy makers in government and business to ensure that they have the long-term interests of their citizens and the planet in mind." Another aspect is the integration of local communities, local cultures, and knowledge that can ensure a WSA is implemented successfully and that therefore also "encourage learning beyond the classroom, across groups and ages, creating spaces for diversity that supports the learning process" (Kemper, 2021, p. 7). Integrating local and cultural communities may also help learners understand their local environment and relate their education to it (Kemper, 2021). Integrating cultural and local knowledge or engaging with indigenous communities, for instance, benefits not only local learners but also all who relate to the local environment and can therefore become a "catalyst for action" (Hargis et al., 2021, p.12; cf. Restoule & Chaw-win-is, 2017). Sherp's model of school improvement may be of great benefit, enabling schools to contextualize by providing the flexibility to incorporate current and local concerns and to work toward sustainable learning and actions.

The abovementioned call for greater contextualization could also contribute to this. The idea of contextualizing can also be used to understand the learning needs of individual learners, who are part of the school community. When considering inclusion-oriented ESD at the level of school structures, the Index for Inclusion (Booth & Ainscow, 2019) offers a helpful way to incorporate inclusion within ESD practices and WSAs. The Index for Inclusion, a tool for the development of inclusive schools, has been used to support the review and development of schools and as a resource for promoting a systemic approach to inclusive schools; its tools (questions and indicators) also integrate aspects of sustainability. The Index for Inclusion places three dimensions at the center of school development, namely, "creating inclusive cultures," "establishing inclusive structures," and "developing inclusive practices" (see Table 8.1): "Cultures reflect relationships and deeply held values and beliefs. Changing cultures is essential for sustainable development" (Booth & Ainscow, 2019, p. 23).

Table 8.1 Dimensions of school development (Booth & Ainscow, 2019, Index for Inclusion)

Dimension A: Creating inclusive cultures	Dimension B: Producing inclusive policies	Dimension C: Developing inclusive practices
A1: Building community A2: Establishing inclusive values	B1: Developing the school for all B2: Organizing support for diversity	C1: Constructing curricula for all C2: Orchestrating learning

The questions and indicators of the three dimensions of inclusive school development help schools contextualize their ESD practices, using a WSA to introduce inclusion-oriented ESD and to empower and encourage all learners, regardless of their individual abilities, to take action to move their school community toward sustainability. The principles of inclusive pedagogy can help extend ESD practices and ensure that teaching and learning methods meet the individual needs of all learners as well as taking the learners' (cultural) context into account; they can also empower learners to engage with sustainability issues. The following excerpt of quality criteria as suggested by Reich (2014) can assist with the introduction of principles of inclusive pedagogy (Table 8.2).

Table 8.2 Approximate translation [authors' own] of an excerpt from quality criteria for inclusive pedagogy by Reich (2014)

Interpersonal relationships based on inclusive principles Competence grid with basic qualifications and differentiations Multivalent professional teamwork, personal assessments, and support for special needs Participation and equal opportunities/equity Heterogeneous learning groups and peer-to-peer learning	Learning tasks in learning environments Workshops and elective opportunities Projects and interdisciplinary contexts Diversity of methods Active use of media Differentiated assessment and feedback Evaluation and collegial consultation Qualification and further or continuing education or professional development

A core element of inclusion-oriented ESD in the context of a WSA is establishing an inclusive culture (Index for Inclusion by Booth & Ainscow, 2019) and a culture of sustainability (Mathie & Wals, 2022) in schools. Hence, aspects of an inclusion-oriented WSA should be viewed and questioned through the lenses of both ESD and inclusive education, taking into account all areas and aspects of school—for instance, school buildings, school environment, classroom settings, management, all stakeholders (learners, educators, parents, social workers, and principles), and, of course, the teaching and learning culture. However, the question remains as to how WSAs take account of inclusive education.

8.4 Research Design

The research takes an exploratory approach to finding answers to the following questions: how can WSAs contribute to the implementation of inclusion-oriented ESD? What are the key success factors and what are the greatest challenges? To identify answers, leading question-based expert interviews were conducted based on Gläser and Laudel's interview guide (2010), gathering data about the experts' understanding of and expertise in inclusion-oriented ESD in the context of WSAs.

Prior to data collection, the methodological procedure for the expert interviews was approved by the University of Vechta Ethics Committee. The experts were selected according to the following criteria: published research in the field of inclusion-oriented ESD and/or WSAs; teaching experiences in formal education in the field of inclusion-oriented ESD and/or WSAs; and institutional engagement, for instance in teacher training institutions or education ministries, in the field of inclusion-oriented ESD and/or WSAs. We approached 13 experts from three groups via email. First, the results of a scoping literature review on inclusive ESD (Rončević & Rieckmann, 2024) enabled us to identify experts who had already published on the broader notion of inclusion-oriented ESD and/or WSAs. Second, we approached the international multi-stakeholder ESD Expert Net, which has members from India, Germany, Mexico, and South Africa (one author of this chapter has been a member of the ESD Expert Net since 2012). Some network members' work takes a critical view of ESD and its more inclusive approaches, for example, in the context of post-colonialism or indigenous knowledge. This targeted enquiry also led to the recommendation of another expert, whom we were able to include in the group of interviewees. Third, drawing on the authors' experiences in the field of inclusion-oriented ESD in Germany, we approached experts on the basis of the criteria described above. Ultimately, 11 experts agreed to participate in the interviews. Of the eleven (seven women, four men), five came from the academic sector (although at least two of these also undertook work for teacher training institutions or education ministries). Two experts worked in an education ministry, a further three in schools (one school principal and two teachers (with one of the teachers also working at an education ministry)), and one in a governmental teacher training institute. The experts came from Germany, India, Mexico, and South Africa; more than 50% of the interviewees were based in Germany. The reason for the overrepresentation of Germany was that the authors of this chapter also conduct most of their research/work on inclusion-oriented ESD in Germany. Additionally, there has been a particularly lively exchange on inclusive ESD in Germany in recent years, which has been presented in at least two special issues on inclusion and ESD in peer-reviewed journals (*Journal for International Educational Research and Development Education* (2020) and *Journal Disability and International Development* (2015)). The expert interviews were conducted in June and July 2022 and recorded using the Zoom video-call program. The interview length varied from 25 to 55 min. A guide was used to conduct the interviews. The Mytrint program was used for transcription. For the purposes of qualitative content analysis, deductive categories were created, using the questions from the guide. The two deductive created categories were understanding of inclusion-oriented ESD and of a WSA, and the challenges and opportunities of a WSA for inclusion-oriented ESD. Next, inductive categories were formulated in seven phases following the approach of Kuckartz and Rädiker (2022):

1. Establishing the assessment category
2. Identifying and coding text passages relevant to the assessment category
3. Assembling coded segments from the assessment category on a case-by-case basis
4. Formulating expressions of the assessment category and assigning reference points, changing the definition and number of expressions if necessary
5. Evaluating and coding all material
6. Simple category-based evaluation
7. Complex qualitative and quantitative correlation analyses, and visualizations

The transcripts were coded by inductive subcategory (phases 2–4). Coding was undertaken by one of the authors and categorizations discussed and adjusted by both authors (phases 5–7). MAXQDA software was used for the purposes of coding.

8.5 Results

The 11 experts (hereafter referred to as E1, E2, E3, … E11) described their understanding of inclusion-oriented ESD and how it should be implemented in schools and explained how WSAs contribute to the delivery of inclusion-oriented ESD. Both the challenges and the opportunities of implementing inclusion-oriented ESD in the context of a WSA were discussed.

The authors identified three different subcategories (Table 8.3), which shed more light on the interviewees' understanding of inclusion-oriented ESD. Building on this understanding, the experts were asked about the key success factors and the greatest challenges at a structural level. Six categories were identified on the basis of their answers.

Table 8.3 Categories of inclusion-oriented ESD

Interview question	Inductive categories/ codings
Describe how, in your opinion, successful inclusion-oriented ESD can be designed for the school context? (X.4.1)	Inclusion-oriented ESD >> All means all >> Participation >> Contextualizing
What are the key success factors for inclusion-oriented ESD in schools, and what are the greatest challenges? (X.4.2).	Whole school approach Participation/ self-efficacy Teacher training Resources (materials and workload) (Education) System Attitude

8.5.1 Understanding of Inclusion-Oriented ESD in Schools

8.5.1.1 All Means All

ESD has been described as an inherently inclusive approach and is very close to inclusion and inclusive education (E2, E5, E6, E7, and E11). And if we take Sustainable Development Goal 4.7 seriously, then "we don't really need to talk about inclusive ESD, because this type of education simply states that all learners should be taken along" (E2). The same interviewee also mentioned that it was important to emphasize again and again that ESD is an inclusive approach (E2). Another interviewee stated that inclusion-oriented ESD was composed of two major fields: inclusion and sustainability, and that education was the binding link (E8). Inclusion-oriented ESD actually means considering the special needs of every child and every adolescent (E8) and trying to get all learners on board, meeting every learner where they are and enabling everyone to participate and to shape a sustainable and fair future (E2).

8.5.1.2 Participation

Participation in the context of inclusion-oriented ESD was also highlighted by the interviewees. One interviewee (E7) pointed out that ESD places a great deal of emphasis on the possibility of making one's own decisions and taking action on one's own behalf, while another interviewee (E3) mentioned that the inclusive aspects tend to emphasize the idea of working together and the concept of "all means all," regardless of the presence of different socio-economic backgrounds and differently abled learners in the same school (E5). This means that all lessons, and school life in general, must be designed accordingly (E1), and support in the development of materials and strategies to promote inclusiveness should be provided in order to engage all and ensure they can participate (E4).

8.5.1.3 Contextualizing

To achieve inclusion-oriented ESD, the most important strands that pedagogic practices must consider are the different contexts of the learners themselves and also the different contexts of education systems (E3), such as "the locality of the school, the life experience of the students, and the region's cultural heritage and indigenous knowledge" (E10). On the one hand, there is the curriculum, and on the other, diverse learners and their individual (life) experiences and contexts. To build bridges and connect the curriculum to learners' lives, "[...] we use our methods and materials, and our understanding of our students' different contexts and spaces, and we've got this curriculum that we need to use to engage them, to prepare them to become these capable individuals, capable, forward-thinking, capable of taking forward our world view, of taking forward our world-orientated learning and curricula [...]" (E3).

8.5.2 Success Factors and Challenges for Inclusion-Oriented ESD

According to the interviewees, there are certain aspects of the WSA that stand out as the most important. These include questions relating to resources, the education system as a whole, participation and self-efficacy, and teacher training programs. These are also the factors that present the greatest challenges for mainstreaming inclusion-oriented ESD in schools.

8.5.2.1 Resources

Teaching and learning materials and workload were identified as resources and highlighted as factors for success and, reflecting on the current status quo, as great challenges for mainstreaming inclusion-oriented ESD in schools.

Interviewees pointed out that it was very important to provide adequate teaching and learning materials, first to support the day-to-day practice of educators implementing inclusion-oriented ESD. Second, teaching and learning materials need to be designed in such a way as to relate to students' lifeworlds (life experiences), take into account the teachers' and students' learning requirements, and develop students' strengths and competences (E2). One interviewee (E5) explained that in her country, "the system is more, you know, textbook-oriented. And so, it's extremely important within the [local] ecosystem to get stuff into the textbooks. And I think this is where the role played by textbook writers and teacher trainers is really important" (E5). So, inclusion-oriented ESD teaching and learning materials are important because they facilitate exploration beyond what is contained within the syllabus (E10). Accessibility and language learning material is another important aspect, for instance, for indigenous learners and non-native speakers. It would be of great benefit if content could also be made available in their (indigenous) languages. Furthermore, inclusion-oriented ESD materials must of course also reflect the special needs of learners, for example, difficulties with cognition, hearing, vision, or mental health (E1). This interviewee made more detailed suggestions, highlighting that for certain learners, if materials include texts, they should be shorter or simplified, or supported with audio tracks, and also be much more action-oriented and descriptive. That would be of great benefit both to educators and to learners. These kinds of materials would enable teachers who are not very familiar with inclusion-oriented ESD to approach the subject matter differently and provide them with support so that they do not have to start from scratch. Inevitably, however, the inclusion-oriented ESD approach would of course require them to adapt materials to their specific learning groups (E1).

8.5.2.2 Workload

Teachers' ability to take inclusion-oriented ESD forward is often constrained. Giving them more time and greater flexibility would support their activities in this area. One interviewee (E6) reported that as a special education teacher, they only had a few working hours at their disposal to devote to individual learners with special needs. Learners with specific special needs hence do not receive full-time support in inclusive learning environments. Teacher shortages and the workload of individual teachers are even resulting in reductions in the support being made available to individual students with special needs. One interviewee indicated that they did not see any chance at all of promoting ESD in the current school environment (E6) due to the intense and heavy workload.

8.5.2.3 Education System

There are many options for implementing inclusion-oriented ESD at the organizational level in schools, be it through procurement, collaborations with extracurricular partners, the involvement of parents, and so on (E2). And still, the overall framework of the education system is simply still too rigid (E7), meaning that the legal requirements on which work in schools is based, such as decrees, are simply still too narrow and therefore only permit creative solutions within a narrow framework (E7). There is a need for policy to enable such creativity "because without that, nothing moves" (E5). The education system puts up barriers to inclusive educational approaches, and of course also to inclusion-oriented ESD, for example, performance assessments and grading (E6, E7, and E8). One interviewee stated that the system was somehow totally out of step with what would be required of them in the future. Another interviewee criticized standardized performance evaluations and the focus on certain subjects and skills, namely, language, communication, and mathematics (E9).

8.5.2.4 Education and Training: Teaching Competences

Teachers play a major role in taking inclusion-oriented ESD forward. Adequate teacher training is therefore key to the promotion of inclusion-oriented ESD, and this was highlighted by most of the interviewees (E2, E3, E5, E6, E7, E8, and E9).

One interviewee stated that this also meant having teachers who were appropriately trained to ensure that the (special) needs of all students were addressed as fairly as possible. This means that in general it is not sufficient to differentiate the training of teachers by types of school (high schools, special needs schools, or others) or by specific impairments and special needs. All teachers must have at least a basic understanding of disadvantage and disability across the board and of gifted learners (E8). In fact, an important focus of teacher education may simply be raising teacher awareness (E2). On the other hand, one interviewee highlighted that when teachers are aware of the contexts in which individual learners live, and in which schools are located, it is easier for this awareness to permeate the school. When teachers are unaware of these aspects, there are more challenges. There is an increasingly clear need for more focus on teacher competences relating to inclusion-oriented ESD.

8.5.2.5 Whole School Approach as a Driver

The whole school has to get on board (E4) with inclusion-oriented ESD, and a lot of different levels need to be considered (E2). Interviewees pointed out that a WSA could provide good guidelines and could also be reflected in everyday school life, across all subjects, and also in lesson design. A WSA can help make ESD inclusive and feasible for all (E2 and E4), while all areas of school must be looked at/into and reflected upon through the lenses of sustainability and inclusion; inclusion-oriented ESD has to be seen through. It's not just an add-on (E4).

8.5.2.6 Participation and Self-Efficacy

To achieve real social transformation, more people are needed than the few who engage in climate activism, for example. A broad social consensus is required, and achieving that requires many more people from very different backgrounds. In order to identify how to reach all learners and enable them to engage in society, we need to address many more people than we have so far (E2). One interviewee stated that ESD placed much more emphasis on the potential to make decisions and to take action on one's own behalf. In the past, this has not really been made clear in discussions about inclusive education (E7). Participation plays a huge role both for ESD and for inclusion (E11). Another interviewee (E4) raised questions that may help to ensure learner participations, such as, for whom is ESD accessible? So, to whom are activities offered, who is involved, to whom are activities adapted, and who is allowed to participate? Inclusion-oriented ESD ensures everyone is allowed to participate so that all learners can experience self-effectiveness and are able to contribute to change in the environment and for fellow human beings (E4 and E6). First and foremost, learning should reflect learners' individual wishes and needs. This would encourage teachers to make lessons more experimental, with more intense learning but maybe with fewer subjects over a longer period (E8).

8.5.2.7 Attitude

Besides the complex and difficult conditions that may hinder the implementation of inclusion-oriented ESD in schools, the fundamental attitude of teachers toward inclusion-oriented ESD is of great importance: "if someone is motivated, I mean, everything falls into place" (E11). One interviewee stated that if the educators have an open attitude and are willing to implement and develop inclusion-oriented ESD, then it might be possible to overcome problems relating to teacher shortages in schools (E6).

8.5.3 WSA for Inclusion-Oriented ESD

A school based on inclusion-oriented ESD would look quite different from what schools often look like (E11). Three different subcategories (Table 8.4) were identified from the interviews to address the questions of how WSAs can help embed inclusion-oriented ESD and whether any aspects of WSAs would need to be expanded to take account of inclusion-oriented ESD. "[…] the whole school approach is a far more inclusive arena. But if you look at the whole school literature, it doesn't extend to include the wider social politics of sustainability and social justice […]" (E10).

Table 8.4 WSA categories regarding inclusion-oriented ESD

Interview question	Inductive categories/ codings
In your opinion, how can the whole school approach help embed inclusion-oriented ESD? Would certain aspects of the WSA need to be expanded to take account of inclusion? (X.4.3)	Whole school in community approach Structural level Participation and self-efficacy

8.5.3.1 Whole School in Community Approach

"They just don't pay enough attention to the socio-cultural dimension" (E10).

To involve all learners, "we need to consider the different contexts within our educational system" (E3). There is a need for teaching to balance science (scientific knowledge) and heritage (heritage knowledge) and to bring it into learning processes. This would engage education, which is culturally situated, which comes from a context, in that actual context (E9 and E10). Interviewees (E3, E9, and E10) thought that bringing in cultural knowledge would enable learners to relate much better to content, making it relevant to their life and improving their engagement with whole school processes aiming to achieve greater sustainability. Including cultural knowledge in learning processes allows school cultures to promote respect for all learners through the WSA. The WSA would monitor inclusion and encourage social justice because "you can't have inclusion without respect."

At the same time, schools can have an impact on the community, acting as "a mentor, a hub for a nearby community maybe" (E5). On the other hand, one interviewee (E2) described the boundaries of WSAs as being where the schools' action ends. This can be due to education systems, which one interviewee (E7) described as rather rigid and to a certain extent slow-moving. Education systems and their policies are often not compatible with the localities in which schools are situated and do not always facilitate cooperation between schools and communities.

8.5.3.2 Structural Level

Interviewees (E1, E4, E5, and E6) described aspects that would need to be incorporated into WSAs at a structural level to enable inclusion-oriented ESD to be mainstreamed. For inclusion-oriented ESD, fundamental systemic changes are also needed in terms of governance. The core factors mentioned by interviewees were an established structure of support groups and a structure for capacity building. Support groups should include teachers with expertise in a wide range of special learning needs and teacher training.

8.5.3.3 Participation and Self-Efficacy

One interviewee (E8) described the WSA and its promotion in general as very difficult because schools operate within a very hierarchical, top-down system. This impedes the democratization of school activities from the ground up and thus compromises true participatory processes. At the same time, participation and self-efficacy—not only of learners but also of educators—were described as important aspects for—and promoted by—inclusion-oriented ESD within a WSA (E4, E5, and E8).

To establish participation as a core method in schools, one would first need to map stakeholders to determine their influence and the various levels of influence for getting the WSA going. Stakeholder mapping should include groups within the school and different external stakeholders, such as from the informal education sector (E9).

8.6 Discussion and Conclusion

This chapter explored the issues of how WSAs can contribute to the implementation of ESD from the perspective of inclusive education and what the key success factors and greatest challenges are. One important finding is that an overall school culture needs to be developed that ensures all stakeholders of a school community are involved so that all fields of activity contribute to the establishment of the values of inclusion and sustainability. A WSA is thus an excellent starting point for mainstreaming ESD inclusively, especially when it considers the participation from learners and their family and the wider community. Nevertheless, some aspects of inclusive education have been identified that are not yet effectively covered by this approach in the context of sustainability discourse. Although diversity and inclusion are often taken into consideration in ESD discourses, the WSA and how it is promoted in ESD literature does not pay sufficient attention to inclusive pedagogy in ESD practices or to school structures. Practices and structures in school development are also highlighted in the Index for Inclusion (Booth & Ainscow, 2019) and are one of the three vectors for developing inclusivity within a school: inclusive cultures, inclusive practices, and inclusive structures. This study shows that a WSA aiming to achieve ESD for all learners (and thus not only for the particularly motivated or sustainability-affine learners, the already particularly competent learners, or those learners with particularly pronounced cognitive abilities, as is often the case in ESD practice to date) needs to give particular attention to inclusion and inclusive pedagogy in teaching and learning practices, as well as to structure. From the present study, a key to the ability of a WSA to promote inclusion-oriented culture in schools is teachers' attitudes toward inclusive values and the sustainability of school processes. In addition to the teachers' attitudes, the need for more attention to the competences of teachers with regard to inclusion-oriented ESD and the WSA itself becomes clear (cf. Vare et al., 2022). Teachers must be enabled not only to take into account the approaches of an inclusion-oriented ESD in their own lesson planning but also to consider other areas of school in the sense of a WSA at the same time so that ESD can develop to its full potential. This requires specific further teacher training.

This study sheds light on key features of WSAs when they include thoroughgoing community awareness. Interviewees emphasized the importance of schools' practices considering individual learners, including their culture, language, beliefs, and environments. Teaching and learning practices need to connect with all learners' backgrounds to make learning relevant for them and for the community. The study endorses the claims of Wals and Benavot (2017) and Kemper (2021) that schools must be embedded much more deeply in their communities and vice versa. To achieve this, we propose that WSAs should see learning processes, including content, as being part of both the community and learners' environments.

Another finding of this study is that inclusion-oriented ESD repeatedly emphasizes the participation of all learners and that this promotes empowerment and self-reliance. The participation of other stakeholders within the whole school in community approach was also touched upon several times during the interviews. Further practice-based research is required to specify how participation of this nature can work when a WSA is paired up with inclusion-oriented ESD, and to reflect on classroom management and teaching and learning methods. A more differentiated examination of the aspect of participation would also appear necessary regarding inclusion-oriented ESD.

Interviewees emphasized that, in their understanding, ESD is by its very nature an inclusive approach that takes diversity into account. However, they also pointed out that it remains necessary to emphasize the inclusive aspect, since accessibility is not a matter of course in ESD practices. For this reason, aspects of inclusive pedagogy need to be added when the whole school in community approach is being applied to ESD. These may include the provision of support groups, and adaptable and flexible materials with easier-to-understand language, among other things. In addition, it will be important to strengthen inclusive school development processes to ensure accessibility across all levels in the school system and, furthermore, that inclusion becomes part of the ethos of a school.

Some of those interviewed pointed out that education systems often resist key aspects of inclusive education. This can be seen, for example, in the case of different types of schools in some regions, including special needs schools, grammar schools, and schools focusing on pupil performance and competence. As one expert stated, the impact of a WSA is limited when it comes to schools' practice in the real world. It remains to be seen what the specific limitations of the WSA approach are, especially regarding inclusive pedagogy. It will be necessary to identify what structures are most effective at enabling change. Questions include the extent to which the WSA is limited from the outset by discrepancies between education systems, policies, and governance, on the one hand, and school practices, on the other. There is a need for further research to evaluate education systems, focusing on how much exclusion they are reproducing, to ensure inclusion-oriented ESD is accessible to everyone involved with schools.

This study covers insights from only a small sample of experts in the field of inclusion-oriented ESD and WSA and can at most serve as a stimulus for future research and for schools' development practices. Even though the experts interviewed came from different regions of the world and brought perspectives from a wide range of professions—as teachers, researchers, or with experience of work within the ministry of education or teacher training institutions—the authors of this chapter are aware of the limitations arising from the small number of interviews undertaken. It is to be expected that there are many more experts from the fields of ESD and inclusive education around the globe who would have further specific suggestions for the truly inclusive implementation of WSAs in this context. However, at present we do not anticipate that such other experts would raise serious objections to the views expressed on WSAs by those interviewed for this study. A further limitation may be that experts from other educational contexts, such as environmental education, transformative education, peace education, or global citizenship education, would be able to contribute further aspects to a broader discussion on WSAs and inclusion-oriented ESD.

References

Ainscow, M. (2007). Taking an inclusive turn. *Journal of Research in Special Educational Needs, 7*(1), 3–7. https://doi.org/10.1111/j.1471-3802.2007.00075.x

Behinderung und Entwicklungszusammenarbeit e.V. (bezev). (2019). *Bildung für nachhaltige Entwicklung inklusiv als Aufgabe der ganzen Schule.* Eine Handreichung mit praktischen Anregungen für Grundschulen.

Böhme, L. (2019). Politische Bildung für Schülerinnen und Schüler mit sonderpädagogischem Förderbedarf. In *Perspektiven Globalen Lernens an Förderzentren*. Wochenschau- Verlag.

Böhme, L., & Führing, G. (2014). Globales Lernen als Katalysator für Diversity und Inklusion. In Verein niedersächsischer Bildungsinitiativen e.V. (Hrsg.). *Die große Globalisierung für kleine Leute - Globales Lernen mit Grundschulkindern* (pp. 13–15). Barnstorf. VNB. https://www.bizme.de/documents/VNB_Broschuere_Webansicht_NRO.pdf. Accessed 28 December 2022.

Booth, T., & Ainscow, M. (2019). Index für Inklusion. Ein Leitfaden für Schulentwicklung. Mit Online-Materialien. Auch für Kindergärten, Hochschulen und andere Bildungseinrichtungen übertragbar. Herausgegeben von Bruno Achermann, Donja Amirpur, Maria-Luise Braunsteiner, Heidrun Demo, Elisabeth Plate, Andrea Platte. Beltz Verlag. Verlag / Amazon / Bibliothek

Cuthbert, R. (2020). Sigurbjorg Stefansson early school: Learning naturally. In C. O'Brien & P. Howard (Eds.), *The living school: Transforming education* (pp. 76–78). ESWB Press.

Engagement Global and bezev. (2017). *Cashew—a global learning challenge. Learning material for inclusive education.* http://www.esd-expert.net/files/ESD-Expert/pdf/Was_wir_tun/Lehr%20und%20Lernmaterialien/Cashew_South%20Africa.pdf. Accessed 18 August 2022.

Gläser, J., & Laudel, G. (2010). *Experteninterviews und qualitative Inhaltsanalyse als Instrumente rekonstruierender Untersuchungen* (4. Aufl.). VS Verlag für Sozialwissenschaften.

Göransson, K., & Nilholm, C. (2014). Conceptual diversities and empirical shortcomings—a critical analysis of research on inclusive education. *European Journal of Special Needs Education, 29*(3), 265–280. https://doi.org/10.1080/08856257.2014.933545

Gough, A. (2005). Sustainable schools: Renovating educational processes. *Applied Environmental Education and Communication Journal, 4*(4), 339–351. https://doi.org/10.1080/15330150500302205

Greenpeace. (2021). *Whole school approach.* Ganzheitlicher Ansatz zur Schulentwicklung. https://www.greenpeace.de/publikationen.pdf. Accessed 18 August 2022.

Grosche, M. (2015). Was ist Inklusion? In P. Kuhl, P. Stanat, B. Lütje-Klose, C. Gresch, H. Pant, & M. Prenzel (Eds.), *Inklusion von Schülerinnen und Schülern mit sonderpädagogischem Förderbedarf in Schulleistungserhebungen*. Springer. https://doi.org/10.1007/978-3-658-06604-8_1

Hargis, K., McKenzie, M., & LeVert-Chiasson, L. (2021). A whole institution approach to climate change education. Preparing school systems to be climate proactive. In R. Iyengar & C. T. Kwauk (Eds.), *Curriculum and learning for climate action* (pp. 43–66). https://doi.org/10.1163/9789004471818_004

Hargreaves, L. G. (2008). The whole-school approach to education for sustainable development: From pilot projects to systemic change. *Policy and Practice: A Development Education Review, 6*(Spring), 69–74.

Hargreaves, A. P., & Shirley, D. L. (2009). *The fourth way: The inspiring future for educational change*. Corwin Press.

Hinz, A. (2002). Von der Integration zur Inklusion—terminologisches Spiel oder konzeptionelle Weiterentwicklung? *Zeitschrift für Heilpädagogik, 53*(9), 354–361.

Hue, M.-T., & Karim, S. (2022). Whole-school approaches to building inclusive classroom and school. In M.-T. Hue & S. Karim (Eds.), *Supporting diverse students in Asian inclusive classrooms: From policies and theories to practice* (pp. 229–243). Routledge.

Ideland, M., & Malmberg, C. (2014). 'Our common world' belongs to 'Us': Constructions of otherness in education for sustainable development. *Critical Studies in Education, 55*(3), 369–386. https://doi.org/10.1080/17508487.2014.936890

Jordt Jørgensen, N., Dahl Madsen, K., & Husted, M. (2020). Sustainability education and social inclusion in Nordic early childhood education. *Zeitschrift für internationale Bildungsforschung und Entwicklungspädagogik, 43*(1), 27–34. https://doi.org/10.31244/zep.2020.01.05

Kemper, J. (2021). Stimulating collective transformative learning experiences with an ESD whole-school assessment tool. *Glocality, 4*(1), 5. https://doi.org/10.5334/glo.48

Klafki, W. (2007). *Neue Studien zur Bildungstheorie und Didaktik. Zeitgemäße Allgemeinbildung und kritisch-konstruktive Didaktik*. Beltz.

Kuckartz, U., & Rädiker. (2022). *Qualitative Inhaltsanalyse. Methoden, Praxis, Computerunterstützung: Grundlagentexte Methoden.* 5. Auflage. Grundlagentexte Methoden. Beltz Juventa.

Kullmann, H., Lütje-Klose, B., & Textor, A. (2014). Eine allgemeine Didaktik für inklusive Lern-gruppen—5 Leitprinzipien als Grundlage eines Bielefelder Ansatzes der inklusiven Didaktik. In B. Amrhein, M. Dziak-Mahrer & Fachdidaktik inklusiv (Eds.), Auf der Suche nach didaktischen Leitlinien für den Umgang mit Vielfalt in der Schule. Lehrerbildung gestalten, Band 3. Waxmann.

Lindmeier, C., & Lütje-Klose, B. (2018). Inklusion. In M. Harring, M. Gläser-Zikuda, & C. Rohlfs (Eds.), *Handbuch Schulpädagogik* (pp. 43–53). Waxmann.

Mathar, R. (2014). A whole school approach to sustainable development: Elements of education for sustainable development and students' competences for sustainable development. In *Schooling for sustainable development in Europe* (pp. 15–30). Springer. https://doi.org/10.1007/978-3-319-09549-32

Mathie, R. G., & Wals, A. E. J. (2022). *Whole school approaches to sustainability: Exemplary practices from around the world.* Wageningen University, Education and Learning Sciences. https://doi.org/10.18174/572267

Mogren, A., & Gericke, N. (2017). ESD implementation at the school organisation level, Part 1–Investigating the quality criteria guiding school leaders' work at recognized ESD schools. *Environmental Education Research, 23*(7), 972–992. https://doi.org/10.1080/13504622.2016.1226265

Mogren, A., Gericke, N., & Scherp, H.-A. (2019). Whole school approaches to education for sustainable development: A model that links to school improvement. *Environmental Education Research, 25*(4), 508–531. https://doi.org/10.1080/13504622.2018.1455074

Neumann, P., & Lütje-Klose, B. (2020). Diagnostik in inklusiven Schulen—zwischen Stigmatisierung, Etikettierungs-Ressourcen-Dilemma und förderorientierter Handlungsplanung. In C. Gresch, P. Kuhl, M. Grosche, C. Sälzer, & P. Stanat (Eds.), *Schüler*innen mit sonderpädagogischem Förderbedarf in Schulleistungserhebungen*. Springer VS. https://doi.org/10.1007/978-3-658-27608-9_1

O'Donoghue, R., & Rončević, K. (2020). The development of education for sustainable development. Materials for inclusive education in South African curriculum settings. *Zeitschrift für internationale Bildungsforschung und Entwicklungspädagogik, 43*(1), 20–26. https://doi.org/10.31244/zep.2020.01.04

Reich, K. (2014). *Inklusive Didaktik*. Bausteine für eine inklusive Schule.

Restoule J. P., & Chaw-win-is. (2017). *Old ways are the new ways forward: How Indigenous pedagogy can benefit everyone* [Paper presentation]. Paper prepared for CCUNESCO Idea Lab, Ottawa. https://en.ccunesco.ca/-/media/Files/Unesco/Resources/Our%20themes/Encouraging%20innovation/20171026_Old%20ways%20are%20the%20new%20way%20forward_How%20Indigenous%20pedagogy%20can%20benefit%20everyone_FINAL.pdf. Accessed 18 August 2022.

Ricken, G., & Schuck, K. D. (2011). Pädagogische Diagnostik und Lernen. In A. Kaiser, P. Wachtel, B. Werner, & D. Schmetz (Eds.), *Didaktik und Methodik* (pp. 110–119). Kohlhammer.

Roberts, J., & Webster, A. (2020). Including students with autism in schools: A whole school approach to improve outcomes for students with autism. *International Journal of Inclusive Education*. https://doi.org/10.1080/13603116.2020.1712622

Rončević, K., & Rieckmann, M. (2024). *Education for sustainable development and inclusive education: a scoping literature review*. In review.

Scott, W. (2009). Judging the effectiveness of a sustainable school: A brief exploration of issues. *Journal of Education for Sustainable Development, 3*(1), 33–39. https://doi.org/10.1177/097340820900300110

Scott, W. (2013). Developing the sustainable school: Thinking the issues through. *Curriculum Journal, 24*(2), 181–205. https://doi.org/10.1080/09585176.2013.781375

UNESCO. (2014). *UNESCO roadmap for implementing the global action programme on education for sustainable development*. United Nations Educational, Scientific and Cultural Organisation. https://unesdoc.unesco.org/ark:/48223/pf0000230514. Accessed 18 July 2022.

Vare, P., Lausselet, N., & Rieckmann, M. (Eds.). (2022). *Competences in education for sustainable development: Critical perspectives*. Springer International Publishing. https://doi.org/10.1007/978-3-030-91055-6

Vierbuchen, M.-C., & Rieckmann, M. (2020). Bildung für nachhaltige Entwicklung und inklusive Bildung. Grundlagen, Konzepte und Potenziale. *Zeitschrift für internationale Bildungsforschung und Entwicklungspädagogik, 43*(1), 4–10.

Wals, A. E. J., & Benavot, A. (2017). Can we meet the sustainability challenges? The role of education and lifelong learning. *European Journal of Education, 52*(4), 404–413. https://doi.org/10.1111/ejed.12250

Wocken, H. (2010). Restauration der Stigmatisierung! Kritik der "diagnosegeleiteten integration". *Behindertenpädagogik, 49*(2), 117–134.

Wocken, H. (2014). Frei herumlaufende Irrtümer. Eine Warnung vor pseudoinklusiven Betörungen. *Gemeinsam Leben, 1*, 52–62.

9 The Whole-School Alignment Model: Facilitating a Teacher Team in Sustainable Entrepreneurship Education

Aksel Hugo and Elisabeth Iversen

Key Message
Any innovative transformation within a whole-school organism implies introducing tensions within a system that is already tense. Finding pathways towards coherence on sustainability education within a whole school is hence not straightforward; it is *no instrumental pathway.* The message from this study is to instead position the question of coherence explicitly within a transformative perspective, building cultures of inquiry that attune and align their processes. Shared inquiry processes that embrace liminality will also cultivate transformative coherency.

9.1 Introduction

In a recent systematic review of whole-school approaches to sustainability in education, Jorrit Holst (2022) points to the key challenge of having continuous and participative organisational learning processes aimed at institutional coherence regarding sustainability. His underlying question is one of finding pathways towards coherence on sustainability education within a whole school. It is the same question we will address in this chapter. Through the entry point of a team of teachers, we explore possible pathways for transformative coherency. As process facilitators, we design and explore a facilitation model to support them *and* the whole school in developing a sustainable entrepreneurship programme at a vocational upper secondary school in Norway. We situate the study within the wider discourse on school–university partnerships (see, for instance, Martin et al., 2011) and particularly approaches for sustainability education (Mathie and Wals, 2022). We hope this study can contribute to, and even lead the way for, similar school-based sustainability projects that coherently align and affect the whole-school context.

We will first briefly frame the research perspective and situate the context and methodology of the study, before presenting the alignment facilitation model and its rationale. Thereafter, we will distil lessons learned from the facilitation process and then finally review the model and contextualise our findings in the crosslight of theories on transformative entrepreneurship education and whole-school approaches to sustainability education.

9.1.1 Tensions and Liminal Spaces in Transformative Entrepreneurship Education

In his analysis of entrepreneurship education, Zhang (2020) emphasises the different tensions that schools and educators necessarily will face

A. Hugo (✉) · E. Iversen
Faculty of Science and Technology, Norwegian University of Life Sciences,
Ås, Norway
e-mail: aksel.hugo@nmbu.no

A. E. J. Wals et al. (eds.), *Whole School Approaches to Sustainability*, Sustainable Development Goals Series, https://doi.org/10.1007/978-3-031-56172-6_9

in developing entrepreneurship programmes. First and foremost are the tensions that come with the fact that only the students can own the entrepreneurial process, which means teachers must step back from directing and controlling, whilst at the same time remaining responsible for the process. Brown (2015) calls this role 'maieutics' and compares the mentor role in entrepreneurship education to that of a midwife. Second, huge tensions arise in having to combine instrumental learning outcomes with this creative process and, moreover, in aligning learning outcomes with assessments (Biggs, 2003). Finally, the development of enterprises often will require more authentic learning environments outside the limited classroom of the school, and managing this implies a whole field of tensions of space and time management. Emphasising the need for authenticity in the entrepreneurial process, Macht (2016) similarly points out how most tensions can be sourced back to a lack of authenticity and proposes 'authentic alignment' as a framework for entrepreneurship education. Not only are teachers challenged in their delivery role in entrepreneurship education, but, in the same way, students are also challenged in their habitual role as educational receivers. Again, the major tension is between what the authentic process of creating their own business demands and the role students are used to, and often comfortable with, as recipients (Haara et al., 2016).

We will use the concept of *alignment* as an important lens in addressing these tensions and designing and discussing our facilitation model. Another important lens will be the concept of *liminal spaces*, as used by Savin-Baden (2020) in her analysis of transformative learning. The task and process of students creating their own business, with all its tensions, would in her view enable them 'to learn how to live in the liminal: a beginning of engagement with risk' (Savin-Baden, 2020 p57). A liminal space is a space where you learn to sustain the unresolved, and it is according to Savin-Baden (2020) to be likened with a tunnel; it begins with a portal or gateway triggered by a threshold or 'disjunction'. The driving forces for movement through the tunnel 'lay in the students' inner motivations for learning, originating from the perceived meaning of the practical experience' (Savin-Baden, 2020 p51). Savin-Badens' (2020) concept of the liminal space is useful because it helps us link the dimensions of tensions inseparably to the processes of transformative education. Thereby, the concept of liminal space will serve as a lens of analysis applicable to the student journey, the teacher team process and the whole-school transformative process.

9.1.2 Attunement and Alignment in Whole-School Transformative Processes

School-based professional development entails intervention and interference with the current state. Any transition within a whole-school organism implies introducing tensions within a system that is already tense. Schröder (2020) found that contradictions and tensions initiate innovative changes and can lead to learning processes. However, to navigate and at the same time facilitate transformation within the school's complex landscape may be challenging for insiders and outsiders. When educational innovation is driven by teachers, or teams of teachers as in our case, it easily creates noise in the school system, resistance from leadership or unintended collegial conflicts. The facilitation framework needs to situate itself *within* this crossfield of tensions and enable us, as facilitators, to help identify and explore them to build direction and the coherence of a common process.

The concept of coherence-making is often used in theories that address whole-school transformative processes (Fullan & Quinn 2015; Holst, 2022; Mogren et al. 2019). It implies aligning and attuning the aims, cultures and structures of the many layers of operations of a school and building shared understanding. Fullan and Quinn (2015) centre their theory of 'coherence-making' in schools around the notion that coherence in essence is shared depth of understanding about the nature of the work. Building on this concept, we argue that it is useful to distinguish clearly between the two perspectives of *alignment* and *attunement*. Attunement is a term sourced in music, which describes the process of tuning and listening in, so the performance is in tune. When used in social science, it operates in a human relational space (McIntyre Latta, 2004). We understand alignment to describe the operational qualities of coherence within and across educational processes, teacher teams and the whole-school organisation. To align tensions and incoherence means, first and foremost, to transform structurally or operationally.

9.1.3 Context of the Study

This study revolves around a development project for sustainable entrepreneurship education at a vocational upper secondary school in a major city in Norway. The school leadership group contacted the continuous teacher education group at our department and jointly applied for funding a project to improve an existing sustainable entrepreneurship programme through a national continuing education scheme. Seven teachers were set to teach the entrepreneurship programme the following school year, and all of them participated in the development project. The project aims were framed as collectively developing a more coherent and sustainability-oriented entrepreneurship education programme, improving collaboration and coherence within the team of teachers and attuning this team process to the whole school. Two project coordinators/researchers facilitated and managed the project from the university sector. One teacher was designated the coordinator of the teacher team, and the school inspector represented the leadership group. The collaboration lasted for two years, where 18 months involved co-working time with the teacher team.

9.1.4 Methods and Materials

Through a collaborative inquiry process, we explored how a facilitation model can support a team of teachers in developing a sustainable entrepreneurship programme as a means for finding pathways in a whole-school context. Design-based research was chosen as a suitable approach since it is 'a systematic, but flexible methodology aimed to improve educational practices through iterative analysis, design, development, and implementation, based on collaboration among researchers and practitioners in real-world settings, and leading to contextually-sensitive design principles and theories' (Wang and Hannafin, 2005, pp6–7).

Design-based research has been critiqued for pacifying and/or excluding participants (Barab and Squire, 2004; Engeström, 2011; Lorentzen, 2017; Wang and Hannafin, 2005). To include the teachers as well in the first phase, we started the project with a shared understanding of our roles and responsibilities in line with the earlier experiences of the second author (Iversen and Jónsdóttir, 2018). This meant that the teacher team, the leadership and the educational researchers clarified and shared explicitly their own goals within the project. We continued with making visible the different actors' responsibilities in being part of the project. For instance, it was the teachers' responsibility to transfer student voices in the project when the teachers undertook certain changes to their teaching. The leadership arranged workshops to define and align visions of this specific project concerning sustainability to the visions of the whole school. Lastly, it was our (the coordinators) responsibility to arrange seven workshops over the course of 18 months. The aims of these workshops were to step-by-step help drive the sustainability entrepreneurship project processes forward and to contribute with our considerations from an outside perspective.

The main data source is a semi-structured focus group interview conducted at the conclusive phase of the project. Here, almost the whole team participated: six of the teachers and the school inspector. The interview revolved around how the teachers experienced being part of this project, if it had made changes to their teaching, and questions related to the content of the WSAM. The last workshop was held in format of a world café on assessment. From this session, the teacher team generated six large note sheets that were put up on a wall. This session was held after the interview and focused more on assessment and the students. Longitudinally, we gathered process supportive data. The supporting data sources are e-mails and preparatory notes before, during and after workshops, questionnaires that the teachers filled out and notes from phone calls.

Concerning data analysis, we started by sorting the data corpus to get an overview of the whole 18-month-long process. The next step was to analyse the interview and find concepts and themes in the data material that could influence the WSAM being explored.

9.1.5 The Whole-School Alignment Model (WSAM)

The purpose of the WSAM was to attune the teacher team's process in two directions: towards the students and their processes on the one hand and towards the leadership group of the school on the other hand (the three figure arrows in Fig. 9.1). We use the concept of attunement here for the vertical dimension because it entails harmonising three parallel processes. *Towards the leadership group* means we need to (1) anchor and dialogue our process and (2) attune it to the structure, culture and aims of the school. *Towards the student process* means we need to anchor and dialogue the programme innovation process to the revisioning and restructuring of the student process.

To start, we agreed to work with the different dimensions of alignment in three steps. Since we were starting in early spring, it made sense to first focus on structural alignment, attuning and anchoring the next yearly plan of the entrepreneurship programme into the school's plan. In addition, we wanted to provide an attunement instrument unifying the teachers and start with a concrete assignment that could quickly yield results. The second step would focus on working out a common understanding of the aims and core competences, and the third step would focus on aligning purpose (competence aims) to corresponding evaluation formats, assessment methods and criteria.

The model was proposed and its applicability discussed with the team of teachers in the initial phase of the project. It seemed to resonate well with all participants as a simple common understanding of what we were going to try out, how it could work as a tool for improvement of the entrepreneurship education programme itself and how it could at the same time serve as integration tool of programme improvement within a whole-

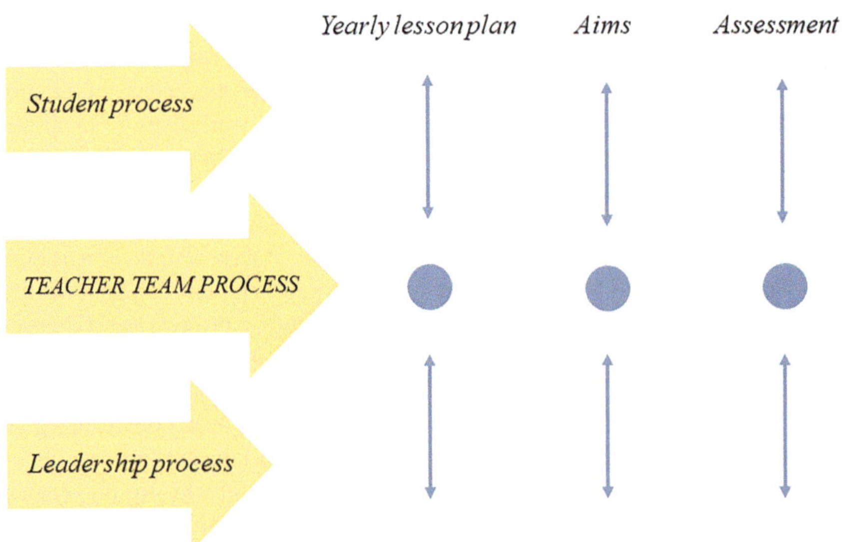

Fig. 9.1 Whole-school alignment model (WSAM): attuning actors and aligning key processes

school sustainability transition context. We could have included the students in the collaborative process but chose not to due to limited resources and time. Their voices speak only indirectly, through what the teachers share.

9.2 The Teacher Team Finding Their Pathways

In the following four sections, we present the teacher team's experiences with their own process of finding their pathways to a shared understanding of the students' entrepreneurship process and attunement of the teacher team's own innovation process with the leadership of the school. We start with their reflections on the first steps of structural alignment, coming together as a team working with the lesson plans. From there, we move to their process of finding a clear shared understanding of the aim or learning objectives of entrepreneurship education and how these learning objectives should align with their forms of assessment. In the third section, we will present their process of renewing the teaching role and, in the fourth, how they experienced the change of the whole school through the project.

9.2.1 The Structural Alignment Process: 'You Need to Start Somewhere'

Reflecting on the first phase of the process, the teachers describe a first major shift that took place in simply 'gathering' and sharing understanding. As Hermod reflects: 'It was that first meeting we had together all of us about this project. It was the first time we teachers met up and gathered our thoughts about how [we see] this entrepreneurship formation path. That was very rewarding'. Idunn notes how 'enduring the chaos' is strengthened by a safe social space: 'both the students and we must own it—and in a way stand securely in it—not feel that we are on shaky ground all the time. So, I think that's very good, to be able to work together on it'. It seems that sharing experiences, mapping tensions, formulating goals and openly discussing the way forward were perceived as appropriate ways to start the project.

During the interview, they were asked how it was to start aligning structures through the school's yearly lesson plan. Frigg answers: 'you get an overview of the process you will go through during the year'. This is echoed by Freya: 'it was a way to structure a common idea'. Tor expresses some concerns: 'There were some bumps in the road with the planning [laughing] since we arranged for many things that, in retrospect, we shouldn't have used any time on'. Idunn extends: '[The lesson plan] is a process of trial and error, and we discover things along the way: "Where do the student end up if we continue this path?"—and then we need to adjust'. Freya adds: '[In starting] with the yearly lesson plan, it was also possible to get everyone involved. Everyone can be on the ball [...] and make a joint effort'. Bringing in the school leadership perspective, Balder, the school inspector, points out that 'the lesson plan is an adaptation to the school year and is the wheel of the school. The whole school needs to adjust the processes within the school year. You need to start somewhere, and you need to stop somewhere as well'. Hermod expresses a counterpoint by stating that 'if I could do this again, I believe it would have been better to start with the students' learning goals [...] as we really didn't know what we were doing'. His conclusion is that, in retrospect, he would have started with getting direction and learning goals in place and then the yearly lesson plan and, lastly, aligned this to assessment.

Both Freya's answer and Balder's summary echo back our intention that the lesson plan helps structure a process integration. However, Hermod does have a valid point: we all clearly saw that there are reasons for starting a process with the learning objectives to create an alignment or shared understanding of objectives and direction for the students' enterprises. Our experience as facilitators during the start-up was that the first phase was perceived as successful because it yielded a 'team feeling of achievement' of something practical and relevant for their process. Also, it made their project and processes visible and anchored in the leadership. The leadership group clearly appreciated the approach of aligning structures onto a common map of activity.

9.2.2 The Process of Clarifying the Aim and Ways of Assessing: 'We See a Goal on the Horizon'

The second process of clarifying the goals of the student journey became more challenging than we expected. It was as if the more the team worked to pin down the aims of the process, the more the essence of the student transformation process slipped. Balder describes their first internal workshop to map the students' core competences of sustainable entrepreneurship as unsuccessful and how it led to a shift in approach: 'We tried first by using a "reversed curriculum method", with no success, and went on to use "understanding by design" and "backwards planning". Then, we came closer to detecting what the goals should be. But we aren't really there yet'. Freya adds: 'a bit of a pain to deal with such things, but very useful. I feel like we're getting a glimpse of a goal there, that we're getting closer to… can't quite see it, but it is getting clearer'. This connotation of 'useful pain' is worth noticing. It translates to a quality of perseverance in the team's inquiry process, a sustaining of the unresolved over time. Freya describes it as a quality of constantly returning: 'We were continuously returning to learning goals and asked: "What are the core competences when working with sustainable entrepreneurship?"'. And she lists some of the questions they have been working on: 'What kind of professional competences should they [the students] end up with?' and 'How can entrepreneurship as a method help the students reach them?' Similar questions trickled down to other work teams at the school and were discussed in regular team meetings. For instance, Frigg shares from the animal science programme that 'in weekly meetings we've been discussing how students, regardless of their company role, may achieve the same competence'.

Another important question the teachers worked with concerning the students' process was 'what makes an enterprise sustainable?' The importance of the social dimension of establishing a student enterprise arose repeatedly during the workshops and again in the group interview. Idunn expresses this core of the process this way: 'It is a social form of teaching... They have roles, they are in real situations all the time, there are conflicts at the social level and directly connected to the work. It's a way of working that gives great opportunities for a lot of collaboration, but [individual] differences become very clear in how the students work within such a frame'. Freya adds: 'Thinking education in relation to social sustainability'. The skill of sustaining collaboration and working together within a long-term group process was seen as a core competence in entrepreneurship education.

The final step of the WSAM is student assessment. During this project year, the teachers changed the assessment method from a final report (used in the previous school year) to a student portfolio. They also made a template for teachers in selected subjects on how to assess the portfolio, based on the core competences they identified in internal workshops and with students. The portfolio was described as something dynamic, and this flexibility was used as a main argument for the change. As Freya exclaims: 'We can adjust the content in the portfolio and its criteria, working with it in a continuous spiral process'.

There was later also a workshop focusing on students' self-assessing their entrepreneurial process and ways of building assessment into various stages of their journey. In the final workshop, the team returned to the unresolved question of student goals and landed a collaborative first synthesis of core competences the students should gain from conducting sustainable entrepreneurship. In hindsight, what we see is a non-linear and meandering process, where student goals and assessments are worked back and forth, always closely linked.

9.2.3 Reversing the Teacher and Student Roles: 'How Much should I Meddle?'

A recurring question that the teachers often discussed in the workshops was the role of the teacher in interaction with the students and their enterprises. Rose expresses: 'I find it hard because it is the students' companies. How much should I meddle and set guidelines? Tor comments: I agree [with Rose]. The students often become so passive. To realize that it isn't me as the teacher who runs this company, it's the students. You can't hold the students' company artificially alive. [Some of] the enterprises would have been bankrupt if they did this on their own. But there is a lot of learning in that'. He explains that what he is really waiting for in the student process is a shift, where 'at last I experience that the student takes ownership of the enterprise'. He continues, 'It's incredibly difficult, that intersection as a teacher to feel that "if I go out now, nothing will happen"—but to go out anyway'. Freya later returns to what Tor was sharing: 'It is very difficult to balance being able to create that ownership, and the feeling that you are an actor that makes a difference, [an actor] that is valuable in this context here. That's perhaps what you achieve then, when you go out the door and leave them behind?' Then, she adds: 'And you're still there!' What Freya is saying here to Tor is that even though you step back, you are still there accompanying their journey. There is a core tension here in how much teachers should meddle in the student enterprise process. We noted that the teachers become truly invested in the students' companies, and they want the companies to succeed, but without taking them over and running them for the students.

9.2.4 Renewing Identity: The Whole-School Transformation Process

In the beginning of the professional development, the teacher team leaned heavily on a model designed by a national NGO (non-governmental organisation) to help teachers conduct entrepreneurship education for youth. In the interview, some of the teachers describe their trajectory. Tor expresses: 'The YE[1] model has been too governing in everything early in the process'. Freya elaborates: 'We dared to take a step away from the YE model, when we shaped that yearly lesson plan to create something that is ours. It's like we're beginning to reach a goal about why we shall [do this] and why [in previous years] we have [done it]. We do not blindly buy into something'. And she comes back to the notion of ownership: 'We are probably not there, you probably won't get there either, but we are well on our way to creating something that is ours. It is starting to sort of come up with an objective about why "we shall", why "we have"—about method and then using it... in contrast to blindly buying a scheme elsewhere, or copying something'. Freya appears to achieve 'ownership' and autonomy around knowing 'why you do what you do as a teacher'. Idunn adds the importance of the ownership that the leadership has taken to the process: 'I think it's important that the work here is intentional. It's anchored both in the college and within the management—and that it's prioritized, that time is set aside'. She adds: 'There are many things that are urgent, that should have been done, but this is a priority, or else nothing would have come of it—this is an important point'. Balder responds: 'Yes, we meet and we talk and patiently believe we'll find the answers to the core competences. I think it' s very important to get there, then. And then you can see what the best ways to get there is. But I think perhaps that in the future, the yearly lesson plan will also have different content'. All in all, what the teachers describe here are dimensions of the school transformation process. What they describe is again a process of taking ownership and a process that leads to renewed identity. At the programme level, Balder describes the renewal as 'moving from YE to SEE'.[2] For

[1] UE (Ungt Entreprenørskap) is an NGO supporting entrepreneurship education in schools in Norway through a process template. Our translation into English, 'young entrepreneurship', abbreviated YE.

[2] BUB is the acronym for Bærekraftige Ungdomsbedrifter

the teachers, this means renewed identity as a collaborative community of practice. Idunn sums it up: 'It's not random or just a feeling, it's more reasoned and more... a bit under the umbrella here, as there are more of us here. We're not just one or two teachers who runs this alone'. At a whole-school level, taking ownership means creating a unifying school vision and profile.

9.3 Narrow Passages and Recurring Motifs

Looking back at the inquiry process, there are some narrow passages that stand out. These are events of unsuccess, leading to new pathways and creating new insights. And there are some recurrent motifs that can signify transformative educational processes. In this section, we will discuss some of these recurring motifs in light of theories of transformative education and 'coherence-making' in schools.

9.3.1 Narrow Passages Through Liminal Space

Our attempted structured facilitation process clearly demonstrates that collaborative inquiry and alignment are in no way a controlled, straightforward pathway. On the contrary, breakthroughs often came as sudden shifts and often as the result of dead ends. In hindsight, we discover that it was Tor's sharing of the pain and 'the essence of what is most difficult in the mentor role' that brought clarity to the essence and crux of the individual student journey. Similarly, it was through a painful failure in the instrumental attempt at pinning down the core competences and a sustained 'churning ahead' that new approaches to an open inquiry process started. Regardless of process or actor—the student journey, the teacher team journey or our own process of testing out a model—through entering liminal spaces, passages allowing insight opened to us. As described by Land, Rattray and Vivian (2014), they are triggered by a disjunction where old ways of seeing, doing, thinking or behaving can no longer be held onto. Savin-Baden (2020 p50) describes how this liminal space is characterised by a stripping away of old identities: 'Learners move through the tunnel of the liminal space and emerge with a shift in learner subjectivity, a discursive shift, or a shift of a conceptual, ontological (e.g. identity shift), or epistemological nature'. The students' painful passage into a position of 'taking ownership', which the teachers described, implies a liminal mentor role 'enabling students to learn how to live in the liminal; a beginning of engagement with risk [...] and uncertainty' (Savin-Baden, 2020 p57). The stripping away of old identities also fits well with their own process, of letting go of the identity of the entrepreneurship programme and building confidence and ownership in their own unique programme profile.

In the course of a passage through a transformative process, not only one's personal identity may change, but also one's social and cultural identity. In his study on transitions in adult learning, Field (2012 p1736) presents the idea of *liminal identity*, which entails that such an identity can be 'shaped through social and cultural processes which are formed and re-formed in dynamic relationships with others'. At this upper secondary school, professional development appears to have influenced how these teachers see themselves as part of the school, contributing to renewing its identity.

9.3.2 Recurring Motifs: Shared Ownership to Shared Questions

In essence, we started out with quite a linear model focusing on coherence-making. Fullan and Quinn's (2015, p30) definition of coherence as the 'shared depth of understanding of the work' proved fruitful, but their instrumental approach to 'coherence-making' in schools gave us only the alignment side. We have seen that

(in English, sustainable entrepreneurship education, abbreviated SEE), which was also the name of the project.

shared understanding of structures, aims and evaluation criteria is important, but more important is the community of inquiry and the values the school, teachers and students are sourced in. The collaborative action inquiry approach invites all involved actors (and also facilitators) into the non-linear and liminal space where coherence means *shared ownership of shared questions*. The core of the entrepreneurial process is that students learn to take common ownership of enterprise innovation. It means sustaining both big and small unresolved questions as they appear for as long as they take to resolve. The same goes for the team of teachers in their 18-month quest to gain clarity and identity of their entrepreneurship programme. They also need to share unresolved questions or tensions in their programme design and mentoring role. This is the very process that yields a depth of a shared understanding, an attunement of colleagueship and an alignment of structure and operation.

The common denominator of innovative cultures of practice is, according to Gillebo and Hugo (2007), that they shape their activities around intentional interaction, dialogue and inquiry with a long-term commitment: 'They develop common tasks by uniting individual commitment to shared questions. And the intensity and ecology of the shared questions build the carrying capacity of the innovation culture' (Gillebo and Hugo, 2007 p1). The endurance and perseverance of the teacher team in their long and winding process fit well with such a description. Their contribution to the process of evaluating and redesigning the facilitation model also proved valuable, as we now shall see.

9.3.3 Reviewing the Model: 'These Arrows Should Perhaps Be Bent'

The teacher team's feedback on the question of sequencing in the model was quite revealing; there are reasons for starting with structural alignment processes, and there are reasons for starting with goals for students and even assessment methods. Towards the end of the discussion of this paradox, after a little pause, Balder exclaims: 'I think that these arrows should perhaps be bent... that this goes in circles'.

Bending the arrows is a good metaphor for moving from a linear to a non-linear understanding of the inquiry process. In hindsight, there was nothing inherently wrong with our linear model. On the contrary, as a tool of 'coherence-making' between the teacher team and the leadership of the school, it did prove fruitful. However, there was a condition for this fruitfulness: that both we as facilitators *and as* teachers allowed ourselves to leave the linear path—and not just leaving it but 'feeling safe and secure in uncertainty', as Idunn specified.

The revised model below is an attempt to capture this non-linear shift, placing the liminal space at the centre of the process of finding pathways. Here, the unifying light green represents the commitment to shared questions, the wheel and process of collaborative inquiry (Fig. 9.2).

On the left, the model still has the key actors and processes of attunement between them, which the small circles signify. On its right side, it still has three fields of alignment to secure institutional coherence. They are no longer in a sequence but entwined. The first field represents the alignment of structural and operational issues, the second the alignment of educational programme design issues (aims and assessment) and the third field the situational alignment (authentic arena and pedagogical process). These fields of alignment correspond to the fields of tensions in entrepreneurship education we discussed initially and which were described by Zhang (2020).

9.4 Conclusion and Implications: Facilitating Transformative Agency

In conclusion, the WSAM we started out with was able to aid a structured facilitation process of pathfinding and to connect our teacher–team process coherently to a whole-school context. Most importantly, it was able to support a collaborative culture of inquiry in which teachers gained ownership of their process and identity. At the same

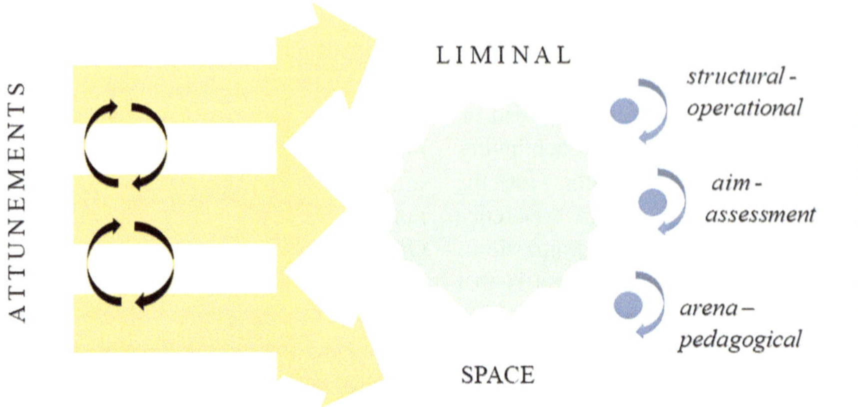

Fig. 9.2 WSAM re-designed: attuning and aligning a sustainable entrepreneurship education project in a transformative whole-school context

time, it clearly revealed the limits of a linear approach to sustainable entrepreneurship education and to the facilitation of institutional coherence on sustainability. A consequence of this realisation is a redesigned WSAM, where the liminal space of collaborative inquiry processes assumes the central role of navigating and attuning inherent tensions and aligning structures, programme design, space and pedagogies to co-create coherency. As a partnership tool, the WSAM has implications for 'navigating the terrain [...] in school-university partnerships' (Martin et al., 2011). The next step would be to explore these implications.

References

Barab, S., & Squire, K. (2004). Design-based research: Putting a stake in the ground. *The Journal of the Learning Sciences, 13*(1), 1–14. https://doi.org/10.1207/s15327809jls1301_1

Biggs, J. (2003). *Aligning teaching for constructing learning*. The Higher Education Academy.

Brown, S. (2015). *Learning, teaching and assessment in higher education: Global perspectives*. Palgrave.

Engestrom, Y. (2011). From design experiments to formative interventions. *Theory & Psychology, 21*(5), 598–628. https://doi.org/10.1177/0959354311419252

Field, D. (2012). Transitions in lifelong learning: Public issues, private troubles, liminal identities. *Studies for the Learning Society, 2*(2-3), 1736–7107.

Fullan, M., & Quinn, J. (2015). *Coherence: The right drivers in action for schools, districts, and systems*. Sage Publications.

Gillebo, T., & Hugo, A. (2007). Sustainable entrepreneurship: Regional innovation cultures in the ecological food sector. *International Journal of Agricultural Sustainability, 4*(3), 244–256.

Haara, F. O., Jenssen, E. S., Fossøy, I., & Ødegård, I. K. R. (2016). The ambiguity of pedagogical entrepreneurship: The state of the art and its challenges. *Education Inquiry, 7*(2), 183–210. https://doi.org/10.3402/edui.v7.29912

Holst, J. (2022). Towards coherence on sustainability in education: A systematic review of whole institution approaches. *Sustainability Science, 18*, 1015–1030. https://doi.org/10.1007/s11625-022-01226-8

Iversen, E., & Jónsdóttir, G. (2018). A bit more than a fly on the wall: Roles and responsibilities in design-based research. *Designs for Learning, 10*(1), 18–28. https://designsforlearning.nu/articles/10.16993/dfl.79

Land, R., Rattray, J., & Vivian, P. (2014). Learning in the liminal space: A semiotic approach to threshold concepts. *Higher Education, 67*(2), 199–217. https://doi.org/10.1007/s10734-013-9705-x

Lorentzen, R. F. (2017). *Lærerens dilemma—mellem ideal og praksis. En virksomhedsteoretisk analyse af progressive undervisning med IT i dansk*. Aarhus University.

Macht, S. A. (2016). "Authentic alignment"—A new framework of entrepreneurship education. *Education + Training, 58*(9), 926–944.

Macintyre Latta, M. A. (2004). Attunement to the creating process in teaching & learning. *Journal of the Canadian Association for Curriculum Studies, 2*(1), 211–226. https://digitalcommons.unl.edu/teachlearnfacpub/13

Martin, S. D., Snow, J. L., & Franklin Torrez, C. A. (2011). Navigating the terrain of third space: Tensions with/in relationships in school-university partnerships. *Journal of Teacher Education, 62*(3), 299–311. https://doi.org/10.1177/0022487110396096

Mathie, R. G., & Wals, A. E. J. (2022). *Whole school approaches to sustainability: Exemplary practices from around the world*. Education & Learning Sciences/Wageningen University. https://doi.org/10.18174/572267

Mogren, A., Gericke, N., & Scherp, H.-Å. (2019). Whole school approaches to education for sustainable development: A model that links to school improvement. *Environmental Education Research, 25*(4), 508–531. https://doi.org/10.1080/13504622.2018.1455074

Savin-Baden, M. (2020). Learning ecologies: Liminal states and student transformation. In R. J. Barnett & N. (Eds.), *Ecologies for learning and practice: Emerging ideas, sightings and possibilities* (p. 50). Routledge.

Schröder, L.-M. U., Wals, A. E. J., & van Koppen, C. S. A. (2020). Analysing the state of student participation in two eco-schools using Engeström's second generation activity systems model. *Environmental Education Research, 26*(8), 1088–1111.

Wang, F., & Hannafin, M. J. (2005). Design-based research and technology-enhanced learning environments. *Educational Technology Research and Development, 53*(4), 5–23. https://doi.org/10.1007/BF02504682

Zhang, J. (2020). Pedagogical alignment for entrepreneurial development. *Entrepreneurship Education (Online), 3*(3), 239–244. https://doi.org/10.1007/s41959-020-00039-z

10 Architecture as Educator for Sustainable Grown-Upness: An ESD Performance Framework for School Habitats

Anna W. Vanderveen, Masi Mohammadi, and Jos J. N. Lichtenberg

Key Message
This chapter addresses the 'educational performance' of physical environments. This is understood as the habitat's gestures to (re)direct pupils' attention to the world so that they may encounter what the world is asking from them. Building on the cultivation and subjectification paradigms in education, this chapter introduces different intentions of building design, either to provoke certain (pro-environmental) behaviours or to educate for 'sustainable grown-upness'. The differences between such *cultivational* and *educational* approaches are discussed extensively.

10.1 Introduction

It requires no introduction that a building can contribute to sustainable development through the way it deals with energy, materials and air quality. Yet, the impact of the physical environment on sustainable development goes far beyond this. The way we shape the physical interface between individuals within their communities and the natural environment is a reflection of larger cultural values and holds a close relationship with our attitude towards the non-human world (Hertzberger, 2008; Upitis, 2015). We shape our buildings based on these values, and afterwards, this influences how we develop and internalise values ourselves. Or as Churchill stated in a speech in 1943 already: 'We shape our buildings; thereafter they shape us'.

Although valid for every building typology, this is particularly important for the locations within and through which education takes place. Schools are in the unique position of being places where the new generation is given the time to meet the world and meet themselves in relation to the world (Biesta, 2021). School habitats, complementary to school habits, play an educational role of their own. Correspondingly, the Italian 'Reggio Emilia' pre-primary schools refer to the physical environment as a 'third teacher', after peers and adults (Strong-Wilson & Ellis, 2007).

10.1.1 School Habitats and Sustainability

In order to study the role of physical environments in education, and education for sustainable development in particular, this chapter builds on a dichotomy set by Veenhoven (2004). Veenhoven

Anna W. Vanderveen, Masi Mohammadi and Jos J.N. Lichtenberg and with the grateful use of the feedback from Hak van Nispen tot Pannerden and Gert Biesta

A. W. Vanderveen (✉)
SME, Utrecht, The Netherlands
e-mail: vanderveen@sme.nl

M. Mohammadi · J. J. N. Lichtenberg
Department of the Built Environment, Eindhoven University of Technology,
Eindhoven, The Netherlands

A. E. J. Wals et al. (eds.), *Whole School Approaches to Sustainability*, Sustainable Development Goals Series, https://doi.org/10.1007/978-3-031-56172-6_10

addresses two kinds of 'qualities' in life, divided into *external* and *internal* qualities. In the first case, the quality is in the (social, physical or otherwise) environment; in the latter, it is in the individual.

The category of external qualities is referred to as 'the liveability of the environment', covering good living conditions. This quality of life is central in the thinking of building professionals (such as architects, building technologists and physicists)—while it is important to stress that the liveability of the environment regards a variety of environments, which are certainly not limited to the physical environment. The second category regards the 'life-ability' of the person, concerning how well one is equipped to cope with the challenges of life, and masters 'the art of living and living well' (Cruz, et al., 2009). This quality of life is central in the thinking of educators.

The division of life chances into 'liveability' and 'life-ability' provides a categorisation to position earlier work on physical school environments in the context of education; see Fig. 10.1. This figure is developed in the study presented in this chapter and visually presents the gap in research and practice it responds to: 'architecture *as* education'.

Research and action on 'green' or 'sustainable' architecture *for* education (left in Fig. 10.1) primarily focus on the enhancement of the liveability of the environment. Much of the work focuses on energy efficiency and material usage, aiming at the creation of a liveable environment at a global scale. Furthermore, research concentrates on liveable environments at a smaller scale, as vast research stresses the effects of the indoor environment on student and teacher outcomes, including health and productivity (Wargocki & Wyon, 2007; Spengler et al., 2001). Also, the physical environment has an influence on how education, didactics and pedagogies are shaped. For example, similar to how open space plans in office environments are penetrating common practice, many schools are introducing 'learning plazas' where pupils can study individually or work in groups, rather than in a classroom with a teacher at the front (Hertzberger, 2008). Hence, the interaction between architecture *for* education and education is multiple.

Meanwhile, architecture *as* education (central in Fig. 10.1) has an educational role of its own. Although increasing numbers of both educators and design professionals are becoming aware of the important role physical space plays (Cleveland & Fisher, 2014), school habitats hardly gain attention in global research on education for sustainable development (ESD), and insight into how physical space contributes to ESD is lacking.

10.1.2 Whole School Approach to ESD

The dichotomy visualised in Fig. 10.1 shows that in the context of ESD we should not only be concerned with the school habitat's *ecological* performance (to improve the liveability of the environment) but also with its *educational* performance (to foster the life-ability of the student). This perspective on physical environments is especially relevant in the context of a 'whole school approach' to education for sustainable development (ESD). A whole school approach (WSA) provides a framework and starting point for a 'whole system redesign' of education, with the aim of realising holistic, systemic and collec-

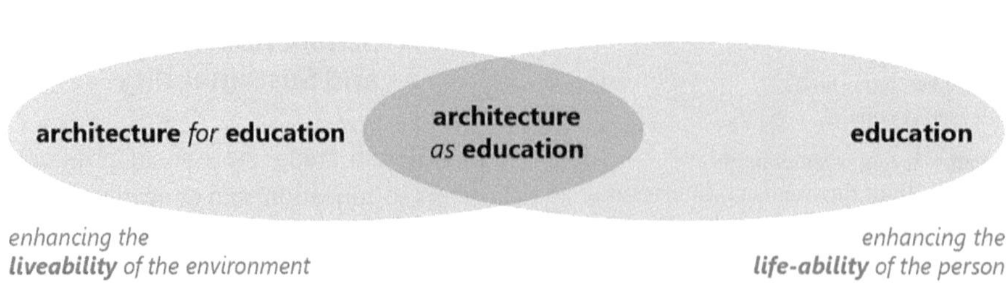

Fig. 10.1 Architecture *for* education and architecture *as* education

tive reflexive efforts towards sustainability as agreed upon by those affected (Wals & Mathie, 2022). Different variations of a WSA exist, but they all point at the multifaceted aspects of anchoring sustainability meaningfully in education and emphasise that the whole is more than the sum of all the parts (Wals & Mathie, 2022). The WSA 'flower' model gives an impression; this is the visual presentation of the conceptualisation of a whole school approach to ESD currently in use in countries like the Netherlands and Norway. It is presented visually in Fig. 10.2.

School habitats are often considered in the context of the WSA element 'institutional practices', because this area deals with practices related to energy, greenery and food, among other things. Approaching school habitats as 'educators' rather than objects invites us to take a broader perspective.

10.1.3 Research Goal and Structure of the Chapter

Following the above, we can conclude that the gap in research and practice regards the common space of architecture and education: architecture *as* education (see Fig. 10.1). Therefore, the goal of this study is to reveal the performance characteristics of school habitats which support or provide education for sustainable development (ESD). The study presented in this chapter elaborates on how the physical environment acts as an educator and presents a framework capturing the ESD performance characteristics of physical school environments. The resulting 'ESD performance framework for school habitats' is presented in Sect. 10.5 and builds on the theory and concepts as described in Sections 10.3 and 10.4. Section 10.3 elaborates on the vision on ESD that

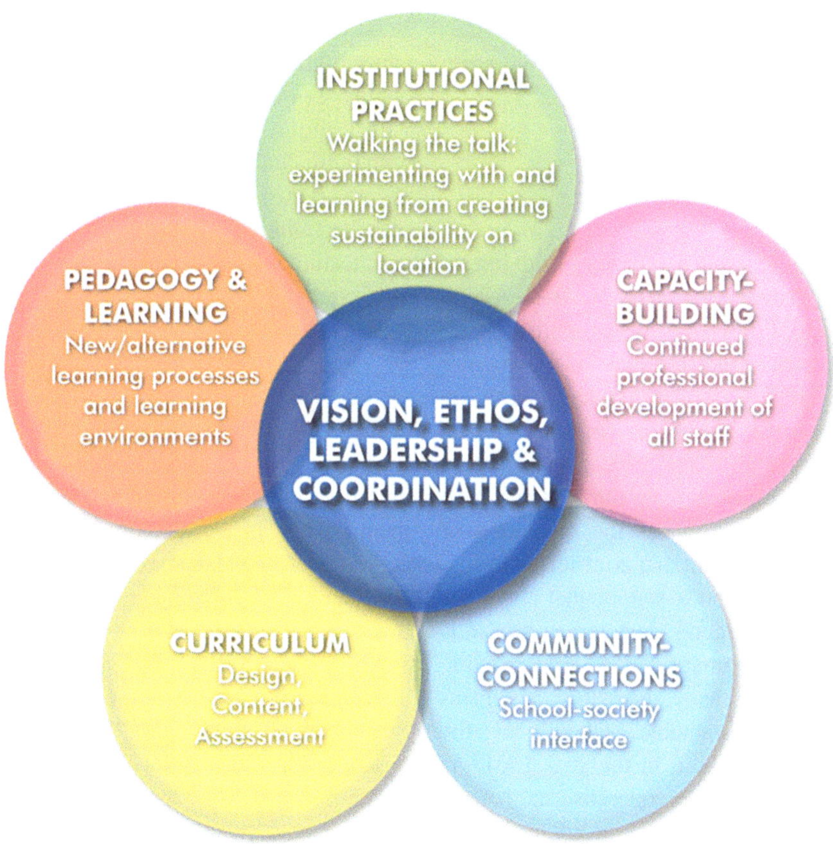

Fig. 10.2 WSA 'flower' model (Wals & Mathie, 2022)

is taken in this study and presents the axioms on which the framework is built. Section 10.4 elaborates on the 'performance concept' for building, which provides a basis to study and present the *educational* performance of physical environments as presented in the framework. Section 10.2 elaborates on the research methodology, and Sect. 10.6 regards the discussion and conclusions of the study as presented in this chapter.

10.2 Research Methodology

In the initial stage of this study, no hypothesis was present with regard to characteristics of the educational performance of school habitats related to ESD. Therefore, an 'inductive' research approach is taken. With an inductive stance, theory is the outcome of research (Bryman, 2012). In other words, the process of induction involves drawing generalisations and ideas out of observations. In order to arrive at an applicable hypothesis, the study began with unstructured literature and case study research.

10.2.1 The Case Study

The case study is used exploratory and instrumentally to distinguish design characteristics. The case study is thus used to accomplish something other than understanding this particular school environment, so the inquiry regards an 'instrumental case study', as opposed to an 'intrinsic case study' (Stake, 1995). In order to be able to draw conclusions from the case, a school has been selected which has the reputation of being a built environment that contributes to sustainability learning and a school that 'breathes sustainability'. The school selected as a case is 'de Sokkerwei', a primary school in Castricum, the Netherlands. By many, the school is considered the first 'sustainable school' in the Netherlands and was built in the year 2001. As the aim of the study is to concretise the role of physical school environments in education for sustainable development, the case study is used exploratory to better understand this phenomenon. To this end, the case is studied by means of unstructured interviews with pupils and staff and the one-site observation of these actors with regard to how the building is used. Moreover, the study involved an exploration of what has been written about the case in order to gain an understanding of why the case is valued by others and a conversation with one of the designers of the building in order to gain an understanding of the ideas behind the design.

10.2.2 The Inductive Process

The literature study is used complementarily to the case study research, to gather relevant theories, models and knowledge both in and outside the domain of building, in order to apply this understanding in the field of physical school environments. The inductive search for an applicable model in this context has been influenced thoroughly by the extent to which it can be understood easily, in order to secure the societal relevance of the outcomes of the study. For input and quality control during this process, a meeting with a group of experts and multiple meetings with individuals have been held. The attendees of the expert meetings were professionals from various disciplines, including building technology, architecture, architectural philosophy, philosophy, anthropology, environmental policy and management, sociology and pedagogy.

10.3 Vision on ESD

Talking about 'education for sustainable development'; what do we actually mean? That is the central question of this section. Education can be considered an 'instrument that can be used to move towards a more sustainable world' (Wals & Jickling, 2002; Buckler & Creech, 2014). Therefore, the UN Decade of Education for Sustainable Development (2005–2014) was supposed to bring substantial changes after the UN's conclusion in 2005 that 'at current levels of unsustainable practice and overconsumption it could be concluded that education is part of the problem' (UNESCO, 2005), since current educational systems still largely build on the paradigm, which has caused the social, ecological and economic problems currently leading. However, an analysis of the decade with worldwide action in theory, politics and practice concluded that it was 'business as usual in the end', primarily because it failed to challenge neoliberalism and foster global citizenship (Huckle & Wals, 2015). As Orr (1991) argued: 'We must not assume that it is education that will save us, or advance us; rather it is education of a certain kind'. Hence, we need to rethink our approach towards education—including the physical environments in which education takes place. Therefore, this section elaborates on the vision on ESD that underlies the study presented in this chapter.

10.3.1 The Hidden Curriculum of School Habitats

In the introduction, we have elaborated on the notion of the physical environment as an educator. This perspective on physical space and place stresses the need to contemplate on *which* lessons we want to teach pupils through education and physical educational settings. According to Orr (1997), educational buildings are typically approached as structures that quietly serve the educational process, but require no care and interaction of those who use it. Orr describes this architecture as the 'hidden curriculum' of the building, which powerfully influences the education given and the learning process of pupils. As Taylor (1995) wrote:

> We expect schools to prepare children for living in a democratic society, yet we provide a learning environment that resembles a police state — hard, overly durable architecture, giant chain-link fences, locked gates, guards, and even guard dogs. ... Such architecture fails to encourage the sense of ownership, participation, or responsibility required for a democracy.

In 1997, Orr even stated that, as commonly practised at the time, educational buildings could easily be converted to use as a factory or prison. Almost twenty years later, Upitis (2015) observed that schools have been built largely as a reflection of the 'factory model' for teaching and learning, meaning that pupils are 'processed' until they are deemed ready to leave school as if they were products to be fabricated.

10.3.2 Education and Sustainable Development

It can be argued that it is not too strange that schools reflect the 'factory model' for teaching and learning. Many of our educational buildings originate from times when the factory model dominated the development of the world. Industrialisation resulted in wealth, and, in most nations, this was accompanied by a rise in average (subjective) well-being (Veenhoven, 2004). Nonetheless, since the 1960s, there has been growing concern about our natural environment. In combination with emerging social challenges around the world, this has given rise to a call for more 'sustainable' development. Since then, enormous efforts have led to remarkable (relative) efficiency gains in many fields, for example, related to energy use (Veenhoven, 2004). However, there is more to sustainable development than efficiency advances. As Wals and Rodela (2014) wrote:

> Popularly stated, sustainability is not only about doing the things we do better (i.e. more efficiently) but also about doing things differently (i.e. developing new routines) and, perhaps foremost, about doing better things (i.e. developing new principles, vantage points and values).

To many, it is an appealing idea to use education as an instrument to shape such a 'better', more sustainable future. However, a genuine approach towards sustainable development in education asks for a careful consideration of the role of education in societal challenges. Considering education an 'instrument' for societal challenges holds the idea that 'others' are in the position to determine what the world and the future should look like, while pupils and students are to commit themselves to the view of others and give substance to it.

Wals and Jickling (2002) explored possible ways of designing education for sustainable development (ESD). On the one hand, 'instrumental views' of ESD are presented, juxtaposed with more 'emancipatory views'. The instrumental approach is described as resulting in an 'eco-totalitarian regime', where rewards, punishments and conditioning of behaviour forcefully create a sustainable society from the perspective of the natural environment. Since this society will almost certainly lack 'justice' and 'happiness', the emancipatory view is proposed, representing 'a very transparent society, with action competent citizens, who actively and critically participate in problem solving and decision making, and value and respect alternative ways of thinking, valuing and doing'.

10.3.3 Cultivation, Subjectification and Sustainable Grown-Upness

One could argue that the concepts Wals and Jickling (2002) describe closely relate to the 'cultivation' and 'subjectification' paradigms described by Biesta. Biesta (2021) describes the worldwide dominant approach towards education as the 'cultivation' of children and young people as *objects* and argues that what seems to be forgotten in education is that pupils are human beings and *subjects* in their own right, who face the challenge of living their own life and of trying to live it well. It is therefore that Biesta suggests that in addition to the widespread idea of education as *cultivation*, we need a different 'paradigm' in education. He describes this 'existential' paradigm as *subjectification*, that is, of the arrival of the 'I' in the world as subject of its own life, not as object of forces or desires from 'elsewhere'. Therefore, Biesta argues that education should be 'focused on equipping and encouraging the next generation to exist "in" and "with" the world, and to do so in their own right'.

This chapter departs from the idea that students should not be approached as objects that will enable a better future, but as subjects in their own right. The distinction between these paradigms of education provides a framework to consider the legitimacy of ESD from an educational perspective, rather than the perspective of a society that asks for a certain kind of citizens while confronting societal challenges. It provides us with the question of whether or not subjectification and sustainable development are compatible intentions of education. Hence, the question is as follows: Can we expect that a genuine approach—that is building on the subjectification, rather than the cultivation paradigm—to education will contribute to sustainable development?

The following citation of Biesta (2021) provides a hopeful perspective on the compatibility of ESD and subjectification:

> This is not to suggest, of course, that existing in one's own right is the same as "just doing what one wants to do." On the contrary, to exist as subject "in" and "with" the world is about acknowledging that the world, natural and social, puts limits and limitations on what we can desire from it and can do with it – which is both the question of democracy and the question of ecology.

Biesta argues that what is at stake in the context of subjectification is the possibility for the child to exist as subject of its own life. In other words: subjectification (which Biesta considers a life-long process) is necessary to be 'free'. Freedom is then understood as the ability to act based on a careful consideration of whether what you want to do, what you desire or what you desire to do will help to live well and coexist well. It is how people 'enact' their own 'subjectness', that is, to not be overwhelmed by impulses and forces from both inside and outside him or herself, but to develop and hold a mature relationship with these forces, which can be understood as 'voluntary self-limitation' (Biesta, 2019; Meirieu, 2007). It holds the challenge to be in the world without imagining ourselves at the centre of the world (Meirieu, 2007).

It is important to stress that this definition of 'freedom' differs from many neoliberal interpretations of the word, which put the emphasis on rolling back 'external' regulations and more choices for individuals (De Lissovoy, 2015). Using this neoliberal interpretation, 'freedom' could be seen as a complicating factor for sustainable development, as it is a major facilitator of the (over)consumption patterns that cause unsustainable development.

On the other hand, interpreting 'freedom' as a kind of 'grown-upness' to enact one's own life, rather than to be driven by forces from outside and within, holds a beckoning perspective. This interpretation of being 'grown-up' has got nothing to do with age. Being 'grown-up' or 'mature' in this context means that we do not solely focus on our own wishes and desires, but constantly ask the question of whether what we wish and desire is good for our own life, our life with others (democracy) and life on a planet with limited possibilities (ecology)—and what democracy and ecology ask from us. This chapter, therefore, interprets ESD as education for 'sustainable grown-upness'.

10.3.4 Relevancy for This Chapter

This vision on education, and on education for sustainable development in particular, is highly relevant for the remaining of this chapter, because it holds great consequences for the study presented in this chapter. Whereas behavioural and (eco-) psychological sciences, among others, provide knowledge about how to induce sustainable, pro-environmental behaviours, a genuine *educational* approach asks for another kind of thinking and doing. We could say that this educational performance primarily regards the extent to which pupils are given the chance to meet the (physical) world in order to encounter what the world asks from them, which is a whole other objective than to induce behaviour. That requires 'the world' to be present, visible and 'readable' in the school environment. The idea of physical environments as educators invites us to explore how school habitats 'speak' to the students as subjects, rather than moving and cultivating them as objects, towards the behaviours considered 'sustainable'.

10.4 The Performance Concept for Building

In order to study how physical place and space can be educational, this section introduces the adoption of the 'performance concept' for building (Szigeti & Davis, 2005). Performance-based building is an approach to the design, construction and evaluation of buildings to meet certain performance requirements. Hence, rather than focussing on building elements and technologies, it offers a way of approaching physical environments as 'performers'.

In order to address this, the study builds on the three domains adopted from the performance concept for building: demand, supply and performance as a bridge between them. These domains can be illustrated by means of the 'hamburger model', constituting two halves of a hamburger bun, which represent the demand and supply side of building (Gielingh, 1988). While this study is interested in the educational role that physical environments play, the *educational performance* is the focus of our interest. Therefore, in this study, education is the demanding discipline and building is the supplying discipline.

This section explores the demand and supply sides of the 'performance concept bun' and provides an introduction on what we can consider the 'educational performance' of school habitats that bridges the gap between demand and supply. The framework as presented in Sect. 10.5 builds on this exploration.

10.4.1 An Exploration of Supplies: School Habitats

The 'supply' side of the 'hamburger model' regards the actual physical environment designed or constructed. There are several examples of schools that have developed their physical environment as 'architecture as education'. For example, famous examples are the Italian 'Reggio Emilia' pre-primary schools (which refer to the physical school environment as a 'third teacher', after peers and adults), the 'Green School' in Bali and 'de Sokkerwei' in the Netherlands. Kong et al. (2014) studied the special design elements and features present in the famous 'Green School' in Bali, which give the school the character of a 'three-dimensional textbook' for environmental education. In a similar manner, Dutt (2013) explored how physical environments play a role in mediating students' relationships with the natural world. Both demonstrate the special design elements and features present in each particular school environment. Yet, since the design of physical environments is highly contextual and dependent on the specific function and location, such design elements cannot simply be copied to other settings. Their studies are typical for much of the work done regarding the impact of the physical environment on education for sustainable development (ESD). They have been conducted primarily in relation to case studies where natural and outdoor environments dominate the built ones and focus on pro-environmental behaviour only.

Performance-based building provides us with a concept to transcend the level of building elements and technologies. What is the special, educational 'performance' that these physical environments deliver? Which lessons can we draw from specific, yet valuable cases? Answering these questions brings us to the next domain of the model adopted, the domain of performance as a bridge between demand and supply.

10.4.2 An Exploration of Performance: Habitats as Educators

As elaborated on earlier, this chapter draws on the existential paradigm of education that Biesta (2021) describes as 'subjectification'. In order to give the next generation a fair chance at their subject-ness, Biesta highlights the importance of *teaching*, not understood as the transmission of knowledge and skills but as a gesture, an *act* of (re)directing the attention of students to the world, so that they may encounter what the world is asking from them. We build on the idea that education, or educating, refers to an *activity*: to something educators *do*. Educating is understood as a form of intentional action, which is something educators do deliberately, including the perhaps slightly odd but educationally important category of intentional nonaction; something educators deliberately do *not* do (Biesta, 2021).

Starting from the notion of the physical school environment as an educator means that its educational performance should lie in its ability to (re) direct the attention of students. The cases discussed above regarding 'good practices' of habitats as educators aimed to conceptualise their results and capture them in design models, aspects, themes or principles in order to provide insight into the influence of physical settings on students. Their inquiries are, however, limited in scope, primarily due to the strong focus on nature connectedness only. This study responds to that by a broader scope, applicable for natural and built environments, in rural and urban settings. But before going in-depth on the educational performance of school habitats in the context of ESD, we will elaborate on the demand side of the performance model, in order to be able to be more specific about the educational performances aimed for.

10.4.3 An Exploration of Demands: Subjectification

As elaborated on in Sect. 10.3.3, the process of subjectification towards 'sustainable grown-upness' is understood as the mature relationship with external and internal impulses and forces, meaning that we are not led by internal or external impulses and forces, but constantly ask the question of whether those forces drive us to actions that we consider good for our own life, our life with others (democracy) and life on a planet with limited possibilities (ecology)—and the ability to act accordingly to the answers we find on these questions. In this respect, it is important to stress that people do not only have an 'outside' of observable behaviour and actions but also an 'inside' of thoughts and feelings that are not visible from the outside (Prange, 2012), but constitute an important driver of action. Hence, subjectification requires another scope than the observable behaviour of people. Therefore, from the perspective of subjectification, the educational performance of a school habitat should not lie in its ability to provoke certain behaviours, but to direct the attention of pupils to things in the world that require their attention, hoping that the educational gesture of redirecting the attention will be able to foster the intentions and motivation of people to act.

How can a physical school environment make such a gesture? How can it 'speak to the student'?

In order to conceptualise this, this chapter adopts the organising principle of 'the heart, the head and the hands' (Sipos et al., 2008). The principle comprises a metaphor of ways to attend and take part in education, through feelings, thoughts and actions. This model is adopted because of its ability to inspire education and building professionals, as it draws attention to a diversity of qualities of physical environments. Previous research has shown that his research is indeed a suitable model to this end (Singleton, 2015).

Architecture as an educator for sustainable development should then have the quality to (re) direct the attention of pupils and therewith provide material for their feelings, thoughts and actions. We will elaborate a bit further on these three domains before moving on to the way school habitats may give substance to this, in Sect. 10.5.

Feelings cover the domain of affection and compassion. One could argue that the current *eco*logical crisis manifests itself as an *ego*logical crisis, that is, as a manifestation of the ego that positions oneself 'above' the world and treats it as composed of objects to fulfil the own desires, rather than as approaching it as subjects in their own right (Biesta, 2019). The domain of 'engaging the heart' builds on this notion. Fostering 'sustainable grown-upness' through the engagement of the heart aims to make pupils feel that the world transcends the level of the individual: that the individual subject is part of an ecosystem with other subjects and to value and respect the needs and integrity of others and the other (Biesta, 2019). As such, engaging the heart regards the intentional action to direct pupils' attention to 'others' and 'the other'.

Thoughts cover the domain of the contents of sustainability and the ability to reflect on issues accordingly (Wals., 2015). Such understanding and reflections require one to adopt an integral, holistic view, in which relationships are understood and connections are seen.

Actions are about what an individual *does* with his/her feelings and thoughts. Feelings and thoughts may result in intentions, motivation and cognition to act in a certain manner. The domain of the 'hands' addresses the actual action taken.

Feelings, thoughts and actions therewith compose three areas of interest when considering the educational performance of physical environments. Furthermore, it should be noted that such categories can be distinguished, but not separated. Reciprocal relationships exist between feelings, thoughts and actions. Rather, the distinction can be used to map different dimensions of the performance of physical school environments, in order to clarify discussions and position earlier research regarding it.

10.5 The ESD Performance Framework for School Habitats

This section holds the theoretical understanding as developed in this study, presented by means of a framework. The framework regards the performance characteristics of physical school environments considered *educational* in the context of education for sustainable development (ESD).

Considering physical environments as educators that (re)direct pupils' attention provides us with the question *whereto* the attention should be directed in the context of ESD. To answer this question, the framework is built on the three domains of the 'hamburger model' of performance-based building (supply, demand and the bridge between them; see Sect. 10.4) and on the three educational domains as elaborated on earlier (feelings, thoughts and actions; see Sect. 10.4.3). As a result, three areas are defined regarding the educational performance of physical school environments: 'engage the heart', 'exemplify to the head' and 'enable the hands'. In each area, performance characteristics are defined. As a result, the ESD performance of school habitats is captured in seven performance characteristics (A1 to C1); see Fig. 10.3. The following sections elaborate on the three focus areas distinguished and the performance characteristics that constitute these categories.

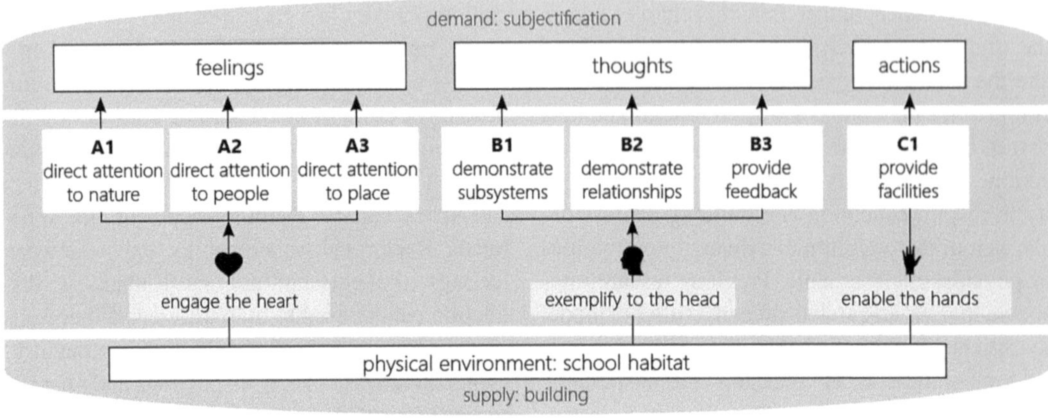

Fig. 10.3 ESD performance framework for school habitats

10.5.1 Engage the Heart

The first domain of educational performance of the habitat is based on the common distinction of three hierarchically situated and dynamically interrelated dimensions of sustainability, namely (1) the biological, geological and climatological substrate and its planetary boundaries; (2) the social relationships between humans; and (3) the human-made structures (Wals., 2015). Together, they capture the domains of 'others' and 'the other'. This leads to three educational performance characteristics in the focus area 'engage the heart'. We will elaborate on these three: the physical environment its ability to (re)direct the attention of students to (1) *nature*, (2) *people* and (3) *place*.

10.5.1.1 Direct Attention to Nature

The design characteristic 'direct attention to nature' (A1 in Fig. 10.3) regards the physical school environment's ability to let pupils encounter non-human life and non-living nature. It is important to stress that from a subjectification perspective, conditioning and inducing behaviour is not the aim of this performance characteristic. Rather, it aims at giving the next generation a fair chance at a mature relationship with nature, by facilitating students meeting the world and encountering what the world is asking of them. The example below illustrates how the presence of and attention towards nature provides the opportunity to engage the heart through education, providing certain opportunities in the physical environment. It also shows that not only the presence of nature but also the experiential interaction with the natural world, including animals, is of great importance (Dutt, 2013). In such cases, one could conclude that the engagement of the heart takes place through the simultaneous involvement of the head and the hands.

> **Example**
>
> A good example is the encounter with a plant. What is interesting about that meeting is that we can think so hard of and about the plant, but the plant will not grow a millimetre faster. Seen in this way, the plant requires a different attitude—one of care and attention. That makes gardening such an interesting topic for the school curriculum. The encounter with an animal does something similar. In that meeting too, something is asked from us, an appeal is made to us—a demand for care and attention and respect for the integrity of the animal.
>
> (Biesta, 2019), *translation by the authors*

A general remark that can be made on the physical environment's ability to educate is that pupils have to open themselves to 'hear' what place and space 'say'. Although this is true for every kind of education, this is particularly relevant in the case of the physical environment.

After all, the 'language' a physical environment is able to use and the gestures it is able to make are quite different from the way we are used to receive education through spoken words and human interactions from and with humans. The presence of nature in a school environment holds interesting possibilities to strengthen the ability of school habitats to (re)direct the attention of students, because it comprises living organisms. They change over time, due to age and seasons, and require suitable circumstances to survive over time. If the needs are not met, living organisms will die, which often is a quite effective way to (re)direct the attention of people.

Example

Seasonal variations comprise an interesting category of how nature can draw pupils' attention. On the school site, this is given substance to through the presence of deciduous trees. As a result of these trees, in the autumn the school grounds are covered with leaves. Almost all toddlers were playing with these leaves, rather than with the facilities provided for play. They specifically mentioned that the leaves were one of the reasons they so much appreciated the fall season. Conifers would not have had the ability to draw the attention of pupils, and to let them meet the specific period the world was in at that time. The leaves invited them to play, but also sweep the school grounds a little to keep it accessible.

Research findings by the authors, through case study work

10.5.1.2 Direct Attention to People

The design characteristic 'direct attention to people' (A2 in Fig. 10.3) regards the way the school habitat makes pupils encounter others. This includes a variety of aspects of built environments. We will elaborate on a few.

An important issue in this respect regards the issue of 'scale'. New technologies have given rise to modern aesthetics, which have drastically changed architectural expressions, such as large parts and spans. As a result, the 'human scale' in the built environment is often being threatened. This also affects schools, where relatively small people are present. In order to engage all of their hearts and to enable autonomous use to all, a variety of scales are thus required. For example, this can comprise indoor windows at child heights, furniture in suitable sizes and places to hide or withdraw in sizes that foster a feeling of safety.

Moreover, the needs and wishes of users of built environments show a wide variety, not only because they vary in length and height, but also because many people have some kind of disability that influences how they (can) use physical space. In many studies, the physical environment has been identified as a dominant factor for a more just and inclusive society in which all needs are understood as integral to society's order and not identified as 'special' (Farrington & Farrington, 2005; Woodcraft, 2012; Sherlawa & Hudebineb, 2015). The example below comprises some reflections on the accessibility of built environments in relation to their *educational* performance.

Example

The building of the school studied was not well accessible for all pupils, including one with a wheelchair. However, listening to the pupils, the situation in this school is not experienced as a discriminating, negative environment. On the contrary, some of them experienced barriers in the environment (such as doorsteps) as an incentive for caring for each other. The physical environment that was limiting (and maybe even felt discriminating) for some, was actually of great educational performance for others.

Research findings by the authors, through case study work

Also, routing is an interesting issue in school habitats. Entrances, corridors and halls powerfully influence who meet each other and who don't and in what manner. For example,

> **Example**
> The routing in the case school is aligned with its circular shape. In the morning, right before the lessons start, two pupils perambulate this circular route through the building with a bell to inform everybody about the current time. Rather than a reprimanding message of an automated system, the ringing bell therewith becomes a gentle reminder of peers to start their activities.
> *Research findings by the authors, through case study work*

Hertzberger (2008) has done extensive work on routes, stairs and human interaction.

10.5.1.3 Direct Attention to Place

Finally, the design characteristic 'direct attention to place' (A3 in Fig. 10.3) regards the physical school environment's opportunities to meet the particular place where the school is located, including human-made structures. The encounter with place provides the students to encounter what the place asks from them. For example, research has shown that participation in the maintenance and operation of school habitats can generate a sense of commitment and responsibility, which can be interpreted as a mature relationship with the needs of place (Ramkissoon et al., 2012). Attention to place therefore facilitates students a grounding in the history, culture and ecology of their surrounding environment, which is necessary to encounter what the place asks from them.

10.5.2 Exemplify to the Head

The notion of a school habitat as a 'three-dimensional textbook' (Spengler et al., 2001) leads us closer to the idea that physical place and space can 'exemplify to the head'. The way a school habitat can make this educational gesture can be structured by means of a metaphor regarding a system of gear wheels. By means of this metaphor, three design characteristics can be explained, leading to the notion of educational performance in the focus area 'exemplify to the head' as the physical environment its ability to (1) *demonstrate subsystems*, (2) *demonstrate relationships* and (3) *provide feedback*.

Firstly, in order to function well, each gear of the system needs to be in order. Hence, one needs to understand the 'subsystems' within the larger system (B1 in Fig. 10.3). For example, the presence and recognition of a rain barrel may direct the attention of pupils to the presence of water in their physical environment. Yet, specialised knowledge on the individual gear wheels is not sufficient. The interaction and relationships between the wheels (B2 in Fig. 10.3) are of importance. 'Systems thinking' is required to understand how those things, which may be regarded as (sub)systems, influence one another within a larger system. For example, the physical environment could show the route of the rainwater from the roof, through the downspout, to the rain barrel, and a nearby watering can may foster the understanding of pupils that rainwater is useful and necessary for the plants in the school garden.

> **Example**
> During their evaluation, pupils concluded that they often misuse the waste bins provided, because they do not have proper understanding regarding the different waste flows. Therefore, they proposed to illustrate on the bins which kind of products they can throw in that bin in order to enhance their understanding.
> *Research findings by the authors, through case study work*

Finally, understanding subsystems and relationships requires an additional step to appeal to pupils. For example, one can be motivated to contribute to the prevention of further climate change and understand that one's energy use is relevant to this end, but if one is not acquainted with one's own energy use, it will be hard to make effective changes. The last characteristic distinguished in this respect therefore is 'provide feedback' (B3 in Fig. 10.3).

Example

The roof of the primary school is covered with about sixty solar panels. Due to the slope of the roof, these panels are clearly visible from the school grounds. Hence, these "subsystems" are demonstrated. However, the pupils showed no understanding of what these panels are for, that is how they relate to the bigger "system." Also, no feedback is provided how the energy production relates to the energy consumption of the school. It remains unclear whether there is an appeal to the school community, and pupils in particular, to rethink the way they deal with energy.

Research findings by the authors, through case study work

Regarding all three design characteristics in this area, an important note should be made. The rise of 'sustainable' efforts in the building industry is leading to many product developments to improve building performance. For example, measures are taken to decrease energy use. Simple examples regard heating and lighting systems, which turn off and on automatically, based on schedules or sensors. If working well, these kinds of systems should be able to enhance the *liveability* of the environment (see Fig. 10.1). However, from an educational perspective such measures may be considered undesirable. One could say that the pupil's chance of meeting the world and encountering what the world is asking from them is diminished; the call to act as a subject in the world is eliminated from the physical environment. From an educational perspective, we could therefore also plead for 'intentional nonaction' (Biesta, 2021) as an educational gesture, non-automation in buildings, to foster interaction and let students meet the appeal the physical environment is making.

10.5.3 Enable the Hands

This final focus area for the educational performance of school habitats 'enable the hands' requires a careful introduction. The framework presented in this section is built on the subjectification axiom, focussing on the student to enact his/her own 'subject-ness'. The concept of subjectification puts emphasis on the importance of *intentional* action, that is not on whether we should consider certain behaviours as either 'good' or 'bad' and how to induce it, but on the question of whether one acts as a subject of his or her own action rather than as an object of forces driving him or her. Until now, we discussed educational gestures a physical environment can make in order to foster such 'sustainable grown-upness' of pupils, by (re)directing the attention to the world in order to encounter what the world asks from them. Whether or not a student opens himself, encounters the world and takes actions to respond to the appeal encountered is up to the student—that is the key issue at stake.

Hence, this focus area should not be understood as provoking or inducing pro-environmental behaviour. Rather, 'enabling the hands' is about making actions *able*. It is about providing the facilities necessary to take the actions aimed for at the individual level. Therefore, 'provide facilities' (C1 in Fig. 10.3) comprises the performance characteristic distinguished in this focus area. Facilities are of crucial importance in order to be able to enact one's own 'subject-ness' and 'sustainable grown-upness', since the key issue at stake with regard to subjectification is what one *will do* with what he or she has learned and how he or she has developed (Biesta, 2021). However, when we dive into the idea of 'facilities' a bit deeper, it turns out to be a complicated and interesting concept in the context of subjectification. We will elaborate on a few attention points when working with this performance characteristic.

10.5.3.1 Educational?

Although 'facilities' may sound as a neutral term, the way facilities are shaped may serve as external incentives or disincentives for actions. External stimuli aim at changing the circumstances under which behavioural choices are made by stimulating or repressing certain opportunities (Steg & Vlek, 2009; Max-Neef, 1991), treating people as objects to exhibit certain behaviours. As such, they can be considered highly *cultivational*. Hence, the question is whether it is possible to provide facilities without taking a cultivational approach. Through incentives and disincentives, certain actions can be provoked or induced, without these actions being *intentional*. This can be done either by limiting people in their freedom of choice or by making certain options more attractive or easy than others. For example, 'nudge theory' and 'fun theory' describe concepts of influencing actions *without* limiting people in their freedom of choice (Thaler & Sunstein, 2008). Thaler and Sunstein (2008) explain that putting fruit at eye level counts as a nudge while banning junk food does not. If the customer already had the intention to buy fruit, 'nudges' may help to realise his intentions. In that case, external stimuli can help develop new habits and routines, due to which those can be maintained and reinforced over time (Defra, 2008). External stimuli for action can then be experienced as supportive with regard to internal motivations for action.

Although one can hardly be against healthy food consumption, it is important to stress that reaching this effect through cultivational 'interventions' lacks to be *educational* and hardly contributes to 'sustainable grown-upness', since the actions are not *intentional*.

10.5.3.2 Self-Limitation and Self-Activation

'Sustainable grown-upness' is understood as not being led by internal or external impulses and forces, but constantly ask the question of whether those forces drive us to actions that we consider good for our own life, our life with others (democracy) and life on a planet with limited possibilities (ecology). Acting according to the answers we find on these questions demands something from us. 'Voluntary self-limitation' (Meirieu, 2007) and 'voluntary self-activation' regards the things we deliberately do and not do because we consider it 'good'. It emphasises that living a 'good' moral life is not something that comes to us, but requires our care and commitment. The example below illustrates this through a simple everyday situation.

Example

Each classroom has its own facilities for separated collection of plastic, paper and residual waste. However, several pupils concluded that they hardly used the plastic bin and threw their plastic waste in the residual waste bin instead, because they had to walk too far to the first-mentioned.

Research findings by the authors, through case study work

Following the above, developing the art of 'voluntary self-limitation' and 'voluntary self-activation' could be considered a key aspect of education for sustainable development. The school could serve as a place for this personal development. Hence, making 'sustainable behaviour' easy or the default actually eliminates this educational aspect from the physical environment. This is not meant as a pledge for unsustainable environments, but highlights the importance of considering the intended impact (i.e. either ecological, educational, cultivational or psychological) when designing physical environments.

10.6 Discussion and Conclusions

10.6.1 A Whole School Approach and Teacher Literacy

The work presented in this contribution departs from the idea that a physical environment is able to (re)direct the attention of students to the world; this is considered the *educational performance* of a school habitat. The case study on which this chapter is built suggests that much of the possibilities present will only be seized if a 'whole-school approach' is taken. The findings of this study primarily stress the importance of teachers. Educators have the important task of making use of the possibilities provided in their physical environments. The case study has shown that without the commitment of teachers, a large discrepancy will exist between the actual and intended use of buildings. This is strongly related to the findings of Martin (2008) and Cutter-Mackenzie and Smith (2003), who found that teacher literacy is the 'missing paradigm' in environmental education. Without their expertise and attention, it is very likely that the ability of a school habitat to educate 'gets lost'. This counts as long as a genuine definition of 'education' (i.e. education as subjectification) is taken, since a more cultivational influence will remain present.

10.6.2 A Moral Agenda in Education?

Alongside aims related to subjectification, whole school approaches to ESD typically also pay comprehensive attention to the *cultivational* performance of physical environments. For example, Shallcross et al. (2006) and Wals and Mathie (2022) emphasise the importance of action-orientated sustainability education that ensures what is taught in the classroom is echoed in the everyday actions of the whole school community. Such statements depart from a normative agenda (i.e. sustainable development) in education, and the appropriate behaviour of pupils. Important to add in the context of physical environments is that such a normative agenda is always present. If the school does *not* depart from an agenda of sustainability, it automatically departs from an agenda of *unsustainability*. This regards the 'hidden curriculum' of (un)sustainability that the school habitat exposes.

In this contribution, we have elaborated on such 'everyday actions' a bit under '9.5.3 Enable the hands'. The framework includes a performance characteristic related to the provision of facilities in order to make sustainable actions able. More cultivational characteristics of physical environments to foster pro-environmental behaviour of students are beyond the scope of this chapter.

10.6.3 Conclusions

This chapter presents the 'ESD performance framework for school habitats'. The performance characteristics presented in the framework of this contribution capture the *educational performance* of school habitats related to sustainable development. They comprise 'educational gestures' a physical environment can make with regard to the domains of the heart, the head and the hands of pupils. Education, in this respect, is understood as an empowering concept that strengthens 'subjectification' of pupils. This entails respecting the fact that pupils are human beings and *subjects* in their own right, who face the challenge of living their own life and of trying to live it well, rather than *objects* that should be 'cultivated' to expose certain behaviours. Therefore, this contribution builds on the notion of 'sustainable grown-upness', meaning that we do not focus on our own wishes and desires, but constantly ask the question of whether what we wish and desire is good for our own life, our life with others (democracy) and life on a planet with limited possibilities (ecology)—and what democracy and ecology ask from us. Cultivational characteristics to foster pro-environmental behaviour are not included in the framework, although the provision of facilities to enable actions can be cultivational.

The framework can provide guidance in practice. In diverse stages of building, such as planning, programming, design, construction and occupancy, the diverse performance characteristics can be used as criteria for building design. The framework is useful for professionals in the fields of education and building design. The examples provided illustrate that this approach to physical environments can be implemented on diverse scale levels, from the site (school grounds), to structure, building skin (facades and roofs), building services, the space plan and 'stuff' (Brand, 1994). The cultivation and subjectification paradigm, as discussed extensively, are relevant for designers, because they provide a dichotomy of different intentions of building design, either to provoke certain (pro-environmental) behaviours or to educate for 'sustainable grown-upness'.

References

Biesta, G. (2019). Duurzaamheid in het onderwijs: agenda of principe? – Een pleidooi voor realiteitszin. *Waardenwerk: Journal of Humanistic Studies, 78-79*, 59–64.

Biesta, G. (2021). *World-centered education: A view for the present*. Routledge.

Brand, S. (1994). *How buildings learn: What happens after They're built*. Viking.

Bryman, A. (2012). *Social research methods*. Oxford University Press.

Buckler, C., & Creech, H. (2014). *Shaping the future we want – UN decade of education for sustainable development (2005–2014) final report*. UNESCO.

Cleveland, B., & Fisher, K. (2014). The evaluation of physical learning environments: A critical review of the literature. *Learning Environments Research, 17*, 1–28.

Cutter-Mackenzie, A., & Smith, R. (2003). Ecological literacy: The 'missing paradigm' in environmental education. *Environmental Education Research, 9*(4), 497–524.

De Lissovoy, N. (2015). Neoliberalism and the contradictions of freedom: Ideology, subjectivity, and critical pedagogy. *Texas Education Review, 3*(2), 45–49.

Defra. (2008). *A framework for pro-environmental behaviours*. Department for Environment, Food and Rural Affairs.

Dutt, I. (2013). Built environments: Green spaces as a silent teacher. In D. Zandvliet (Ed.), *The ecology of school* (pp. 85–104). Sense Publishers.

Farrington, J., & Farrington, C. (2005). Rural accessibility, social inclusion and social justice: Towards conceptualisation. *Journal of Transport Geography, 13*(1), 1–12.

Gielingh, W. (1988). *General AEC reference model (GARM)* (pp. 165–178). Construction Informatics Digital Library.

Hertzberger, H. (2008). *Ruimte en leren* (p. 010). Uitgeverij.

Huckle, J., & Wals, A. (2015). The UN decade of education for sustainable development: Business as usual in the end. *Environmental Education Research, 21*(3), 491–505.

Kong, S., Rao, S., Abdul-Rahman, H., & Wang, C. (2014). School as 3-D textbook for environmental education: Design model transforming physical environment to knowledge transmission instrument. *The Asia-Pacific Education Researcher, 23*, 1–15.

Martin, P. (2008). Teacher qualification guidelines, ecological literacy and outdoor education. *Australian Journal of Outdoor Education, 12*, 32–38.

Max-Neef, M. (1991). *Human scale development: Conception, application and further reflections*. The Apex Press.

Meirieu, P. (2007). *Pédagogie: Le devoir de résister [Education: The duty to resist]*. ESF Éditeur.

Orr, D. (1991). What is education for? *The Trumpeter*, 99–102.

Orr, D. (1997). Architecture as pedagogy II. *Conservation Biology, 11*(3), 597–600.

Prange, K. (2012). *Erziehung als Handwerk. Studien zur Zeigestruktur der Erziehung*. Ferdinand Schöningh.

Ramkissoon, H., Weiler, B., & Smith, L. (2012). Place attachment and pro-environmental behaviour in national parks: The development of a conceptual framework. *Journal of Sustainable Tourism, 20*(2), 257–276.

Shallcross, T., Robinson, J., Pace, P., & Wals, A. (2006). Creating sustainable environments in our schools. In *Education and learning sciences* (p. 205). Trentham Publishers.

Sherlawa, W., & Hudebineb, H. (2015). The United Nations CRPD: Opportunities and tensions within the social inclusion and participation of persons with disabilities. *European Journal of Disability Research, 9*, 9–21.

Singleton, J. (2015). Head, heart and hands model for transformative sustainability learning: Place as context for changing sustainability values. *Journal of Sustainability Education, 9*, 1–16.

Sipos, Y., Battisti, B., & Grimm, K. (2008). Achieving transformative sustainability learning: Engaging head, hands and heart. *International Journal of Sustainability in Higher Education, 9*(1), 68–86.

Spengler, J., Loftness, V., Bayer, C., Bradley, J., Earthman, G., Eggleston, P., et al. (2001). *Green schools: Attributes for health and learning*. The National Academies Press.

Stake, R. (1995). *The art of case study research*. SAGE Publications.

Steg, L., & Vlek, C. (2009). Encouraging pro-environmental behaviour: An integrative review and research agenda. *Journal of Environmental Psychology, 29*(3), 309–317.

Strong-Wilson, T., & Ellis, J. (2007). Children and place: Reggio Emilia's environment as third teacher. *Theory Into Practice, 46*(1), 40–47.

Szigeti, F., & Davis, G. (2005). *Performance based building: Conceptual framework*. CIB (PeBBu) General Secretariat.

Taylor, A. (1995). How schools are redesigning their space. In A. Meek (Ed.), *Designing places for learning* (pp. 67–76). Association for Supervision and Curriculum Development.

Thaler, R., & Sunstein, C. (2008). *Nudge: Improving decisions about health, wealth, and happiness*. Yale University Press.

UNESCO. (2005). *Guidelines and recommendations for reorienting teacher education to address sustainability*. UNESCO, Division for the Promotion of Quality Education.

Upitis, R. (2015). School architecture and complexity. *Complicity: An International Journal of Complexity and Education, 1*(1), 19–38.

Veenhoven, R. (2004). Sustainable consumption and happiness. In *Driving forces and barriers to sustainable consumption*. University of Leeds.

Wals, A., & Jickling, B. (2002). "Sustainability" in higher education: From doublethink and newspeak to critical thinking and meaningful learning. *Higher Education Policy, 15*(2), 121–131.

Wals, A., & Mathie, R. (2022). Whole school responses to climate urgency and related sustainability challenges. In *Encyclopedia of educational innovation*. Springer.

Wals, A., & Rodela, R. (2014). Social learning towards sustainability: Problematic, perspectives and promise. *NJAS – Wageningen Journal of Life Sciences, 69*(1), 1–3.

Wals. (2015). *Beyond unreasonable doubt: Education and learning for socio-ecological sustainability in the Anthropocene (inaugural address)*. Wageningen University.

Wargocki, P., & Wyon, D. (2007). The effects of outdoor air supply rate and supply air filter condition in classroom on the performance of schoolwork by children. *HVAC and R Research, 13*(2), 165–191.

Woodcraft, S. (2012). *Social sustainability and new communities: Moving from concept to practice in the UK*. ASIA Pacific International Conference on Environment-Behaviour Studies. .

11 Developing a Whole (Pre)school Approach to Sustainability: Insights from Global Citizenship and Early Childhood Education Across Nordic Countries

Farhana Borg, Ingrid Pramling Samuelsson, and Niklas Gericke

Key Message
Wholeness in Nordic preschool education is defined through four perspectives: *The greening of the whole preschool, The whole child's learning, Wholeness in preschool teaching,* and *Thematic approach as a perspective on a whole preschool.* These perspectives, derived from global citizenship education, serve as a basis for implementing a whole school approach to sustainability in preschool education. However, the lack of societal engagement in early childhood education must be addressed, given its significance for children's learning for sustainability.

11.1 Introduction

This chapter focuses on early childhood education, which in many ways differs from school education. Early childhood education often takes a holistic perspective on children's learning and the preschool in general, making it a suitable arena in which to implement a whole school approach to sustainability. The chapter discusses this from the perspective of the preschool system in the Nordic countries, namely Sweden, Denmark, Norway, Finland, and Iceland. In the Nordic countries, preschool education has taken inspiration from global citizenship education (Pramling & Pramling Samuelsson, 2011), which we argue makes it even more possible to align preschools with a whole school approach to sustainability. Global citizenship can be defined as an awareness of the wider world and the development of an understanding of how the world functions socially, economically, culturally, and environmentally, as well as active engagement in community and international issues to make the world a more equitable and sustainable place (Twigg et al., 2015). There are important connections between global citizenship education and a whole school approach to sustainability, and this is stated by the United Nations (2023): "Connected within the Target 4.7, Global Citizenship Education (GCED) and Education for Sustainable Development (ESD) are recognized as mutually reinforcing approaches"—the intention being to build a sustainable world.

The world is becoming increasingly global and connected, and the concept of "citizenship" has been extended beyond the traditional views

F. Borg (✉)
Dalarna University, Falun, Sweden
e-mail: fbr@du.se

I. P. Samuelsson
Gothenburg University, Gothenburg, Sweden
e-mail: ingrid.pramling@ped.gu.se

N. Gericke
Department of Environmental and Life Sciences, Karlstad University, Karlstad, Sweden
e-mail: niklas.gericke@kau.se

A. E. J. Wals et al. (eds.), *Whole School Approaches to Sustainability*, Sustainable Development Goals Series, https://doi.org/10.1007/978-3-031-56172-6_11

of country and continent. Global citizenship education includes local issues, such as the living conditions of people, and connects those conditions with global issues (Wals et al., 2017). For children to become global citizens, they need first to become citizens in preschool[1] by learning to stand up for their rights, taking responsibility for their actions, respecting and valuing diversity and other cultures, and learning to plan together and resolve common problems. Global citizenship education helps children find their place in the international community (Borg & Pramling Samuelsson, 2022). To create a sustainable and just society for all, education needs to emphasise "a more relational and caring way of being and becoming" rather than knowledge and literacies (Wals & Mathie, 2022, p. 1).

The whole school approach to sustainability embraces a holistic and participatory educational philosophy that aims to enhance the potential of the school environment to function as an authentic and meaningful place of learning. The whole school approach has been described in different ways in the literature (e.g. Henderson & Tilbury, 2004; Mathie & Wals, 2022; Mogren et al., 2019). In a review of the literature, the integration of three lines of action is frequently visible in all whole school approach models: first, the environmental management ("greening") of the school; second, the establishment of ongoing partnerships with the broader local community to address issues of social-environmental sustainability; and third, the incorporation of sustainability in the curriculum (Gericke, 2022). In this chapter, we specifically investigate the ways in which these three lines of action can be identified at the preschool level while addressing global citizenship.

Global citizenship education shares several educational characteristics with the whole school approach to sustainability, as pointed out in the above quote from the UN. Both approaches provide a framework for reorienting and redesigning education for future generations in which a holistic perspective on the learner and the school is taken (Borg & Sporre, 2023; Gericke, 2022). Moreover, both approaches incorporate the learning goal of building a sustainable world based on collaboration with society beyond the school. However, the literature has shown the difficulties of implementing a whole school approach to sustainability in primary and secondary schools and, furthermore, provides no examples of a whole school approach to sustainability in preschools (Gericke, 2022). This indicates that the whole school approach to sustainability remains unexplored in early childhood education. In this chapter, we argue that preschool in the Nordic context could be a promising arena in which to develop a whole school approach to sustainability based on existing teaching practices that build on traditions from global citizenship education. Indeed, this chapter aims to show and critically reflect on how global citizenship education in preschool, from a Nordic perspective, can illuminate the possibilities and barriers to developing a whole school approach to sustainability.

11.2 Global Citizenship in Early Childhood Education and a Whole School Approach to Sustainability

Children live in the here and now. How they experience and view the world can differ considerably from how adults experience and view it (Sommer et al., 2010). The concept of global citizenship considers the shared and interdependent global experience and includes values such as diversity, interdependence, and empathy (WEF, 2021). To find their place in the international community, children should be engaged in local and global sustainability issues with consideration given to their age and stage of development (Borg & Gericke, 2021). Global citizenship is closely related to education for sustainability (EfS), which is about caring for oneself, for other people, and for the world. EfS emphasises the importance of local and global issues, stressing

[1] Preschool refers to early childhood education and care, and it is a synonym for kindergarten, nursery, and pre-primary education for children who have yet to enter compulsory school. In this paper, we use both preschool education and early childhood education.

that all education should promote vital values, necessary knowledge, and required skills that can contribute to a sustainable future (United Nations, 2015). EfS is often described as being a foundation for a whole school approach to sustainability (Henderson & Tilbury, 2004). A whole school approach to sustainability has the overall aim of transforming the world into a more sustainable place by way of both direct actions, such as greening of schools and collaboration with society, and indirect actions in teaching, such as EfS.

11.3 Linking a Whole School Approach to Sustainability with Early Childhood Education Research

In this chapter, we argue that a whole school approach to sustainability embraces many approaches to teaching that are similar to those described in global citizenship education. The whole school approach to sustainability holds a holistic and participatory educational philosophy that aims to function as an authentic and meaningful place of learning to orient the world in a sustainable direction. As pointed out earlier, a whole school approach to sustainability generally proposes three lines of action: environmental management ("greening") of the school; establishment of ongoing partnerships with the wider local community to address social-environmental sustainability; and incorporation of sustainability in the curriculum (Gericke, 2022). To some extent, these actions overlap because children's learning and the curriculum are often central in preschool. In this section of the chapter, we specifically investigate the ways in which these three lines of action can be identified at preschools in previous early childhood education research.

When it comes to the first line of action, namely greening, literature reviews on preschool education indicate that there is a long tradition of being in the outdoors to work with greening issues (Bascopé et al., 2019; Davis, 2009). In the Nordic countries, outdoor education and experiences in nature are embedded in early childhood education. The most common activities in preschool are gardening, being in nature, and picking up litter in forests—that is activities closely related to greening (Ärlemalm-Hagsér & Sundberg, 2016). For example, in Sweden in the 1980s, the outdoor association *Friluftsfrämjandet* established a chain of preschools called *I Ur och Skur* ("Come Rain or Shine"), a major educational aim of which was for children to spend the whole day outside to become environmentally aware and enjoy natural environments (Gericke et al., 2020). Based on the eco-school movement, there are also two types of eco-certifications available in Sweden: one is "Green Flag" by the Keep Sweden Tidy Foundation, and the other is "Preschool for Sustainable Development" by the Swedish National Agency for Education. The purpose of these eco-certifications is to promote and implement EfS from a holistic perspective as suggested in a whole school approach to sustainability. Currently, there are many preschools in Sweden and the other Nordic countries that are actively engaged in sustainability issues from a holistic perspective (Bascopé et al., 2019; Borg, 2017).

In terms of the second line of action, which is about collaboration with society at large, there is less advancement, yet there is potential for preschools to develop in this specific area. In the literature, there is little about studies on collaboration between preschools and society at large. Preschools need to extend their collaboration with a broader segment of society that is more than just about greening issues. From a formal perspective, this shift has already occurred. In the new national Swedish curriculum for preschool (Lpfö 18), active participation in the society of the individual child is emphasised. It gives the following learning outcomes for preschool children:

- A growing responsibility for and interest in sustainability and active participation in society
- An ability to assume responsibility for their actions and for the environment in the preschool
- An understanding of democratic principles and the ability to cooperate and make decisions (Skolverket, 2018)

There are some positive findings from recent research that investigated the professional development of preschool teachers in projects relating to the implementation of a whole school approach to sustainability. As a result of their participation in those projects, preschool teachers were found to be able to include sustainability from a holistic perspective in their teaching. They felt confident in their ability to extend their collaboration to the local community and parents, as well as children and teachers from preschools in other countries (Borg, 2019; Borg & Gericke, 2021).

The third line of action is about the incorporation of sustainability into the curriculum. Preschools in Sweden and the other Nordic countries have a holistic perspective in which education connects to children's everyday experiences from both within the preschool and beyond the preschool in their everyday lives. In preschool education, play, learning, and care are necessary for a wholeness in childhood and all are considered essential (Broström, 2017; Pramling Samuelsson & Asplund Carlsson, 2008). Furthermore, preschool teachers commonly collaborate with other members of staff, such as childcarers, kitchen staff, and caretakers, such as is suggested in a whole school approach to sustainability. For example, preschool teachers and childcarers often plan their work together with the children, and the principal is often involved as well. Preschool teachers in Sweden share mealtimes with the children, and the children often collect their food themselves from the kitchen staff. They are also all active in greening efforts, such as composting (Ärlemalm-Hagsér & Sundberg, 2016). Many preschools aim to reduce food waste and offer more locally produced food (Engdahl et al., 2021). It could be argued that the preschool is closer to the whole school approach to sustainability ideal than schools that differentiate more strictly between teaching and other school activities. McKeown and Hopkins (2007) identified four educational levels that are important in the work towards sustainable schooling. These are disciplinary, whole school, educational system, and international policy. At the preschool, disciplinary boundaries are absent and not quite applicable. There are no school subjects to consider, and as such, teaching takes place at the (whole) school level or the unit level. Because of this, preschools do not face the same problems of reductionism as are commonly experienced in primary and even more so secondary education (Gericke et al., 2020). Instead, the teaching approach is holistic and aims to connect new knowledge with the children's life world.

To conclude, preschool teaching is only loosely regulated, and at preschool, it is easier to work holistically and more flexibly. Furthermore, in recent years, the Swedish preschool curriculum (Skolverket, 2018) has become more prescriptive than it once was when it comes to teaching, and there has been more emphasis on aspects of sustainability (Ohlsson et al., 2022). In that sense, the curriculum supports preschools in their work to address sustainability issues from both an educational and a greening perspective.

11.4 Wholeness in Preschool Education

Preschool has a different perspective on education compared to compulsory school since it aligns more with the whole school approach to sustainability. This is because the view on learning differs in relation to the physical and emotional development of children, which is determined by their age. Furthermore, teaching is not organised by school subject but rather by theme so that children are aware of the relationship between what they are working with and their everyday experiences (Pramling & Pramling Samuelsson, 2011). In the Nordic countries, preschool is well-established and is for children between the ages of one and five. Preschools in the Nordic countries are subsidised by the government and the municipalities; they work according to a national curriculum; they are regulated by school law; and they employ university-educated teachers. Upon a review of policy texts and research in the Nordic countries, we were able to identify four aspects of existing preschool education that each has a basis in the global citizenship tradition of viewing the child and the preschool from a wholeness perspective (Wagner, 2006). Together they form a foundation on which, we would argue, it is possible to build when operationalising a whole school approach to sustainability. We describe these four perspectives on wholeness in Nordic preschool education under four headings: *The greening of the whole preschool; The whole child's learning; Wholeness in preschool teaching;* and *Thematic approach as a perspective on a whole preschool.*

11.4.1 Greening of the Whole Preschool

In Sweden, the issue of greening in preschools has increased in importance in recent decades. For example, there has been a move to rid building environments of toxic materials, such as harmful plastic, and to make sure that cleaning fluids are non-toxic (Borg, 2019). A recent study on sustainable preschools showed that many preschools procure eco-food or local produce, serve vegetarian food once or twice a week, and grow vegetables on school grounds (Engdahl et al., 2021). An increasing number of new preschools are climate-neutral. It is also common for Nordic preschools to have recycling boxes so that children learn from an early age to sort things into different containers and cupboards where parents can leave used toys and clothes for other children (Engdahl et al., 2024). Such examples demonstrate how the greening aspect of the whole preschool approach to sustainability has been significantly addressed in preschool education practice and research in Sweden. However, as mentioned earlier, there is little research on how to transfer the greening aspect from the preschool into society beyond the preschool. Therefore, more developmental work and research are needed that can investigate how preschools can interact with society on issues of sustainability in studies on the whole school approach to sustainability.

11.4.2 The Whole Child's Learning

Traditionally in preschool education, the child is the focus, and the importance is on how they develop as individuals. The aim of early childhood education is to support children to develop cognitive, social, and motor skills, as well as—not least—values and ethics that are central to sustainability (Broström, 2017; Pramling & Pramling Samuelsson, 2011). These are important aspects when talking about a whole (pre) school approach to sustainability. A learner-centred perspective views the child as a whole in global citizenship education and is an important baseline when addressing a whole school approach to sustainability in preschool education. It is important for children to develop language competence so that they can become literate citizens. The journey towards the whole child's learning is important in a whole school approach to sustainability because children need to be able to use and apply knowledge and make decisions according to their validated knowledge. This is important for them as citizens and is, consequently, a goal of a whole school approach to sustainability. As such, preschool is fundamental in terms of the lifelong learning of global citizens.

11.4.3 Wholeness in Preschool Teaching

All elements of the preschool day provide children with an education—that is to say, they are being educated even at times when the teacher is not teaching explicitly according to the curriculum. This distinction between spontaneous teaching and planned teaching is clear in the new Swedish curriculum (Skolverket, 2018), which is very different from teaching in compulsory school in which all teaching is steered by the teacher. Play, learning, and care are integral components of holistic preschool education. Teachers can play with children, look out for their well-being, and challenge and support their ideas and understanding of the world around them. Another way to describe preschool education in the Nordic countries is to say that its formal educational components are in essence informal (Wagner, 2006). Furthermore, the literature presents teaching strategies that have been developed to help teachers with children's learning and that describe the importance of communication for young children (Doverborg et al., 2019). These teaching strategies can be used to further develop whole preschool approaches for sustainability because they start from the authentic experience of the child rather than the prescribed content of the curriculum.

When it comes to teaching about and for sustainability, it is important both to recognise the interconnectedness of environmental, social, and economic perspectives and to link facts, stories, and imagination and play with real-life issues (Eilam & Trop, 2010). In a preschool setting, teaching in line with global citizenship aspirations can be viewed as a shared, interactive experience, which means that rather than being instructional, teaching in the preschool setting is more about making somebody aware of something (Björklund & Pramling Samuelsson, 2020). The curriculum should be open and flexible, and it must provide opportunity for thought, reflection, and invention—indeed, opportunity for the child to be creative. Preschool education does not separate play from learning since the job of the preschool is to facilitate children's learning and development in a form of teaching where play is central (Pramling & Wallerstedt, 2019). What the teacher can do is to point out a direction—to engage children in activities. In this way, children can be involved in decision-making in preschool teaching since they are to perceive themselves as having agency so that they can consider themselves active citizens, both locally and internationally (Borg & Gericke, 2021). This is also in line with a whole school approach to sustainability and differentiates between preschool teaching practices and those of compulsory school.

The predominant idea is that no one (the preschool teacher in this case) can construct teaching; rather, participants—in this case, the child and the preschool teacher—contribute jointly to learning. As such, the aspects of preschool teaching strategies that centre on wholeness in education are an existing tradition within preschool education in Sweden and the other Nordic countries that can inform the development and initiation of whole preschool approaches (Bascopé et al., 2019; Doverborg et al., 2019).

11.4.4 Thematic Approach as a Perspective on a Whole Preschool

In preschool education in the Nordic countries, thematic work has traditionally been central to the organisation of the curriculum and teaching. Thematic teaching integrates different subject areas under one theme and over the years has had many synonyms, among them *centres of interest* and *projects*. The official term in the curriculum is *thematic work* (Doverborg et al., 2020). What this means is that learning is not organised by school subject such as is the case in formal education at school. Thematic work integrates various topics: for example, the farm, the forest, water, the city, the seasons, and friendship. To make children aware of aspects of content areas, different subjects are used: for example, mathematics and language are put into a context that makes them meaningful. Thematic work in Swedish preschools can last from one week to six months (if not longer) so that teachers can incorporate and combine different subjects while focusing on a specific topic or theme. This is another aspect of the Nordic early childhood education tradition that fits well with the agenda for a whole school approach to sustainability.

Since children's participation and influence are important for democracy and relate to citizenship, their voices and opinions are central. This relates to the UN Convention on the Rights of the Child (1989) as a source in the sustainability work to enable children to develop agency and to dare to make decisions and develop civil disobedience instead of simply being followers. According to Struthers (2016), children can learn about their rights so long as adults afford them the opportunity to exercise them by allowing them—the children—to be heard and have influence. Studies show that it is not easy to implement this aspect of global citizenship (Theobald et al., 2023).

11.5 Lessons Learned from Global Citizenship to a Whole School Approach to Sustainability

In this chapter, we discuss the potential of preschool education in Sweden and the other Nordic countries in relation to whole school approach to sustainability. Based on our analysis of early childhood literature, we identified four characteristics of current preschool education in the Nordic countries, mainly Sweden, that can facilitate further development of a whole school approach to sustainability: *The greening of the whole preschool; The whole child's learning; Wholeness in preschool teaching;* and *Thematic approach as a perspective on a whole preschool* (Gericke, 2022). We conclude that there is a great opportunity to promote a whole school approach to sustainability in preschool education in Sweden and the other Nordic countries today, much due to the fact that preschool education in the local context has been influenced by global citizenship education. We would argue that the current organisation, practices, and policy documents of the Nordic countries provide good incentives for preschools to work from a whole school approach to sustainability perspective based on their preschool education traditions. There are many preschools that have come a long way when it comes to establishing a whole school approach to sustainability without explicitly using the term or knowing what the term whole school approach means (Borg, 2019). One aspect of a whole school approach to sustainability in Nordic preschools that we identified as possibly weak is collaboration with society at large: here, there is a lack of both good examples and research, which calls for more research and development work in this area.

A more general conclusion of this chapter is that the whole school approach to sustainability benefits when consideration is given to the issue of globalisation and alignment to global citizenship education with a whole school approach to sustainability. If Agenda 2030 is to be achieved and preschools are to be agents of change for a more sustainable future, partnerships are needed not only at the local level but also at the global level. We, therefore, suggest that efforts to implement a whole school approach to sustainability must be aligned with efforts to achieve global citizenship education. For this to happen more systematically in efforts to achieve a whole school approach to sustainability in preschools, it is not enough to offer professional development. We would suggest that perspectives from a whole school approach to sustainability be included in early childhood teacher training programmes. Preschool teachers need to develop their skills and competence in this area so that they can work with both curriculum issues as well as greening and cooperation with society at large since their role is crucial for children's learning of sustainability (see Borg, 2019; Borg & Gericke, 2021). In this way, we would argue that the preschool has the potential to be both a direct and an indirect agent of change that fosters children's action competence for sustainability (Sass et al., 2021).

References

Ärlemalm-Hagsér, E., & Sundberg, B. (2016). Naturmöten och källsortering-En kvantitativ studie om lärande för hållbar utveckling i förskolan. *Nordic Studies in Science Education, 12*(2), 140–156.

Bascopé, M., Perasso, P., & Reiss, K. (2019). Systematic review of education for sustainable development at an early stage: Cornerstones and pedagogical approaches for teacher professional development. *Sustainability, 11*(3). https://doi.org/10.3390/su11030719

Björklund, C. & Pramling Samuelsson, I. (red.) (2020). Innehållets didaktik i förskolan. Liber. http://libris.kb.se/bib/bnx9136982xc9xbf

Borg, F. (2017). *Caring for people and the planet: Preschool Children's knowledge and practices of sustainability*. (Doctoral thesis). [Elektronisk resurs]. Umeå universitet.

Borg, F. (2019). A case study of a green flag-certified preschool in Sweden. *Hungarian Educational Research Journal, 9*(4), 607–627. https://doi.org/10.1556/063.9.2019.4.52

Borg, F., & Gericke, N. (2021). Local and global aspects: Teaching social sustainability in Swedish preschools. *Sustainability, 13*(7), 3838.

Borg, F., & Pramling Samuelsson, I. (2022). Preschool children's agency in education for sustainability: The case of Sweden. *European Early Childhood Research Education, 30*(1), 147–163. https://doi.org/10.1080/1350293X.2022.2026439

Borg, F., & Sporre, K. (2023). Children's empowered inclusion in early childhood education for sustainability. In D. Stacy (Ed.), *International perspectives on educating for democracy in early childhood: Recognizing young children as citizens* (1st ed., pp. 241–260). Routledge. https://doi.org/10.4324/9781003229568

Broström, S. (2017). A dynamic learning concept in early years' education: A possible way to prevent schoolification. *International Journal of Early Years Education, 25*(1), 3–15. https://doi.org/10.1080/09669760.2016.1270196

Doverborg, E., Pramling, N. & Pramling Samuelsson, I. (2019). *Att undervisa barn i förskolan* (2:a upplagan). Liber.

Doverborg, E., Pramling, N., & Samuelsson, I. P. (2020). *Att arbeta med tema i förskolan - en metodikbok*. Liber.

Eilam, E., & Trop, T. (2010). ESD pedagogy: A guide for the perplexed. *The Journal of Environmental Education, 42*(1), 43–64. https://doi.org/10.1080/00958961003674665

Engdahl, I., Pramling Samuelsson, I., & Ärlemalm-Hagsér, E. (2021). Swedish teachers in the process of implementing education for sustainability in early childhood education. *New Ideas in Child and Educational Psychology, 1*(1), 3–23. https://nicepj.ru/articles/article/9634/

Engdahl, I., Pramling Samuelsson, I., & Ärlemalm-Hagsér, E. (2024). How different choices people make in everyday life can contribute to sustainable development in Sweden. In J. Davies & S. Elliot (Eds.), *Young children and the environment*. Cambridge University Press.

Gericke, N. (2022). Implementation of education for sustainable development through a whole school approach. In G. Karaarslan-Semiz (Ed.), *Education for sustainable development in primary and secondary schools*. Springer.

Gericke, N., Manni, A., & Stagell, U. (2020). The green school movement in Sweden—Past, present and future. In A. Gough, J. C. Lee, & E. P. K. Tsang (Eds.), *Green schools globally: Stories of impact on education for sustainable development* (pp. 309-332). Springer.

Henderson, K., & Tilbury, D. (2004). *Whole school approaches to sustainability: An international review of sustainable school programs*. Report prepared by Macquarie University for the Australian Government Department of the Environment and Heritage.

Mathie, R. G., & Wals, A. E. J. (2022). *Whole school approaches to sustainability: Exemplary practices from around the world*. Education & Learning Sciences/Wageningen University. 62 pages. https://doi.org/10.18174/566782

McKeown, R., & Hopkins, C. (2007). Moving beyond the EE and ESD disciplinary debate in formal education. *Journal of Education for Sustainable Development, 1*(1), 17–26. https://doi.org/10.1177/097340820700100107

Mogren, A., Gericke, N., & Scherp, H.-Å. (2019). Whole school approaches to education for sustainable development: A model that links to school improvement. *Environmental Education Research, 25*(4), 508–531. https://doi.org/10.1080/13504622.2018.1455074

Ohlsson, A., Gericke, N., & Borg, F. (2022). Integration of education for sustainability in the preschool curriculum: A comparative study between the two latest Swedish curricula. *Journal of Childhood, Education & Society, 3*(1), 12–27. https://doi.org/10.37291/2717638X.202231130

Pramling, N., & Pramling Samuelsson, I. (Eds.). (2011). *Educational encounters: Nordic studies in early childhood didactics*. Springer.

Pramling Samuelsson, I., & Asplund Carlsson, M. (2008). The playing learning child: Towards a pedagogy of early childhood. *Scandinavian Journal of Educational Research, 52*(6), 623–641. https://doi.org/10.1080/00313830802497265

Pramling, N., & Wallerstedt. (2019). Lekresponsiv undervisning – ett undervisningsbegrepp och en didaktik för förskolan. *Forskning om undervisning och lärande, 7(1)*, 7–22.

Sass, W., Boeve-de Pauw, J., De Meyer, S., Gericke, N., & van Petegem, P. (2021). Actions for sustainable development through young students' eyes. *Environmental Education Research, 27*(2), 234–253. https://doi.org/10.1080/13504622.2020.1842331

Skolverket. (2018). *Curriculum for the preschool 2018. The Swedish National Agency for Education.* Skolverket.

Sommer, D., Samuelsson, I. P., & Hundeide, K. (2010). *Child perspectives and children's perspectives in theory and practice. International perspectives on early childhood education and development* (Vol. 2). Springer Science & Business Media.

Struthers, A. E. C. (2016). Breaking down boundaries: Voice and participation in English primary education. *The International Journal of Children's Rights, 24*(2), 434–468. https://doi.org/10.1163/15718182-02402011

Theobald, M., Busch, G., Mushin, I., et al. (2023). Children as citizens of a global society: Learning together in culturally and linguistically diverse class-

rooms. In S. DeZutter (Ed.), *International perspectives on educating for democracy in early childhood: Recognizing young children as citizens*. Routledge.
Twigg, D., Pendergast, D., & Twigg, J. (2015). Growing global citizens: Young Children's lived experiences with the development of their own social world. *International Research in Early Childhood Education, 6*(1), 79. https://doi.org/10.4225/03/58100c68eed90
United Nations. (2015). *Transforming our World: The 2030 agenda for sustainable development*. United Nations. Retrieved from http://www.un.org/ga/search/view_doc.asp?symbol=A/RES/70/1&Lang=E, December 3, 2015.
United Nations. (2023). *Global citizenship education*. United Nations. Retrieved from http://www.un.org/ga/search/view_doc.asp?symbol=A/RES/70/1&Lang=E, December 3, 2015.
Wagner, J. T. (2006). An outsider's perspective: Childhoods and early education in the Nordic countries. In K. M. Borman (Series Ed.), J. Einarsdottir, & J. T. Wagner (Vol. Eds.), *International perspectives on educational policy, research and practice* (pp. 289–306). Information Age Publishing.
Wals, A. E. J., & Mathie, R. G. (2022). Whole school responses to climate urgency and related sustainability challenges: A perspective from Northern Europe. In M. Peter & R. Heraud (Eds.), *Encyclopedia of educational innovation* (pp. 1–8). Springer. https://doi.org/10.1007/978-981-13-2262-4-263-1
Wals, A., Weakland, J., & Corcoran, P. (2017). Preparing for the Eocene: Envisioning futures for environmental and sustainability education. *Japanese Journal of Environmental Education, 26*(4), 71–76.
WEF (World Economic Forum). (2021). *What is global citizenship?* World Economic Forum. https://www.weforum.org/agenda/2017/11/what-is-global-citizenship/, June 11, 2021.

Part 2

Practices

12 Activating the Petals in the 'WSA Flower Model': Critical Factors for Sustaining Schools' Pathways Towards a Whole-School Approach

Ingrid Eikeland and Astrid Sinnes

Key Message

Our key message in this chapter is to acknowledge individual schools' paths towards a WSA to sustainability, a journey that will look different depending on the local context of each school. We therefore recommend future stakeholders—internal or external to the school—to depart from and build on already existing initiatives when embarking on a path towards a WSA to sustainability. At the same time, we stress the need within schools to activate more petals in the WSA flower model to sustain the schools' sustainability focus.

12.1 Introduction

Although working towards a whole school approach (WSA) is widely recognised in both research and policy as crucial for developing the sustainability competences of students and staff, there are few examples of schools that have integrated this approach in an extensive way (e.g. Wals & Mathie, 2022). Moreover, the examples that do exist seem difficult to replicate or scale up, since schools are complex entities and closely linked to local changing policies and conditions (Augenstein et al., 2020). Trying to identify critical factors that cause schools to move from 'ordinary' to WSA to sustainability schools is therefore complex. Nevertheless, to release the full potential of educational institutions' contributions to sustainable development, the question remains crucial within the sustainability education discourse.

Research on schools' development towards sustainability often describes processes where schools move from individual teachers' engagement within their subjects to a more integrated approach where sustainability increasingly becomes part of the school ethos (Scott, 2013). The teachers and/or school leaders often seem to be crucial actors in developing and institutionalising a WSA (Mogren et al., 2019; Scott, 2013).

In this chapter, we describe four Norwegian upper secondary schools' paths towards a WSA. Since 2017, these schools have been part of a university–school partnership consisting of four schools, a school owner and a teacher education. The aim of this partnership has been to develop the learning institutions' sustainability education. We describe how the four schools took different paths towards achieving a WSA. We conclude that schools may follow different paths towards sustainability, yet the crucial factor seems to be

I. Eikeland
Faculty of Science and Technology, Norwegian University of Life Sciences, Ås, Norway
e-mail: ingrid.eikeland@nmbu.no

A. Sinnes (✉)
University Managment, Norwegian University of Life Sciences, Ås, Norway
e-mail: astrid.sinnes@nmbu.no

A. E. J. Wals et al. (eds.), *Whole School Approaches to Sustainability*, Sustainable Development Goals Series, https://doi.org/10.1007/978-3-031-56172-6_12

that more than one petal in the WSA flower model (Wals & Mathie, 2022) needs to be activated to sustain a school's journey towards a WSA. We also acknowledge the 'third space' (Zeichner, 2010) that developed through the university–school partnership as being crucial for both activating the petals and keeping them alive.

12.2 Background

To understand a school's path towards a WSA to sustainability, several researchers have worked with frameworks built around 'stage responses', whereby some responses are more in line with promoting a WSA than others (e.g. Scott, 2013; Sterling, 2004). For example, these types of frameworks describe weak and strong responses to sustainability education, where weak responses have more of a 'bolt-on' quality, leading to more fragmented and sporadic initiatives. A strong response to sustainability is described as more 'built in' and transformative, where sustainability principles are integrated between the different aspects of the school system. These types of frameworks are useful tools for schools to reflect upon and guide their path towards a WSA to sustainability.

To further conceptualise what a WSA to sustainability entails, the WSA flower model identifies overarching levels and initiatives (petals) that are important for schools to emphasise and activate to move towards a WSA. The petals are as follows: institutional practices (walking the talk and experimenting with and learning from creating sustainability on location), capacity building (continued professional development of all staff), pedagogy and learning (new/alternative learning processes and learning environments), curriculum (design, content and assessment), community connections (school–society interface) and vision, ethos, leadership and coordination. These petals are described in detail by Wals and Mathie (2022) and have been central in this chapter's analysis of the upper secondary schools' paths towards a WSA.

Both the WSA flower model and the stage responses can be seen as conceptualisations of a WSA that focuses on scaling up, activating and connecting several sustainability initiatives in the school. However, Augenstein et al. (2020) highlight several challenges—including more stakeholders and more initiatives—from simply scaling up sustainability initiatives without acknowledging and embracing the complexity and struggles of being in a change process. Thus, the development process needs to be context sensitive and inclusive and involve all actors in the process. These perspectives are also emphasised by other sustainability education and WSA researchers, who argue that nurturing supportive relations, both internal and external, is an important aspect to bring about change. Moreover, in democratic decision-making, ideas should be constantly explored and negotiated to create a shared and praxis-oriented vision to guide the change process (Mogren et al., 2019; Scott, 2013; Sterling, 2004; Verhelst et al., 2020).

With these perspectives in mind, we were interested in exploring four upper secondary schools' journeys towards a WSA to sustainability. These schools have all been supported by a university–school partnership. This partnership formed a third space in which the schools, together with a teacher education, could mutually explore and develop the sustainability activities in the learning institutions. Within the field of teacher education, third space theory describes a non-hierarchical relationship between partners, with the aim of creating new learning opportunities for both the teacher education and the schools (Robson & Mtika, 2017; Taylor et al., 2014; Zeichner, 2010). In these types of collaborations, careful attention must be given to the needs, perspectives and interests of all participants involved (Daza et al., 2021). A third space describes a symbolic space where boundary crossing becomes the norm and binary attitudes (such as teacher vs. student or theory vs. practice) are abandoned (Grudnoff et al., 2017; Norwegian Ministry of Education and Research, 2020). The aim of the third space that is described in this chapter has been for schools and the teacher education to mutually develop a whole institution approach to sustainability. Although the partnership is mutual, with the aim of also developing the sustainability focus in the teacher education, the focus of this chapter is on the four schools that were part of the partnership.

12.3 About the University–School Partnership

In 2017, the teacher education, together with the county council (school owner), initiated a university–school partnership with four upper secondary schools in the local area. Establishing partnerships between a teacher education and schools is part of the national strategy for teacher education institutions in Norway, introduced by the white paper 'Teacher Education 2025 – National Strategy for Quality and Cooperation in Teacher Education' in 2017 (Norwegian Ministry of Education and Research, 2017). With these partnerships, the intention is to create a stronger connection between the teacher education and schools, institutions that often act as independent forums, to equally strengthen the teacher education and schools' educational quality, innovation and relevance (ibid.). In this specific partnership, the aim was to promote and develop a WSA to sustainability.

The four upper secondary schools in this university–school partnership were chosen in dialogue with the teacher education and school owner and were based on the schools' interest to promote sustainability education and motivation to participate in a collaborative partnership with the teacher education. In addition, the selection process was also driven by an interest in recruiting different types of schools; one school was college preparatory, two schools were a mix of college preparatory and vocational school and one school was a vocational agriculture school. The schools varied in size, spanning from 500 to 2000 pupils. The initial dialogues with the schools went through the principal, followed by the establishment of a coordinator position both in the schools and in the teacher education to follow up the partnership on a day-to-day basis. The coordinators in the schools were either teachers, department heads or both. The coordinator in the teacher education (the first author of this chapter) was a science educator. A research group was also established in the teacher education to follow up and support the collaborative partnership, led by the second author of this chapter.

Among the activities in the partnership has been a reference group meeting every half-year to discuss the status of and ways forward for the partnership and also to discuss how to approach a WSA in our different institutions, based on both practical and theoretical inputs. In addition, seminars and meetings were carried out with the different participants: seminars and meetings within the research group; seminars and study trips with the reference group; coordinator meetings; and specific research and development projects that were initiated between the teacher education and the schools. The partnership is equally funded by the teacher education and school owner. Although the schools were selected as partners based on their interest in developing a WSA, the schools were still at the initial stages in their sustainability work when the project started. The schools can all be considered as representative of typical Norwegian upper secondary schools.

12.4 Data and Analysis

The empirical data for this chapter are based on written material from the partnership: signed agreements between the different parties, presentations and minutes from meetings and seminars, a midway report in 2020 and school observation notes. In addition, a master's thesis and field notes from PhD students carrying out research at the schools have been used as data (see a full overview of the written material in Table 12.1). As both the first and second authors have been directly involved in the partnership since the beginning as project coordinator and research leader, respectively, we also have a unique insight into the context behind the written material. The described cases in this chapter have been shared with the school coordinators for reliability assurance.

Wals and Mathie's (2022) WSA flower model has been used as an analytical tool to describe the four schools' paths towards a WSA. We use the model to identify what sustainability initiatives (petals) were active in the different schools at the beginning of the partnership in 2017 and what petals were activated as the partnership evolved.

Table 12.1 An overview of the empirical data for this chapter

When	What	Empirical data
Autumn 2017	Signed agreement between the teacher education and school owner	Written agreements
Spring 2018	Half-year meeting	Presentations and minutes
	School visits	Observation notes
Autumn 2018	Signed agreements between the teacher education and the individual schools	Written agreements
	Half-year meeting	Presentations and minutes
	Coordinator meeting	Minutes
Spring 2019	Half-year meeting	Presentations and minutes
	Coordinator meeting 1	Minutes
	Principal meeting	Presentations and minutes
	Coordinator meeting 2	Minutes
	Study trip	Presentations and observation notes
Autumn 2019	A two-day seminar with representatives from the different parties	Presentations and minutes
	Coordinator meeting	Minutes
Spring 2020	Half-year meeting	Presentations and minutes
	Coordinator meeting	Minutes
	School visits	Observation notes
Autumn 2020	Half-year meeting	Presentations and minutes
	Coordinator meeting	Minutes
Spring 2021	Half-year meeting	Presentations and minutes
	Coordinator meeting	Minutes
Autumn 2021	Half-year meeting	Presentations and minutes
	Coordinator meeting	Minutes
	Continued education workshops for teachers and school leaders (part of PhD study)	Presentations and observation notes
Spring 2022	Principal meeting	Presentations and minutes
	Coordinator meeting	Minutes
	Continued education workshops for teachers and school leaders (part of PhD study)	Presentations and observation notes

We first collected the relevant data to be analysed. Data consisted of notes written by participants in the project, minutes of meetings and data collected by researchers and students in the project (described in Table 12.1). Subsequently, the two authors individually went through the data critically and marked sustainability initiatives in the schools according to what petals in the flower model they each represented. We then came together and separated the initiatives according to how they had developed throughout the different phases of the project period. Together, we identified some central activities from each school that we see as representative of the development of each school's WSA journey. After doing this, we reread the documents to double check that we had not missed out on any central initiatives. We then wrote a narrative for each school's development process. It is beyond the scope of this chapter to write extensively about all the activities that were carried out by the schools during the project period. The cases must therefore be read as instrumental case studies (Yin, 2014) that illustrate different schools' journeys towards sustainability, not as complete descriptions of all activities in the schools.

12.5 Describing the Cases

12.5.1 School 1

School 1 is a college preparatory school situated in a small city 20 minutes out of the capital. The school has around 700 students.

The focus on sustainability at School 1 started with a group of three teachers wanting to collaborate to develop the sustainability focus in their school. To get inspiration, the teachers went on an excursion with a group of researchers from the teacher education to Globala Gymnasiet in Stockholm. Globala Gymnasiet has a particular focus on sustainability and can be considered a model school for a WSA to sustainability. The trip was inspiring and led to the teachers creating a transdisciplinary project on consumption in their school. The project became an annual event. Initially, it was carried out as a 7-week project before being reduced to 5 weeks. The project received much attention from policymakers, politicians and researchers and was ultimately awarded the best sustainability project of the year from the Teacher Union. With this success, the leadership at the school decided that the transdisciplinary project should be carried out for all students in their first year of upper secondary school. This meant that all teachers had to be involved. However, not everyone was happy about being forced into working in transdisciplinary teams, and the three enthusiastic teachers were dissatisfied about having to pull the weight of the teachers who were resistant to be involved. The three enthusiastic teachers who had initially started the project lost their motivation for the project, and the transdisciplinary project ended after being carried out for 3 years.

Despite the transdisciplinary project ending at School 1, this project inspired two of the other university schools (Schools 3 and 4) to start up similar initiatives but on a larger scale and spanning a whole school year. There will be more on this in the description of the other cases.

The story of the rise and fall of this transdisciplinary project is probably typical for many sustainability projects in schools. For various reasons, enthusiastic teachers who develop exciting projects can burn out after some years. However, the sustainability focus at School 1 blossomed in other areas.

As part of the university–school partnership, School 1 was committed to developing a WSA to sustainability. Thus, the sustainability focus was in many ways manifested in the school's vision. A sustainability coordinator had also been appointed at the school, and he continued following up the sustainability initiatives. Another factor contributing to the sustainability focus continuing was the new curriculum being implemented in Norway in 2020, requiring more interdisciplinarity and a stronger focus on sustainability. The school was also in a process of being environmentally certified.

All in all, several of the petals in Wals and Mathie's (2022) figure were activated: pedagogy and learning, curriculum, institutional practices and vision, ethos, leadership and coordination. A continued professional development programme was initiated from the teacher education, involving teachers from different disciplines in the school. This initiative activated yet another petal, capacity building. Currently, teachers involved in the course are developing a new transdisciplinary study at the school, like the full-year initiatives at Schools 3 and 4. The sustainability coordinator and the principal have also initiated a project that aims to develop a sustainable cantina. A process has begun in order to make the cantina the sustainable heart of the school and to develop the cantina as a learning arena for exploring and developing sustainable food choices (Brox, 2022).

12.5.2 School 2

School 2 is a vocational preparatory school situated in the countryside, 50 minutes out of the capital of Norway. The school has around 500 students.

While the sustainability focus of School 1 started with enthusiastic teachers developing an innovative pedagogy initiative, School 2 started their focus on sustainability by developing sustainable institutional practices. Being an agriculture school, School 2 had worked to promote and implement sustainable and climate-neutral farming into their farming practice long before this partnership started. The school has also been environmentally certified for several years. These initiatives have to a large degree been initiated and run by the leader group at the school. Thus, the sustainability initiatives have mainly revolved around changing the institutional practice of the school without necessarily affecting the pedagogy and learning and teachers and students at the school.

Although the leader group of School 2 has had a vision for a sustainable school, the leader group has struggled to get teachers involved in developing their pedagogy with a stronger focus on sustainability education and transdisciplinary teaching. In addition, the leader group has wanted to create a stronger link between the institutional practice initiatives and the learning process of their students, whereby all the students (not just the agriculture students) could be more involved in the farming practice. For example, students could have been involved in running the farmers' market that exists at the school, where products like flowers and vegetables are sold to the local community. However, to create good collaborations between the daily practice at the farm with the students' lesson plan demands a high level of coordination.

In 2020, a PhD student from the teacher education started a collaboration with a teacher at School 2 to develop an educational programme on the sustainable use of water. From this collaboration, they created a student-led project. Students would explore how the school could use their water resources more sustainably and develop prototypes through engaging with the technical personnel and farmers at the school and local businesses in the area. This initiative activated both the petals of pedagogy and learning, curriculum, institutional practice and community connections. The project has also sparked interest from other teachers, who have been inspired to carry out similar projects in their own teaching.

Also within the period of the university–school partnership, the leader group at School 2 has to a larger degree focused on carrying out professional development initiatives for all teachers. The aim has been to create a culture around sustainability education among the teaching staff. The process is not rushed, and the leaders want to take small steps spanning several years. The capacity-building petal was also activated through the continued professional development programme offered from the teacher education to teachers in the university schools. Three teachers from School 2 spent time evaluating the sustainability focus at the school using the WSA flower model as a starting point to develop new sustainability practices. Based on this analysis, the teachers developed a transdisciplinary teaching plan. The teachers decided to apply for funding to buy a pasteurisation machine. The idea was that the students should learn to use milk from the school's own cows to develop products that could be sold in the school's cantina. The purchase of this equipment generated enthusiasm in the school and opened a new perspective on the school's own teaching (for more details, see Mathie & Wals, 2022).

A continuing professional development programme for the school leaders was also offered by the teacher education parallel to the teachers. At a joint workshop for school leaders and teachers, the principal at School 2 enthusiastically declared that he would change the school's greenhouses from focusing on growing one type of plant for sale to becoming a learning arena for sustainable inquiry teaching.

Although both the vision/leadership petal and the institutional practice petal had been activated for years at School 2, activating the capacity building and pedagogy and learning petal in addition seemed to be crucial to stimulate the rest of the petals towards a whole school focus on sustainability.

12.5.3 School 3

School 3 is a mixed vocational and college preparatory school, situated 30 minutes out of the capital. The school has around 600 students.

Before the university–school partnership was established, the leader group at School 3 developed a strategic document stating the vision for the school, and one of the main points was for the school to contribute to the green shift and sustainable development. Largely with the enthusiasm of two teachers, School 3 had already established beehives on the school roof in 2016, along with a greenhouse. Since then, despite the established vision, School 3 has experienced frequent changes in principals and leadership, a situation that has interrupted the implementation of the sustainable vision at the leadership level. However, through the university–school partnership, a sustainability coordinator was hired at the school. The sustainability coordinator carried out an internship at Globala Gymnasiet, and after this, a new transdisciplinary educational programme was established at the school, focusing on inquiry and sustainability. The new programme was popular, and after a year, it expanded from one to two classes. In the third year, however, the number of applicants to this study programme decreased. Because of this, the new leadership at the school decided to put the programme on hold, a decision that disappointed the teachers involved.

In this school, the sustainability focus has existed in the pedagogy in several of the education programmes. For example, one teacher who is also a Ph.D. in the teacher education is working to develop inquiry-based teaching related to renewable energy that aligns with WSA perspectives (for more details, see Mathisen & Johansen, 2024). The pedagogy initiatives have mainly been individual initiatives by various teachers without an overall understanding of sustainability throughout the school. However, the sustainability coordinator has had an important role as a stabilising factor to keep up the continuity of the sustainability focus at the school. For example, the coordinator led the process of the school becoming environmentally certified and established a development group of 10 teachers, representing all departments at the school, to follow up the sustainability focus in their daily practice.

In School 3, the main drivers of the sustainability focus seem to be the teachers and the coordinator. Although the principals have changed, the commitment to the university–school partnership has also been a stabilising factor in maintaining the sustainability focus in the school. Teachers from School 3 also took part in the continuing professional development course offered by the teacher education. This course seemed to have activated several teachers to commit to and collaborate more on developing the sustainability focus at the school.

12.5.4 School 4

School 4 is a mixed college preparatory and vocational school, situated 40 minutes out of the capital. It has around 1300 students.

For a long time, this school has collaborated with the university, particularly through the science centre located on campus related to a sustainable food project. This collaboration also involves School 3. More specifically, the project is organised by and carried out at both the schools and the science centre, with the mutual aim of increasing students' awareness about sustainable food choices. Through the university–school partnership, School 4 has also experimented with a transdisciplinary educational programme in one class, inspired by Globala Gymnasiet and the initiatives in the other university schools. However, partly due to the significant size of the school in a Norwegian context with its 1300 students and a large staff, most of the sustainability initiatives at this school seem to have existed as initiatives of individual teachers and have not necessarily influenced the school's overall sustainability practice. Thus, as also seems to be the case for all three other university schools, this school has had several initiatives focusing on sustainability, particularly within the pedagogy and learning petal, without necessarily being connected and integrated practices as part of a WSA to sustainability.

However, in 2021 a new principal was appointed at School 4. With a strong interest in sustainability, she has focused on bringing the group of leaders on board. In the continuing professional development programme, the teacher education organised for school leaders, School 4 participated with a team of six leaders. This led to School 4 developing an extended plan for the following school year, where it was illustrated how the sustainability focus would be part of the teaching and school practice throughout the year. School 4 has also hired a new chef to cook sustainable food in the cantina. The sustainability coordinator and two other teachers from the school also participated in the continuing professional development course and subsequently developed a transdisciplinary teaching module on sustainable textiles.

In School 4, it, therefore, seems that the sustainability initiative started with the community connection and the curriculum (sustainable foods) petals through the collaboration with the local science centre. The pedagogy and learning petal was activated through the transdisciplinary pedagogy being developed in the school and was strengthened by the introduction of the new national curriculum where the focus on interdisciplinarity was prominent. The continuing professional development programme and the new principal activated yet two more petals and hence brought the school a step further towards the aim of a WSA to sustainability.

12.6 Summary

The short descriptions of how the sustainability focus has developed in these four schools show that they have had different starting points for their paths towards developing a WSA to sustainability (see overview in Table 12.2). While the work at School 1 mainly started within the pedagogy and learning petal by individual teachers developing a transdisciplinary sustainability module, School 2 started their path through the leader group changing the school's institutional practice by developing a sustainable farming practice. School 3 started their path towards sustainability with a leader-initiated strategy to promote sustainability, while the point of departure for School 4 was primarily community connections through teachers collaborating with the local science centre.

Even though the schools had different starting points on their sustainability journey, all the schools have since activated more petals as the university–school partnership has evolved (see Table 12.3). School 1 has especially activated the institutional practice petal, in combination with capacity building, through the process of establishing a sustainable cantina in the school. School 2 has activated their pedagogy and learning and capacity-building petals through competency building for all staff and experimenting with new pedagogies. School 3 has especially activated the pedagogy and learning and capacity-building petals by experimenting with transdisciplinary classes and following up on their green vision with all staff. School 4 has activated the vision/

Table 12.2 A simple overview of the petals that were active in the initial stages of the university–school partnership

Start of the university–school partnership	School 1	School 2	School 3	School 4
Institutional practices		✔	✔	
Pedagogy and learning	✔			
Curriculum	✔			✔
Capacity building				
Community connections				✔
Vision/ethos/leadership/coordination		✔	✔	

Table 12.3 A simple overview of the petals that were active 5 years into the university–school partnership

5 years later	School 1	School 2	School 3	School 4
Institutional practices	✔	✔	✔	
Pedagogy and learning	✔	✔	✔	✔
Curriculum	✔	✔	✔	✔
Capacity building	✔	✔	✔	✔
Community connections				✔
Vision/ethos/leadership/coordination		✔	✔	✔

ethos/leadership/coordination and pedagogy and learning petals through experimenting with transdisciplinary classes and the leaders working to make sustainability an overarching focus through the school year.

From this study, we argue that schools may take different approaches to sustainability, yet a crucial factor is that more than one petal in the WSA flower model (Wals & Mathie, 2022) needs to be activated. For example, in School 3, there was a rapid change in leadership, a factor that could have inhibited the sustainability focus of the school. Nonetheless, the 'green' vision still existed, and the university–school coordinator also played an important role in keeping the momentum. Furthermore, the school has become environmentally certified, and a teacher group has been established that follows the sustainability focus and pedagogical innovations going on. Thus, the leadership petal was not the only activated petal in the school, making the sustainability focus less vulnerable. Still, there were initiatives within the capacity-building, pedagogy and learning, institutional practices and vision/ethos/leadership/coordination petals to drive the sustainability focus forward. Thus, activating more petals will strengthen the response to sustainability (Sterling, 2004; Scott, 2013) by securing a robust and long-lasting path towards a WSA that can handle changing conditions both within and out of school. This point seems especially crucial in 'ordinary' schools like the university schools without a specific sustainability focus to secure momentum in the development process.

Furthermore, we want to emphasise the value of all initiatives, although some of them did not last, such as the transdisciplinary initiatives in Schools 1 and 3. Initiatives like these have inspired new initiatives within the school, across the university–school partnership and beyond, as both teachers, pupils and others become aware of the value of transdisciplinary projects. Such initiatives should therefore not be seen as failures but rather as door openers for future sustainability initiatives. This aligns with Augenstein et al. (2020), who highlight the importance of sparking innovative and experimental processes to contribute to learning, rather than measuring initiatives based on the success of efforts to continue and to be 'scaled up'.

We also want to acknowledge the fact that the university–school partnership has provided a critical third space by connecting individual initiatives throughout the schools, thereby stimulating more petals to be activated. Through the half-year meetings, seminars and daily relational contact and eventually with the continual professional development course for teachers and leaders, we as a partnership have constantly developed our understanding of a WSA, supported each other through ups and downs and inspired each other to work towards a common goal. We acknowledge that the third space ideas of equality between the stakeholders, the freedom for participants to choose how to act on a local level and the focus on supporting local practices, followed by the longitudinal nature of the collaboration and predictable funding, have been successful factors in the partnership. Thus, the university–school partnership has provided a fruitful third space for keeping the petals activated and hence supported the development of the four schools' approach towards developing a sustainability focus.

All the schools still have a way to go to be described as stage-four responses to sustainability (according to Sterling, 2004). During the project period, all the schools have lifted the sustainability activities from being about individuals to being more an aim for the entire school. Nevertheless, all the schools have a way to go before the sustainability focus is integrated fully into the ethos of the schools and before they work systematically with the community on sustainability. Also, although the schools are more integrated in terms of leadership, they are still vulnerable to changing leaders and teachers. We do, however, consider the schools, where all petals of the flower model to a certain degree are activated, to have a promising potential for further development towards a WSA to sustainability.

References

Augenstein, K., Bachmann, B., Egermann, M., Hermelingmeier, V., Hilger, A., Jaeger-Erben, M., Kessler, A., Lam, D. P. M., Palzkill, A., Suski, P., et al. (2020). From niche to mainstream: The dilemmas of scaling up sustainable alternatives. *Gaia, 29*(3), 143–147.

Brox, L. (2022). *Kan kantina være en læringsarena for å utvikle elevers kompetanse til å ta ansvarlige valg og handle miljøbevisst?* [Master's thesis, the Norwegian University of Life Sciences].

Daza, V., Gudmundsdottir, G. B., & Lund, A. (2021). Partnerships as third spaces for professional practice in initial teacher education: A scoping review. *Teaching and Teacher Education, 102*(1), 103338.

Grudnoff, L., Haigh, M., & Mackisack, V. (2017). Re-envisaging and reinvigorating school–university practicum partnerships. *Asia-Pacific Journal of Teacher Education, 45*(2), 180–193.

Mathie, R. G., & Wals, A. E. J. (2022). *Whole school approaches to sustainability: Exemplary practices from around the world.* Wageningen University.

Mathisen, Ø., & Johansen, G. (2024). Students' deliberation on 'greening' the school's energy supply: A case discussed from a WSA perspective. In A. Wals, B. Bjønness, A. Sinnes, & I. Eikeland (Eds.), *Whole school approaches to sustainability*. Springer.

Mogren, A., Gericke, N., & Scherp, H. (2019). Whole school approaches to education for sustainable development: A model that links to school improvement. *Environmental Education Research, 25*(4), 508–531.

Norwegian Ministry of Education and Research. (2017). *Teacher education 2025 – National strategy for quality and cooperation in teacher education.* https://www.regjeringen.no/contentassets/d0c1da83bce94e2da21d5f631bbae817/kd_teacher-education-2025_uu.pdf

Norwegian Ministry of Education and Research. (2020). *Partnerskap i lærerutdanningene – et kunnskapsgrunnlag Delrapport 1. Faglig råd for lærerutdanning 2025.* https://www.regjeringen.no/contentassets/9969c3f46f0c4a4f95ed9eee70b3ed19/partnerskap-i-larerutdanningene%2D%2D-et-kunnskapsgrunnlag.pdf

Robson, D., & Mtika, P. (2017). Newly qualified teachers' professional learning through practitioner enquiry: Investigating partnership-based mentoring. *International Journal of Mentoring and Coaching in Education, 6*(3), 242–260.

Scott, W. (2013). Developing the sustainable school: Thinking the issues through. *Curriculum Journal, 24*(2), 181–205.

Sterling, S. (2004). Higher education, sustainability, and the role of systemic learning. In P. B. Corcoran & A. E. J. Wals (Eds.), *Higher education and the challenge of sustainability* (pp. 49–70). Springer.

Taylor, M., Klein, E. J., & Abrams, L. (2014). Tensions of reimagining our roles as teacher educators in a third space: Revisiting a co/autoethnography through a faculty lens. *Studying Teacher Education, 10*(1), 3–19.

Verhelst, D., Vanhoof, J., Boeve-de Pauw, J., & Van Petegem; P. (2020). Building a conceptual framework for an ESD-effective school organization. *The Journal of Environmental Education, 51*(6), 400–415.

Wals, A. E., & Mathie, R. G. (2022). Whole school responses to climate urgency and related sustainability challenges. *Encyclopedia of Educational Innovation*, 1–8.

Yin, R. K. (2014). Case study research: Design and methods (applied social research methods) (p. 312). Thousand Oaks, CA: Sage publications.

Zeichner, K. M. (2010). Rethinking the connections between campus courses and field experiences in college- and university-based teacher education. *Journal of Teacher Education, 61*(1–2), 89–99.

Piloting a Self-Assessment Tool for a Whole School Approach to Sustainability in Mongolia

13

Erdenebayar Shinetsetseg, Genendorj Battsetseg, Orsoo Oyuntungalag, Lundaajantsan Temuujin, and Mita Nangia Goswami

Key Message
A key message from this chapter is the transformative role of a WSA self-assessment tool in promoting ESD in Mongolian schools. It highlights the development of a tailored checklist that equipped 30 schools with essential skills in management, planning, collaboration, and problem-solving. The initiative empowered schools to devise specific action plans addressing identified challenges, with leadership playing a crucial role. Implementing the tool required efforts and time, particularly in reorienting the view of the school community as fixed and given to a dynamic and shapeable one.

13.1 Introduction

Addressing global sustainability challenges through achieving Sustainable Development Goals (SDGs) is the foremost agenda for many governments worldwide. Target 4.7 of SDG 4 on Quality Education lays out the knowledge, skills, values, and attitudes needed for every citizen to live responsibly within the finite boundaries of the planet and take an active part in addressing local and global challenges through Education for Sustainable Development (ESD) and Global Citizenship Education (GCED). UNESCO (UNESCO, 2015) has urged its member states to focus on human rights, peace, gender equality, sustainable development, and health while meeting target 4.7. The Government of Mongolia (GoM) has put forward the goal of becoming a middle-income country while preserving ecological balance and stable democracy, as articulated in its Sustainable Development Vision-2030 policy document (GoM, 2016). It reflects the aforementioned global agenda in the national-level policies and operations.

Mongolian schools strive to focus on a meaningful contribution to a more sustainable world. The questions of how to nurture students as global citizens, organize school management, and adopt the most effective way of overcoming sustainability challenges within the school walls are of great importance for schools today. To find the answer to the question, we initiated the development of the WSA as a self-assessment tool to help schools reorient toward sustainable development. We piloted the tool in 30 selected schools in Mongolia under a donor-funded project. This paper describes the initial stage of developing and using a WSA self-assessment tool, which we believe will be refined and amended with usage.

E. Shinetsetseg (✉) · G. Battsetseg ·
O. Oyuntungalag · L. Temuujin · M. N. Goswami
Information and Training Center for Nature and Environment (ITCNE), Ulaanbaatar, Mongolia
e-mail: ceo@feemongolia.mn

A. E. J. Wals et al. (eds.), *Whole School Approaches to Sustainability*, Sustainable Development Goals Series, https://doi.org/10.1007/978-3-031-56172-6_13

13.2 Why WSA

WSA integrates sustainability principles into all aspects of school life and enables students to live what they learn and learn what they live. The integration of the WSA in schools is a new concept in Mongolia. WSA is a cohesive and collaborative action by a school community aiming to raise quality and standards across the entire school. Every school has different starting points and needs. Using a WSA might help a school community find optimal solutions to address their needs and challenges. According to UNESCO International Bureau of Education (UNESCO, 2023), it involves addressing the needs of learners, staff, and the wider community not only within the curriculum but across the whole school and learning environment. On the other hand, it promotes a new school management culture in Mongolia as a school community tends to follow school principals' decisions which are always top-to-down.

The first initiative to pilot the WSA in Mongolia was operationalized within the Education for Sustainable Development-Phase II (ESD-II) project implemented by the GoM in partnership with the Swiss Agency for Development and Cooperation in 2019–2022. This project aimed to transform education policies and schools toward achieving sustainability goals under the SDGs and national development goals. The local implementing partner of the ESD-II project, the NGO, Information and Training Centre for Nature and Environment (ITCNE, 2020), coordinated the entire process of piloting WSA and introducing the self-assessment tool to the 30 pilot schools in Mongolia under this project implementation.

Piloting the WSA as a self-assessment tool in the selected schools was a timely response in contributing to the education reforming process in Mongolia. School management and school-level planning have been considered among the weakest areas in the Mongolian education system since its transition to democracy in the 1990s. This has been due to the weak capacity of the schools to plan their activities as per their needs and situation and the lack of a systemic approach in school planning by school leaders. The government has been paying more attention to this area by revising its policy directions. The WSA self-assessment tool development and piloting intended to address this school-level planning capacity deficit since school governance is one of the salient pillars of the WSA.

13.3 How We Have Started

The GoM formed a project consulting team comprising leading education researchers from national education research institutions and specialists with school planning and management background. The team reviewed the government rules and related policy direction, such as the Ministry of Education's (MES) criteria for school performance assessment and the State Inspection Agency's criteria for school inspection. However, these criteria were only related to monitoring school activities and did not investigate the issue of a holistic approach to school management mentioned earlier. Schools had never conducted a self-assessment of their management. The WSA, for the first time, ushered in a school-level planning tool through the self-assessment exercise involving everyone inside and outside the school: teachers, the school's administrative and other staff, students, local communities, and parents while harnessing local wisdom.

A prominent aspect of rolling out the WSA in Mongolia started with selecting schools that would become pilots of the approach. The GoM, the MES, and the Ministry of Environment and Tourism (MET), both partners in the ESD project, nominated a number of Mongolian secondary schools from various geographical locations through an open announcement. The aimag/district local education departments administered the school selection process.

The expression of interest to participate in the WSA tool piloting activity was delivered to all nominated schools to submit their short proposal explaining why they need to adopt a WSA. On receiving proposals from the interested schools, the local education departments reviewed them, assessing them against the prepared criteria. The school selection criteria prepared by the consulting team included the school's willingness to participate in the pilot and their expectations, the school's previous activities in ESD, and the school's motivation to look at their present situation for further improvement.

Based on the review results, the local education departments selected the 30 schools (24 rural and six urban schools) in six aimags and two districts of Ulaanbaatar. The reason for picking up the locations is that SDC's project activities were implemented before in those places, so WSA tool piloting would complement the already invested areas to maximize the project impact there.

The next step was to formulate the actual WSA self-assessment tool suitable for Mongolian school conditions considering their specifics and needs. The project consulting team and ITCNE worked on this demanding task. They used UNESCO's (2014) conceptual framework as the foundation for the design of the self-assessment tool. According to UNESCO (2014), the WSA is an approach that covers all aspects of school governance, organizational structure, operations, research, human resource development, curriculum, school infrastructure, students' participation, partnership with parents and public relations, organizational culture, and school values. In Mongolia, the WSA was considered an appropriate framework for assessing what exists, creating a plan to address gaps, and identifying measures of progress as it includes:

- An overarching supportive, safe, and inclusive school culture
- School-based programs in and out of the classroom that focus on social-emotional learning and respect for differences
- School policy that prioritizes, monitors, and reports on progress for all students, with a focus on eliminating disparities
- School partnerships to bring community programs and resources into the school setting to augment school capacity and address families' needs beyond the classroom

In addition, the consulting team examined and reviewed the national school inspection criteria and standards and the experience of schools using the WSA in countries such as Canada, the United States, the United Kingdom, and India. Besides, the documents on key outputs and the experience of previous donor-funded projects in Mongolia related to supporting school management approaches ("Rural school" project by DANIDA, "Open school" project by Open Society Foundation, and "Eco school" project by Swiss Development Cooperation) have been reviewed.

After reviewing the existing documents and international experiences, the consulting team created a draft version of the WSA tool applicable to Mongolian schools. The tool has 10 handbooks designed to support schools in adapting the tool to their activities.

13.4 The WSA Self-Assessment Tool

The WSA self-assessment tool was developed and described in detail in the following segments. The WSA Self-assessment tool is a series of handbooks for schools. A compendium of 10 handbooks was developed, unpacking each domain of the WSA, ensuring that they were understandable for schools and easy to use for everyone, as the schools were expected to go through the handbooks by themselves.

WSA tool integrates all aspects of school life, focusing more on sustainability challenges. Schools are supposed to review their current status in various aspects of school development and identify the right solutions to address their challenges. The core principle is to involve the entire school community in the school development planning process. The whole school is supposed to learn how to plan together by themselves instead of hanging around a plan prepared by their school administrators only.

The Mongolian WSA self-assessment tool is a comprehensive model that encompasses all aspects of school life:

1. 5 domains and 24 subdomains
2. 4 core themes related to ESD
3. 5 maturity levels of each domain
4. Principles for each domain

13.4.1 Five Domains and 24 Subdomains of the WSA Self-Assessment Tool

The first component of the WSA Self-assessment tool has five domains or the different strands of the WSA and subdomains. Domains are expected to cover all aspects of school life. It has the following five domains, encompassing every aspect of a school: (i) governance; (ii) capacity building and human resource development; (iii) curriculum, teaching, and learning; (iv) school and its surrounding environment; and (v) partnership and collaboration. Each of these five domains is unpacked into 24 subdomains. Table 13.1 presents the five domains and the corresponding 24 subdomains.

Table 13.1 Five domains and 24 subdomains of the WSA Self-assessment tool

Domains	Subdomains
Governance	(a) policy, planning, (b) budgeting, (c) implementation, (d) performance, (e) leadership
Capacity building and human resource development	(a) organizational structure, operational setting, (b) leadership, (c) knowledge, (d) follow-up responsibility
Curriculum, teaching, and learning	(a) curriculum, (b) teaching, (c) learning, (d) learning support actions, (e) evaluation
School and its surrounding environment	(a) outdoor landscaping, safety, security (b) interior furniture, equipment supply, safety, hygiene, (c) use of materials and resources, waste management, (d) building, use of land and protection, e) health, communication, and attitude, quality of social service, accessibility
Partnership and collaboration	(a) social issues: child protection, (b) economic issues: decent employment for youth, (c) environmental issues: ecological education and maturity of youth and children, (d) tradition and culture: nomadic culture, the global Mongolian with heritage, (e) integration of the three pillars of sustainable development

13.4.2 Four Core Themes of the WSA Self-Assessment Tool

The second component of the WSA Self-Assessment tool has transversal core themes. These are as follows: (i) sustainable development; (ii) human rights; (iii) gender equity, and (iv) global citizenship. These have been adapted from UNESCO's Global Education Monitoring Report (2016), where countries were reviewed for the implementation of target 4.7 of SDG 4. It states:

> By 2030, ensure that all learners acquire the knowledge and skills needed to promote sustainable development, including, among others, through education for sustainable development and sustainable lifestyles, human rights, gender equality, promotion of a culture of peace and nonviolence, global citizenship and appreciation of cultural diversity and of culture's contribution to sustainable development (UNESCO, 2020).

13.4.3 Five Levels of Maturity of Each Domain

A third component of the tool is the maturity level which presents the level at which a school is in. We assume that during each round of using the WSA Self-Assessment tool by schools, the maturity level of the domains will change as per the school's needs and efforts. It will also aid the school in tracking and monitoring its progress. The five levels of maturity for each domain are explained in Table 13.2.

13.4.4 Principles for Each Domain

Bringing the principles related to each domain and subdomain would be pertinent here. The principles for each domain have been identified by the consulting team based on the review of existing literature. Each domain's principles required proper justifications as follows:

(i) The principles for the domain of "governance" were adapted from the Handbook *Governance Approach* by SDC (2017).

Table 13.2 Meaning of maturity levels for each domain

	Maturity level meaning				
Five domains	1st level	2nd level	3rd level	4th level	5th level
School governance	No changes	Start changing	Some progress observed	Clear progress achieved in real life	Reaching the objectives of sustainability
Capacity building and human resource development	Collection of data	Data coding and recording	Data processing	Data storage and application	Changes in attitudes and behavior
Curriculum, teaching, and learning	Nonstructured	Start up	Application	Integrating	Transforming
School and its surrounding environment	No changes	Sensing the need for changes	Making efforts for progress	Accomplishing and achieving results and conditions improved	Meeting certain requirements and standards
Partnership	Inform	Consult	Involve	Collaborate	Partnership

(ii) The principles for the domain of "capacity building and human resource development" were selected to ensure their alignment with the ESD core competencies.

(iii) The principles for the domain of "curriculum, teaching, and learning" were taken from the core principles and teaching methods for ESD identified globally.

(iv) The principles for the domain of "school and its surrounding environment" were aligned with the standards of ISO 41001 on "facility management".

(v) The principles for the domain of "partnership" were aligned with the principles of the Global Humanitarian Platform prepared by the International Volunteers' Association (Platform, 2007).

As each domain has different principles applied, let us present one example using the domain of "governance." For the domain of "governance," the following six principles have been identified (see Table 13.3). These principles were taken from handbook *Governance Approach* by SDC prepared under the donor-funded project, and also these principles were previously defined by SDC recommendations (SDC, 2017).

Table 13.3 shows that the domain of "governance" should be aligned with the following six principles: (i) effectiveness; (ii) participation level; (iii) transparency; (iv) accountability; (v) equality; and (vi) response to legal requirements. These principles are supposed to be used by school staff when they identify the current school governance challenges. It would give us more information, for example, on whether school activities involve everyone and whether the school's budget is transparent for everyone. In other words, this domain will disclose whether the school activities administered by the principal meet the principles. On the other hand, this study evaluates the six principles in the school context. The school handbooks have described these principles in detail to make them understandable.

13.5 How Schools Learned to Use the Tool

The ITCNE designed a comprehensive and step-by-step training plan to reach everyone at the targeted schools. Training modules covered the content of 10 handbooks on the WSA. The training plan was a cascading model preparing the first six representatives from each school as the master trainers through Training of Trainers (ToT). At the beginning of the first ToT, a pretest was carried out among the participants to gauge their understanding of ESD concepts and their willingness to introduce the WSA at the school level. At the end of the project in 2023, these

Table 13.3 An example of the six principles for the domain of governance

Subdomains	**Principles**					
	Effectiveness	*Participation*	*Transparency*	*Accountability*	*Equality*	*Legal requirements*
Policy, planning	✓	✓	✓	✓	✓	✓
Budgeting	✓	✓	✓	✓	✓	✓
Implementation	✓	✓	✓	✓	✓	✓
Performance	✓	✓	✓	✓	✓	✓
Leadership	✓	✓	✓	✓	✓	✓

participants will be asked to write a posttest report to document the changes that have occurred. The ToT was conducted during the COVID pandemic, so the training content was prepared and transacted using EdTech solutions. The training package included video lectures, reading materials, and workbooks.

The second step in the cascade training was the training of the other participants of schools by the master trainers. The ITCNE recently interviewed a few school principals of targeted schools who were a part of the WSA piloting process. They highlighted that delivering information or making every teacher understand the sense of WSA in the Mongolian context was challenging. Delivering important messages to all teachers usually takes on a face-to-face mode in Mongolia. Especially for bigger-sized urban schools, it was not easy to convene all the teachers and make everyone understand what was expected to be done. Because it is a self-assessment to be done for the first time involving the entire school community at the school, everyone's participation was important. As a school principal mentioned, for a smaller rural school, it was much easier to reach every teacher. Also, participants mentioned that a general understanding of the importance of ESD and SD was weak at the project's beginning stage, but now this understanding has been enormously improved during the process of piloting the WSA tool. After reviewing the WSA handbooks and receiving a series of training, the target schools administered the first round of self-assessment. Each school's entire community gathered and gave scores to each WSA domain. The scores for each domain were entered into an Excel worksheet that ITCNE developed. This Excel worksheet provided them with the maturity level of each domain automatically. The use of the tool empowered the school community with a clear understanding of their school's strengths and drawbacks.

The actual process of administering the first round of WSA Self-Assessment went through the following stages:

- Analyzing the current situation of school life and identifying the challenges.
- Identifying the solutions to address the current problems.
- Determining the actions to be taken to meet the expectation.

The assessment process required the participation of not only the school teachers and students but also parents and local communities, including local government administrators. Additionally, it was a new exercise to conduct school-level self-assessments by themselves because, in Mongolia, external parties or upper-level authorities usually do assessments. During the first round, a total of 14,000 participants from selected target 30 schools participated in the assessment exercise. 14.9% were from local communities who are sub-national-level administrators, local education authorities, and local service providers in education and health sectors. 84.7% were beneficiaries of the education sector, such as school students, their parents, school teachers, school support staff, and school administrators.

13.5.1 Challenges

There have been some challenges identified. The biggest challenge was the issue of how to involve everyone. A recent interview with a few school principals (one urban school and three rural schools) who were a part of the first round highlighted that it was not easy to invite everyone into this process because some people did not understand why they were doing it. Moreover, the process seemed overwhelming for them, initially covering all aspects of their life.

After the first round, it became simpler, and the second time round, they were able to do the exercise much more confidently and easily. They were also motivated now because they were able to find the solution to the difficult question of how to plan school life and how to make their schools more welcoming and relevant for the learners. Another important lesson was to learn to identify their problems. Schools became aware that they were using a "magnifying glass lens" to assess where they were, what their issues were, and how they would best be able to nurture their teachers and students to become active contributors to sustainable development.

The participants understood that it was an exercise to involve everyone at the school level toward the next step of addressing their problems based on the self-assessment findings. It was a new and challenging goal for schools. It was a whole process of changing the attitude and minds of people at the school, asking them to participate in school-related decision-making, which was a completely new culture in the case of Mongolia. Earlier, school decisions were usually made by only one person, namely the school principal. With the introduction of WSA, a new culture of everyone's participation in decision-making was created and piloted. It was time-consuming and an entirely new way and intimidated the schools in the beginning.

Some schools assessed lower in the governance domain tried to introduce student-led governance involving students in all school affairs. Now, the whole school enjoys this culture. But they shared that it was not easy to change the students' minds. They always waited for their teacher to lead their actions. They learned in this culture gradually.

Having completed the self-assessment checklist, the majority of schools reported in their annual reports to ITCNE that this exercise opened opportunities for further improvement of the school situation, addressing their drawbacks while identifying the causes of the difficulties they encountered. It was an exercise to fully understand the importance of collaboratively planning and monitoring school affairs toward integrating ESD principles. It was, without a doubt, time-consuming and demanding, but it got the whole school to come together and work as a team and ushered in a cultural shift in the way the schools functioned.

13.6 Analysis of the Findings from the First Round of Assessment

The 30 target schools under the project piloted the first round of the WSA Self-Assessment using the newly developed 10 handbooks as the beginning stage of introducing WSA. Schools entered their scores for each domain and subdomains into the Excel worksheet. The ITCNE compiled all the school data into one database for a thorough analysis to see the general trend of the schools across five domains.

To analyze all schools' data on five domains, the Leadership Practices Inventory methodology (LPI) (Kouzes and Posner, 2013) was used in the discrepancies in the level of maturity among schools. This method highlights the voices of stakeholders and indicates key focus areas and underlying conditions rather than simply ranking the end results into poor and good ones.

All school data for each domain and subdomain have been aggregated, producing the national-level average points taken by schools. It can present the facts about where the domains are (see Fig. 13.1).

The highest score schools are supposed to get is five. Figure 13.1 illustrates the average score schools assessed themselves on each of the 24 subdomains, and five domains ranged from 2.31 to 2.75. This may conclude that schools assessed themselves below the medium level in all domains. In terms of maturity levels, this score is at the second level of the five levels of maturity. The average score for the governance domain was 2.31, the youth employment of partnership domain was 2.41, and the organizational structure and operational mechanism of the human resource development domain scored 2.75. The school campus, security, and sanitation of the school and learning environment scored 2.73, whereas the average for the school building, land use, and protection was 2.71. Importantly, these findings can provide rich data for the school community for evidence-based planning and serve as a baseline to measure the outcome of the follow-up Action Plans.

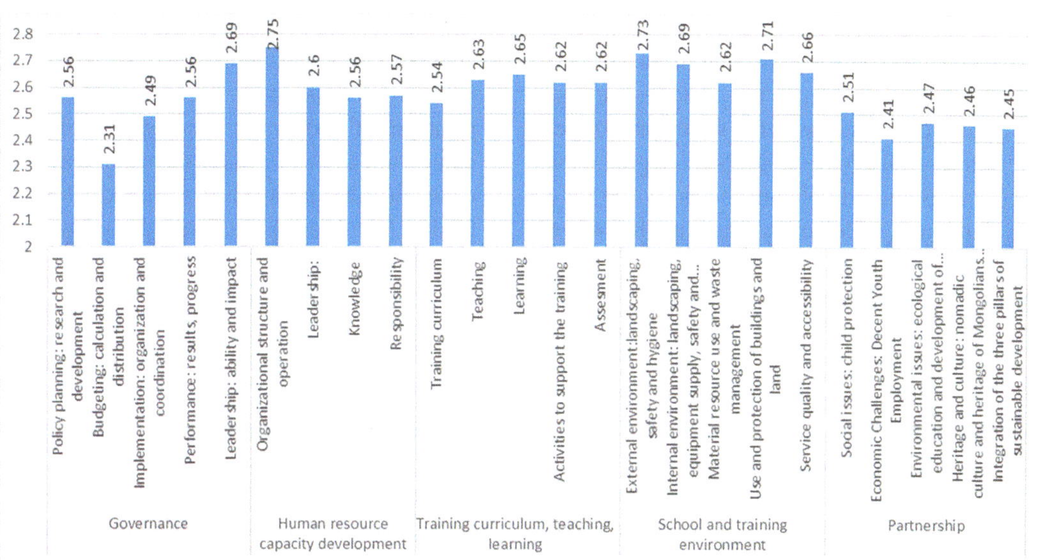

Fig. 13.1 Average maturity level by five domains and their components

13.7 Observations so far

Based on the findings of the Self-Assessment Checklist, schools could identify their weak areas and problems of further attention. During the recent interviews, a few school principals (one urban and three rural schools were invited) said that this beginning stage of planning was not easy in terms of making everyone toward one goal and one activity. Because school-level planning took place differently, by only a few people or the principal alone, all schools could face this challenge. No collaboration in planning any school-related activity became a culture that significantly impacted the introduction of the WSA tool in school activities.

Schools prepared the Action Plan to adapt the WSA tool in their activities considering the findings of the Self-Assessment exercise. During the project implementation period, the project supported schools with small grant funding to implement their Action Plan activities. Funding was provided due to the gap in the government budget allocations to schools. Schools always face a funding gap to implement quality-related activities according to their needs and specifics.

The project team paid monitoring visits to schools twice a year in cooperation with the local support teams to provide mentoring guidance and assistance. Schools also sent progress reports presenting their results of the WSA tool adaptation. The project team reviewed the school progress reports and reached the following conclusions:

(i) Principles of good governance were embedded in school-level planning. The process of involving joint efforts and collective decision-making to arrive at solutions was a step-by-step process and not easy. But now, many schools reported that collaborative decision-making has become school culture. They especially enjoy when students take an active part in this process. This whole approach improved effectiveness in school governance, as they told.
(ii) Each school's plan of action reflected their pressing needs incorporating various ESD elements into the teaching and learning process. This has taken the learning process beyond the cognitive area and has put more focus on social and behavioral competencies, creating a learning environment that connects classroom learning to life and practice. Many schools focused on the domain of school surrounding to provide a

favorable learning environment. They created school outside gardens to be used for science experiments. The school's outdoor and indoor environments were transformed into eco-labs for testing ESD-integrated lessons. Some schools introduced waste recycling and collecting culture at the school level and family and soum levels.

(iii) The ESD themes have penetrated well into school activities. The examples are the schools where the entire school community was bringing down its ecological footprint by saving water and electricity and through better waste management. Schools learned to identify a solution to save energy and recycle. But the challenge was that soums and aimags do not have facilities to collect recycling goods and waste. No proper system is there. However, schools learned to recycle and manage waste.

(iv) An innovative partnership model was created between schools, local authorities, and communities. Schools tried to involve local communities in addressing their problems by seeking support from them. A local mentoring team called Sub-TACG has been formed to support schools in ESD implementation. Besides, some schools tried to do advocacy activities on themes like waste management to influence local administration and communities.

In summary, the results of the inception stage of the WSA adoption in 30 Mongolian schools show that the schools have made progress in strengthening their leadership role in ESD by introducing a new management culture and by upholding common human values despite the various challenges mentioned earlier. The key activities observed in target schools can be summarized below by each WSA tool domain:

1. *Governance:* Improved leadership of school principals and managers was witnessed; they learned how to change school management according to the principles of WSA and were able to influence the community, change attitudes, unite for a common goal, and gradually become a model school for ESD. To become an ESD school, the whole school was involved in decision-making in accordance with the principle of cooperation and partnership and democratization of school management. One of the 30 ESD pilot schools (School Number 28 in Ulaanbaatar) encouraged student participation in the decision-making process by making students members of the school's procurement committee. This school is now named the "School Governed by Students," a first in Mongolia.

   This study is in line with the review results of the study by Henderson and Tilbury (2004). The review results demonstrated that a committee/working group (with management, staff, students, and stakeholder participation) should be formed to decide on actions and review the school planning and operation progress. Plus, it shows that democratic decision-making and meaningful participation of all stakeholders are at the center of WSA.

2. *Capacity building and human resource development:* Schools learned how to integrate the ESD pillars into all school activities by piloting unique approaches to building their capacity. All school staff, including school cooks, learned in their work areas. The cooks became proficient in preparing nutritious and healthy food for children. At the same time, students from the eco-clubs of a school in Arkhangai province trained the support staff about the ecological footprint. They supported them in improving the energy efficiency of their school.

   ESD-integrated lessons were prepared by teachers of different subjects per the "Mutual Learning" principle. As stated earlier, traditionally, cooperation is weak among the teachers, but their attitude changed, and they supported each other, sharing their experiences and planning lessons together. Many schools piloted "ESD Integrated Project-based Lessons," which is a step toward bringing training closer to reality, saving teachers' time, and using available resources to improve the quality of education.

3. *Curriculum, teaching, and learning:* "School-based curricula" has been a new approach toward making learning closer to real-life practices integrating SD pillars. Schools continue testing various approaches and methodologies supported by the national expert team. Its results can feed the process of the new national curriculum cycle to be started in 2024 as best practice examples. To date, over 10 schools have developed and tested school-based curriculum models. This approach helps schools to introduce innovations such as EdTech while using place-based learning. To continue learning under challenging conditions, such as during the COVID pandemic, and promote blended and online teaching methods in the future, 30 percent of the targeted schools have tested the curricula creating a modern blended learning classroom.
4. *School and its surrounding environment:* Schools have made enormous efforts to create a favorable learning environment for students through learning new principles of sustainable use, such as saving water and electricity, growing vegetables in greenhouses, segregating and managing waste, and converting unused spaces into study spaces for the students, popularly known as "corridor libraries." Some schools became model schools of sustainable consumers by creating new habits and activities to recycle and reuse. Forty percent of the pilot schools initiated a new waste management culture not only at the school level but also at the community level. Recycling and collecting waste have become an everyday habit for many people as a result of the school initiative. Another new activity that some schools could promote was creating an eco-environment. One of the pilot schools (School Number 15 in Orkhon province) could create a new garden with plenty of plants where students now do science experiments and plant trees. In addition, this school built a winter greenhouse to provide vegetables for its students (primarily children of nomadic herders) living in the dormitory.

   Our findings show some similarities to Ireland's Green School research (O'Mahony & Fitzgerald, 2001). It indicates that students are less likely to drop less garbage while being more likely to conserve water and energy and think about the environment.
5. *Partnership:* The birth of a local coordinating mechanism comprising the local authorities, community members, and members of the local media to support the ESD schools was an unexpected result for the ESD-II project. These local coordination groups have emerged as champions of ESD. Thus, an innovative partnership model has been created between schools and local organizations, and schools are learning advocacy activities to influence local administration and communities. This partnership has proved effective in many areas: (i) many schools could raise extra funding from local governments, a new initiative in the case of Mongolia. The school in Khyalganat of Bulgan province created a new summer camp where students can explore science experiments and vegetable planting. This school could raise funding of USD 140,000 from the local government and communities. The governor of the Khovd province has earmarked a budget of USD 90,000 for the further development of ESD in the schools of his province; (ii) schools piloted various activities to introduce sustainable consumption like the waste management system in their localities and promoted initiatives to revitalize the local heritage considering that the local community is a live carrier of cultural heritage and a source of learning. Our study findings are supported by the study by Henderson and Tilbury (2004) (p.39), which highlights the importance of the WSA process to create and strengthen links and partnerships with the community. Examples of the par"ners'ip include active participation in projects outside the school boundaries and equal and reciprocal partnerships.

13.8 Summing Up

This pilot of utilizing a WSA self-assessment tool in school activities in Mongolia has shown to be an example to be scaled up in the future. It was a learning process to improve everyday schooling adding more new insights related to school-based ESD development. Implementing the Self-Assessment Checklist also gave schools new knowledge, insights, and skills in school management, school development planning, identification of the challenges in collaboration, and finding solutions in a partnership covering all areas of school practices. This has ushered in a new progressive culture in Mongolian schools and will inspire more schools to follow their example. Moreover, the GoM is keen to scale up the WSA model to all schools in Mongolia. Through this project, schools could develop school-specific action plans to take concrete actions to address the challenges identified from the checklist, and each school's plan was specific to its unique context. Leadership by school administrators was a critical aspect of the successful experiment with the tool. The whole process in 30 schools was not easy to implement right away. It took some time and effort from everyone. Reorienting the entire school community toward collaborating for the school development using the new tool called the WSA self-assessment tool was a key challenge for project implementers and school administrators. Change of attitude and understanding of the school community, from something that is fixed and a given, to something that is dynamic and can be shaped, was the biggest challenge in this process. The new culture of involving everyone in school management and the decision-making process has been a steep and long learning curve for everyone.

References

Global Education Monitoring Report. (2016). *Education for people and planet: Creating sustainable futures for all.* https://unesdoc.unesco.org/ark:/48223/pf0000246230

Henderson, K., & Tilbury, D. (2004). *Whole-school approaches to sustainability: An international review of whole-school programs.* http://aries.mq.edu.au/projects/whole_school/files/international_review.pdf

Information and Training Centre for Nature and Environment (ITCNE). (2020). *Handbooks on WSA and school self-assessment tool.* https://www.esd.mn/

Kouzes, J. M., & Posner, B. Z. (2013). *Great leadership creates great workplaces.* New York: John Wiley & Sons.

Unified Legal Information System of Mongolia. (2016). *Sustainable development vision-2030 policy of Mongolia.* https://legalinfo.mn/mn/detail/11725

O'Mahony, M. J., & Fitzgerald, F. (2001). *The performance of the Irish Green Schools program.* An Taisce Publication. https://www.eltis.org/sites/default/files/case-studies/documents/ireland_research_report_2001_3.pdf

Global Humanitarian Platform. (2007). *Principles of partnership.* http://globalhumanitarianplatform.org/

Swiss Agency for Development and Cooperation SDC. (2017). *SDC's approach to governance.* https://www.shareweb.ch/site/DDLGN/Documents/SDC_Approach_To_Governance-01.pdf

UNESCO. (2014). *UNESCO roadmap for implementing the global action program on education for sustainable.* https://unesdoc.unesco.org/ark:/48223/pf0000230514

UNESCO. (2015). *Incheon declaration and framework for action for the implementation of SDG 4.* http://uis.unesco.org/sites/default/files/documents/education-2030-incheon-framework-for-action-implementation--of-sdg4-2016-en_2.pdf

UNESCO. (2020). *Education for sustainable development: A roadmap (ESD for 2030).* https://unesdoc.unesco.org/ark:/48223/pf0000374802

UNESCO International Bureau of Education. (2023). *Whole school approach.* http://www.ibe.unesco.org/en/glossary-curriculum-terminology/w/whole-school-approach

14 Critical Events in the Systematic Work at an Organizational Level Towards a Whole School Approach to Sustainability in a Swedish Municipality

Annika Manni, Eva Knekta, and Erika Åberg

Key Message
When a whole school approach on ESD is aimed for in practice, we recommend:

- A reflective and systematic work process
- Daily dialogue about how we can jointly contribute to ESD
- A shift in focus from the individual to the organizational level
- To include the ESD work in the ordinary educational systematic quality work
- Surveys for teachers and principals in combination with a follow-up dialogue to get an overview of ESD in practice, and as a means for inclusive developmental work

14.1 Introduction

The issue of sustainability is no longer seen as a matter of individual interest, but rather an unavoidable issue at all societal levels striving for a common future (Bokova, 2015; UNESCO, 2021). A general understanding of the urgency and importance of sustainable transformations is established, and therefore also included in the Swedish national curricula, which clearly states that education for sustainable development (ESD) is not only a single subject matter but should permeate all education in compulsory school. The concept of sustainability/sustainable development is, however, also interdisciplinary, complex and an object of interpretations (cf. Jickling & Wals, 2012) and therefore neither easy to operationalize in a subject-oriented school context nor systematically evaluate.

In Sweden, the state governs the school via aims and guidelines in national curricula, while the municipalities are the main organizers for compulsory basic school, upper secondary school and municipal adult education. As such, the municipality is responsible for implementing the curricula and plays a crucial role in organizing and evaluating education for sustainable development. Municipalities are often organized hierarchically, with a top-down chain of steering, with educational politicians at the top and teachers at the bottom. Although municipalities have an important role in the Swedish educational system, educational research at the municipal level is rare, both in general (Alvunger, 2022) and in the context of ESD.

A. Manni (✉)
Department of Applied Educational Sciences, Umeå University, Umeå, Sweden
e-mail: annika.manni@umu.se

E. Knekta
Department of Science and Mathematics Education, Umeå University, Umeå, Sweden

E. Åberg
The Nature School, Municipality of Umeå, Umeå, Sweden

A. E. J. Wals et al. (eds.), *Whole School Approaches to Sustainability*, Sustainable Development Goals Series, https://doi.org/10.1007/978-3-031-56172-6_14

There is a history of individual teaching efforts within environmental and sustainability education in Swedish schools, and many previous studies on ESD have focused on teaching approaches and educational outcomes, as well as the complexity of classroom teaching and learning (cf. Manni, 2018; Sandell et al., 2005; Sund & Gericke, 2020; Wickenberg, 2004). Over the years, researchers have argued that, beyond individual teaching efforts, there is a need for a holistic and systematic approach to successfully implement ESD into all education (Mathie & Wals, 2022; Rieckmann, 2018; Sterling, 2004). Despite the need for a widened perspective, studies on ESD have only in recent years started to integrate the specific characteristics of sustainability within whole school organization theories (cf. Holst et al., 2020; Mogren et al., 2019). Furthermore, a recent study shows that transforming educational work on sustainability issues from individual teaching efforts to include a whole school has proven to be a crucial, but not easy, task (Mathie & Wals, 2022).

Thus, due to ESD being an ambiguous curricular goal and at the same time a complex concept, with a lack of knowledge about organizational methods or tools, local school leaders at a municipality level might feel they have been left alone to perform the urgent, but somewhat vague, task.

In the study underlying this chapter, we have analysed the comprehensive processes of evaluating and improving the work with ESD in one Swedish municipality. The municipality had identified that, despite several good examples of individual work with ESD, it was hard to get it permeated within all school units. As a first step towards developing their work with ESD in the whole municipal school organization, they decided to systematically evaluate their ongoing work and to build a basis for further improvements. Critical events during the process, and unexpected dimensions within a school development model, are here described and analysed with the hope of contributing to others wanting to implement ESD in a whole school organization.

14.1.1 Aim of This Study

The aim of the study was to gain a rich understanding of the work in practice with ESD at a school organizational level. The questions we seek to answer are:

- What role do different aspects of a school organization play in a successful implementation of ESD in a whole school organization?
- What events can be critical in the process of implementing ESD in a whole school organization?

14.2 School Developmental Work

In Sweden, school developmental work is generally understood and referred to as 'School improvement cycles' (Swedish National Agency for Education, 2022), describing transformative processes as circular and systematic, where action, experience and reflection interact in the ongoing developmental work. The understanding of developmental processes stems from experiential learning theories, where learning continuously includes experiences from external events, reflections (as internal events) and actions in a transformative process for an individual or a group (Dewey, 1966; Kolb, 1984). When analysing the characteristics of such processes, critical events within the phases of the process are key to a deeper understanding of why and how things are entangled.

In practice, school developmental cycles of work start with the question 'Where do we stand?', followed by 'Where do we want to go?' and 'How do we go there?', ending with a reflective question on the activities of choice, 'How did it work out?'. It is common for practitioners to structure and manage their professional tasks this way, often referred to as systematic qualitative work. Furthermore, these circular processes at a school organization level emphasize a dialogic and collaborative approach between all actors in practice, like the idea of professional learning communities (cf. Stoll et al., 2006).

In other words, there is a common understanding of developmental processes continuously going on at individual as well as collective levels in all areas of education. In this study, we focus mainly on the collective level; educational actors in a whole municipality, and in their work with ESD, which will be further described below.

14.2.1 The Whole School Approach and Education for Sustainable Development

In our study, we acknowledge the importance that all staff, all students and all activities are involved in the work with ESD. This is also how we understand the concept of a *whole school approach (WSA)*. Scherp and Scherp's (2007) model for school development in general, also called the whole school approach model, is here used as an analytical framework to understand important areas involved in a school organization. Their model includes four dimensions of a school organization that influence student outcomes. The dimensions are: *Holistic idea, Routines and structures, Professional knowledge creations and Pedagogical practice*. The four dimensions represent a school organization's everyday practice, jointly considered as driving forces of school improvement on a general level (2007).

Mogren et al. (2019) have worked with the Scherps' whole school approach model in an ESD context. They define that the *holistic idea* refers to 'the degree that schools have an articulated, holistic vision of their aims concerning student outcomes, and the pedagogic methods and perspectives that should be applied to realize the vision' (p. 511). *Routines and structures* refer to the degree to which there are established routines and structures at the school that create stability and facilitate teachers' teaching (ibid). *Professional knowledge creations* refer to whether there is an openness to changes and innovation that stimulates teachers' professional development (ibid). Lastly, *Pedagogical practice* concerns what happens in the classroom, the teaching and the learning (ibid).

The results from the Mogren et al. (2019) study showed that the holistic idea is the most essential element as it influences the three other dimensions and 'thus is the primary connection between the Scherp model and the ESD whole school approach' (Mogren et al., 2019, p. 513). Furthermore, the character of the holistic idea of ESD differs from other areas of education in the sense that ESD explicitly aims for the common good (Lotz-Sisitka, 2017; UNESCO, 2021), while education in subject-specific areas aims for the individual's increased learning. This is an important aspect to consider when working with ESD in a whole school organization.

In summary, we examine critical aspects in the work towards a whole school approach to ESD within a municipality school organization by considering school developmental work as circular and systematic processes (Dewey, 1966; Kolb, 1984; Swedish National Agency for Education, 2022), including educational dimensions defined in Scherp and Scherp's (2007) model for school development.

14.3 Method—Narrative Case Study

This case study reports from 4 years of school developmental work in one municipality involving about 100 preschools and 50 schools, including their leisure-time centres. As a case study with qualitative features, it seeks to reach deeper understanding of a specific case through thick descriptions of the situated context (Flyvbjerg, 2011; Yin, 1994). The data material gathered and analysed is comprehensive, including formal documents, meeting notes, field notes, surveys, interviews, photos, etc. In total, more than 2000 teachers, and 100 school leaders participated through surveys, seminars and interviews. Meeting notes ($n = 40$) with school leaders on a monthly basis, and multiple field notes organized in chronological order, are also part of the data we have gathered and used for this case study.

Considering the long-term process of the developmental work of this case, we have chosen to analyse and report it as a narrative case study, following the principles of narrative methodology (Webster & Mertova, 2007). Within such an approach, participants' stories are considered to contribute to valuable and relevant knowledge, forming a narrative case study over time. In particular, the analytical work and structure of reporting the results have been led by a narrative approach (Connelly & Clandinin, 1990; Sonday et al., 2020), which for this case study includes the following stages:

1. *Setting the scene*—data describing the specific context of this case were gathered and analysed in order to present a rich contextual background.
2. *Telling the story*—based on comprehensive data material, the 4-year story of the school developmental process was summed up in chronological order and critical events transforming the process in different ways were identified. Critical events are categorized as external events, reflections and actions according to experiential learning theories (Kolb, 1984).
3. *Analysing the plot*—finally, the story and its critical events were analysed and interpreted in relation to the dimensions in Scherp and Scherp's (2007) model for school development.

14.3.1 Ethical Considerations

Research participation in the school developmental process has undertaken ethical vetting with approval (Swedish Ethical Review Authority, Dnr: 2019–03239). Participants in the surveys, interviews and meetings have been informed in both oral and written form and agreed to be part of a research study. Ethical guidelines have been followed, paying respect to participants' integrity and security, although no personal or sensitive data have been handled (Swedish Research Council, 2017).

14.4 Setting the Scene of this Case

During the last 4 years, one municipality in Sweden has been engaged in a continuous school development process, involving school leaders, teachers and researchers, to systematically strengthen the ongoing work and achieve a whole school approach to sustainability. A municipality is a geographical and administrational district, in charge of organizing public preschools, schools and leisure-time centres in accordance with national guidelines and curricula. This municipality's developmental work stems back to many years of intense and engaged educational work regarding environmental and sustainability issues. Through the so-called Nature school, competence development in various forms has been offered, and a network of representatives from all preschools and schools has met four times a year. At an earlier stage, the academic year also began with joint conferences focusing on ESD for educators and principals. Furthermore, there was also a decision in the political mission plan of 2013 that activities should be developed in accordance with the National Agency for Education's award 'School for sustainable development', and here, the municipality was at the forefront, nationally. Although the municipality had worked intensively with ESD, including many successful individual examples, a general implementation in all schools and preschools had not yet been achieved, which was the driving force of the developmental work underlying this case.

14.4.1 The Story of the Developmental Process for Implementing ESD as a WSA

14.4.1.1 Year 1

The stepping stone for the developmental work that began in January 2019 was a decision by the Municipal Educational Board that resulted in a mission plan that aimed to implement ESD in all preschools and schools between grades 1 and 9.

The intention was that ESD should reach all pupils, regardless of individually engaged teachers. From this political goal, a more detailed action plan was developed and signed by the head of the municipal school department, guiding the work for 4 years (2019–2022). The Nature school together with a reference group on ESD consisting of school leaders were assigned as supporting actors and in charge of the implementation work. Through network structures, organized by the Nature school, a variety of activities were offered to teachers and individual schools on an annual basis. Two researchers were engaged in the project to follow, study and support the implementation. In the beginning, the project work was organized in a top-down network approach, with school leaders and teachers who were interested in supporting network activities (Fig. 14.1).

One of the major critical events in the initial process was the decision to start the developmental work with a survey mapping the current work across the whole municipality, not just including those who had previously chosen to engage in ESD activities. During the first months of the project, the survey was developed in cooperation with researchers, school leaders and teachers. The survey construction was theoretically based on, and structured according to, the general didactic principles asking why, what and how (Uljens, 1997), and were related to descriptions of ESD in national and international steering documents (Swedish National Agency for Education, 2016, 2018, 2019; UN General Assembly, 2015; UNESCO, 2014). The survey consisted of 40 Likert scale items and seven open-ended items and took about 20 minutes to complete. The purpose of the closed questions was to get an overall picture of the teachers' perceptions of current aims (*I work with ESD because I personally find it important)*, content (*I work with the social goals*), methods (*In my teaching, I use learning environments in the local community)*, and perceived needs (*I need more knowledge about the content of the global goals*) related to ESD in their daily teaching practice. Although the closed questions were grouped into three different main areas (why, what and how), the purpose was not to measure three underlying constructs; instead, each question was considered to have an individual value. The purpose of the free-response questions was to gain a deeper understanding of the educators' perspectives, especially with regard to the need for competence development and support, and to capture aspects that may not have been included in the closed questions. For a detailed description of the survey, see Manni and Knekta (2022). The reasons for focusing on teachers in the survey were to map what kind of ESD teaching was offered to all students and children in the municipality in terms of equal ESD education, as well as aiming to direct the ESD seminars and activities towards the needs or desired content that would increase relevance for the teachers.

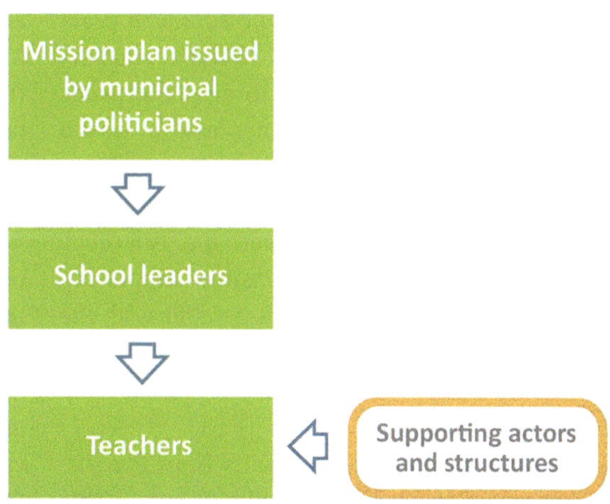

Fig. 14.1 Initial ESD organization (illustration by Matz Glatz). Top-down structure with supporting actors acting in parallel with the ordinary steering chain

The survey was distributed and analysed for the first time in Year 1 (September 2019). Analysis of the data showed that most teachers found ESD important, but fewer regularly prioritized this in their teaching. A comparative analysis between the different educational actors, that is pre-schools, leisure-time centres, grades 0–6 and 7–9 showed ESD was dealt with differently by the different actors and age groups of the children. The results indicated a need for individual in-service training and a lack of routines and structures for ESD work at the individual school level. The open-ended items confirmed and nuanced the need for in-service training. They also gave a rich description of organizational obstacles. Time constraints and a lack of common understanding were two prominent examples of organizational obstacles. Furthermore, the overall organizational structure for ESD was questioned since it implicated a linear top-down structure and not a process-based structure that invited dialogue and feedback. Thus, the obstacles related to ESD were found in the organizational structures and support, not only in the individual needs concerning teachers' knowledge and teaching skills within ESD (as had been anticipated by the Educational Board when planning the survey). Analysing these results was the major critical event derived from the survey, and reflecting on this event within the reference group resulted in several important actions. The main change involved dialogue and feedback between the teachers and the supporting actors (the Nature school and the reference group), which had not explicitly been done before. One action in practice was a change of activities offered to teachers at the ESD network seminars to meet the needs reported in the survey. Another action was a change in the character of network seminars, from lectures to discussions and collegial dialogues. The Nature school and the reference group remained as the responsible supporting actors, continuing to operate in parallel with the ordinary steering chain (Fig. 14.2).

14.4.1.2 Year 2

As a result of the reflected results from the survey, one part of which regarded teachers asking for new ideas on how to work with ESD, documentation of ongoing activities in school practices was sampled. Besides more traditional outdoor environmental activities and artwork with recycled material, new ideas such as a second-hand market and a co-creative playing room were displayed (Manni, 2023). The examples were intended to serve as inspirational examples and to give credit to existing work. Special focus was placed on providing activities adjusted to different target groups to include and empower all actors. The network seminars during the spring of 2020 were organized in the same vein, to meet the needs of the teachers as expressed in the survey.

In the survey, teachers in leisure-time centres reported a greater need for professional development compared to the other teacher categories. This finding led to an in-depth research study focusing on teachers' work with ESD. This external event, which was reported back to the municipality, contributed to shift in focus from routines and structures to also include the holistic idea of ESD (Manni & Knekta, 2022). It became clear that although many understood and valued sustainability as part of teaching and learning, they did not have a common understanding of ESD, and therefore could not really place it within the traditional teaching structures. Furthermore, comparing the results from teachers in leisure-time centres with other groups of teachers, many schoolteachers reported that lack of time to add sustainability to their subject-specific schedules was why they did not teach it. That might indicate a misunderstanding about ESD as a new subject or yet another content to cover in their already full curriculum, rather than an approach and a perspective within all subjects. This finding reflects Johnston's (2009) article about the need to step out of the curriculum box when working with environmental and sustainability issues. In other words, the expressions regarding lack of time to work with ESD also indicate a lack of understanding of the holistic idea behind ESD.

One critical external event in our case that affected not only this developmental work but the entire world was the coronavirus that spread in the spring of 2020. Due to the pandemic restric-

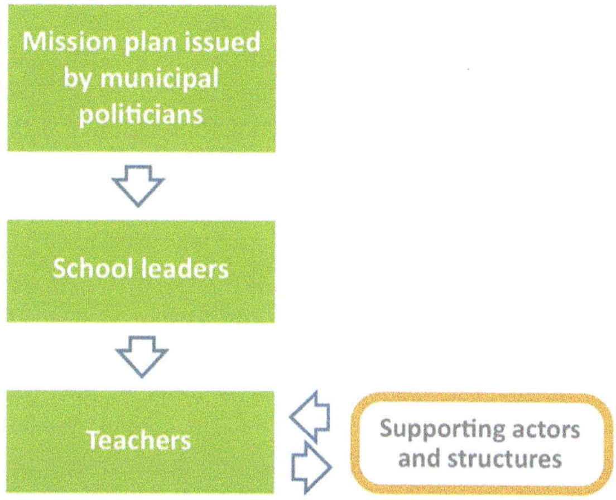

Fig. 14.2 Revised ESD organization model (illustration by Matz Glantz). Top-down structure. Increased focus on dialogue between supporting actors and teachers

tions, all planned activities within the ESD network were put on hold for one semester. Despite the obvious reasons for this pause, it was most unfortunate for the continuous process and dialogue with the participants. During this period of pandemic restrictions, a reflective report on the results from the survey and the follow-up activities was written. One major finding from the report was that the teachers needed more structural support in their ESD work. Further, the report pointed at some disadvantages in assigning two groups outside the ordinary steering chain (i.e. the Nature school and the reference group), responsibility for implementation and support. Being parallel organizations gave them less authority and less opportunity to push through changes within the organization. This event was critical and worked as a turning point for a more conscious process to revise and incorporate ESD within the ordinary steering chain of the whole municipality (Fig. 14.2).

In the autumn of 2020, the pandemic still restricted staff from meeting and working as usual, and new digital forms of meetings and seminars were established, providing space for joint reflective analyses between researchers, the reference group and teachers. At this point, the developmental work was explicitly articulated as a systematic circular process (cf. Swedish National Agency for Education, 2022), not as a linear top-down project. This internal reflection and transformation of how the work was framed and communicated was another of the major critical events in our case, as the new communication process opened doors to communicate with school leaders in general, not only within the established 'ESD-family'. One might say that this critical event was a game changer to addressing ESD as a whole school approach within the general educational organization of this municipality.

At the end of year two, as part of the shift of focus from individual teachers' work to the whole school organization, we decided to develop a new survey directed at school leaders, with the same areas of questioning as in the teacher survey. We will return to the results of this activity as an event in the developmental process.

14.4.1.3 Year 3

After 2 years of developmental work, where the survey mapping had played a significant part in understanding the ongoing work among teachers and the existing school structures, an additional focus was set. This entailed using the survey as a tool not only to provide informative overviews at a municipality level but also for inclusive developmental work at each school unit. Through collaborative work between researchers and the ESD reference group, a 'Support for ESD dialogue' tool for teachers and school leaders was developed. The dialogue tool mirrors the structure of the surveys and provides a simple guide for analysing the survey results and identifying future ESD development.

The developmental processes hence moved between micro and macro levels within the municipality, showing how to pay attention to all levels in an inclusive whole school approach, but also how these levels are aligned. Finally, after some delays due to the pandemic, in May 2021, the 'Support for ESD dialogue' tool was used for the first time in schools with guidance from a researcher and the head of Nature school. This event was critical, as the municipality work was now finally reaching individual schools in the implementation work, i.e. increased agency in the processes, shifting the ownership down from the top (Stoll et al., 2006).

In the autumn of 2021, a new leader in the municipality educational organization, with responsibility for assessing quality, was hired. This turned out to be a door-opener for the school development process, bringing increased engagement from an actor representing the top of the municipality educational organization, who also began to join the ESD reference group meetings.

14.4.1.4 Year 4

At the start of year 4, the developmental process had revealed and resolved several critical aspects in striving for a whole school approach to ESD; the educational organization had improved, and teachers' in-service training was more tuned to their needs, as well as emphasizing dialogue.

But more improvements were yet to come; the results from the survey of the school leaders, developed in late 2020, were analysed and showed that the leaders also needed support in ESD work. The kind of support asked for was on an organizational level but also regarding how to understand the very idea of ESD as a whole school approach. Intensified collaboration and dialogue between the school leaders and supporting actors, an acknowledgement of ESD as important for all, and a need for organizational support were agreed. This event involved discussions regarding supporting school structures, including an emphasis on ESD as a mandatory educational task, instead of optional as before. This change of view was seen in practice when ESD was put on the monthly agendas for principals and ESD seminars to raise competence, specifically among the main school leaders, were asked for. In this situation, the top-down approach was helpful and necessary to engage all levels of the school organization in ESD, not only teachers.

Furthermore, this awareness among the main school leaders included a reorganizational proposal for ESD, now as a task within the ordinary steering chain and not merely reliant on support from external actors (Fig. 14.3). This new structure thus included a shift of responsibility for ESD from the head of the Nature school to the school leaders of the four educational areas within the municipality, that is a shift from a voluntary task to mandatory and participatory-led work. Moreover, the school leaders were also introduced to how to work with ESD as a circular, systematic and dialogic approach to school development (Stoll et al., 2006), with them taking

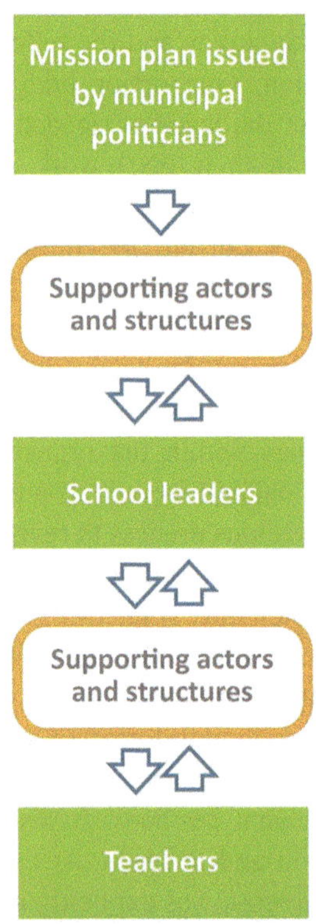

Fig. 14.3 Current ESD organization structure, after 4 years of work (illustration by Matz Glantz). Top-down structure is replaced with dialogue between all actors in the educational organization, and supporting actors are included in the ordinary steering chain

responsibility for the implementation of ESD as a whole school approach.

In this process of change, one critical event proved to be the continuous discussions with the municipal heads of education, school leaders and the supporting actors from the Nature school on what ESD implies as a holistic idea. The opportunity to discuss this issue with colleagues at the same organizational level and other school leaders, as preparation before leading teachers and staff in the ongoing work, was appreciated.

Besides the developmental processes at the school organizational level, seminars and workshops with teachers were also held during this period, mainly with preschools, asking for new ideas and perspectives on ESD and outdoor teaching, and with staff from leisure-time centres wanting to meet colleagues and share ideas about how to work with ESD in their specific practice.

At the end of our 4-year story of ESD development, we conclude that the work in practice has not ended but re-started on a new cycle of improvement. A new survey to teachers and school leaders was administered during late spring of 2022. The tentative results suggest continuous work to deepen the understanding of ESD as a holistic idea and what that might mean as a whole school approach, for teachers as well as for school leaders. Results from the survey will be brought back into the new organizational structure and be part of the future ESD work. School leaders and teachers will also take a more active part in analysing and reflecting on their own results to strengthen the ownership of their developmental work. The next step of further improvements will also include the pupils and their opinions on ESD, in accordance with the initial political mission plan.

14.5 Analysing the Plot and Its Critical Events Using Scherp's Model of School Improvement

In order to analyse the story of this case, we combined its critical events with the dimensions of Scherp's model of school improvement: *Holistic idea, Routines and structures, Professional knowledge creation and Pedagogical practice* (Mogren et al., 2019; Scherp & Scherp, 2007). We differentiated the critical events as either external events, reflections or actions, in line with experiential learning theories (cf. Kolb, 1984). Figure 14.4 displays the critical events in chronological order and how different dimensions of the Scherp model have been uncovered in different stages of the process of our case. Below, we will sum up Fig. 14.4 in words.

When the task of implementing ESD was given in the mission plan issued by the municipal politicians, the formulation in the document had a clear whole school approach as a *holistic idea*—that all units should contribute to all pupils' learning and meaning-making within ESD, although this was not established as a concrete concept. The critical event following the plan was to distribute the survey. Although all areas of the school development model were represented in the survey, the focus came to be on teachers' current work with ESD as well as their need for support or in-service training. Further, as it was only the teachers who were asked to answer the survey, this indicated that they were responsible for conducting ESD. Thus, although the *holistic idea* was prominent in the mission plan, aspects *of professional knowledge creation* and *pedagogical practice* were unconsciously focused on in the beginning of the school developmental process. After analysing the results from the survey, a shift in understanding important factors for successfully reaching a WSA on ESD was seen. The results put focus on *professional knowledge creation* and *pedagogical practice of professionals*, but also on *routines and structures*. The teachers found ESD important but mentioned a lack of *routines and structures* as a main obstacle to its implementation, a result that changed the game of the developmental processes. A new organization model was implemented, and professional development organizations were restructured. In the process of improving professional development for teachers, and in particular for teachers in leisure-time centres, the focus turned back to the *holistic idea*. A *holistic idea* of what ESD is and how it can be implemented, shared by teach-

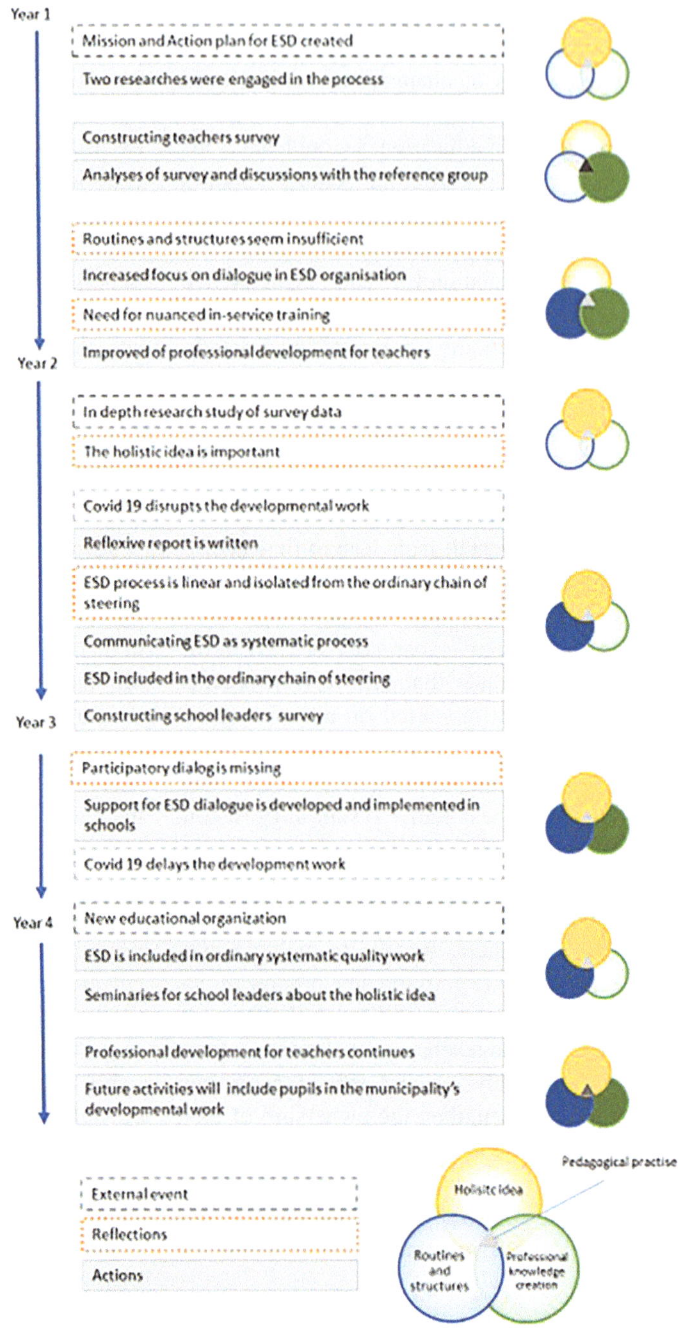

Fig. 14.4 Schematic illustration of the critical events (external event, reflections and actions) in chronological order during the school developmental process (left side), combined with theoretical analysis based on dimensions of Scherp's model of school improvement: holistic idea, routines and structures, professional knowledge creation and pedagogical practice (right side). Filled forms indicate where the focus was at the moment

ers and school leaders, was found to be fundamental to successfully working with the three other WSA dimensions. At this stage of the process, we also put the spotlight on the linearity of the process and how the supporting structures were operating outside of the main organization. These two aspects of *routines and structures* seemed to hamper the possibility of developing a common *holistic idea* and again showed the interconnectedness of the four dimensions of the WSA. Implementation of ESD instead became communicated as a circular process, and the first step towards including ESD in the ordinary quality work in the municipality was taken. *The support for dialogue* for school leaders and teachers was developed as a tool to include all actors in the development work. It aimed to serve as a means for *professional knowledge creation* as well as creating a shared *holistic idea*. This (we believe) will in turn, together with changed *routines and structures*, lay a solid foundation for future *pedagogical practices of ESD*.

14.6 Concluding Remarks

The aim of this study was to gain a rich understanding of the work in practice with ESD at a school organizational level. We searched for knowledge about what role different aspects of a school organization play in a successful implementation of ESD in a whole school organization and what events can be critical in the process of implementing ESD. What might be unique for our case within the context of a whole school approach on ESD, is that the 'whole school' not only includes one school unit but one municipality (the administrational district in Sweden in charge of organizing public preschools, schools and leisure centres), with all its schools, leaders and teachers. Our case should then be seen and understood as an example on a macrolevel. Summarizing the experiences from our case study of the school development work, we finally conclude on some vital and critical aspects that we think are more general, and of interest to others wanting to develop and transform their ESD work by taking a WSA. The first two conclusions regard the process of implementing ESD, while the last one regards dimensions of a WSA to ESD.

14.6.1 The Systematic and Reflective Process of Research-Led Developmental Work for ESD

The characteristics of the process as dialogic, reflective and systematic (cf. Kolb, 1984) have proved to be important for the changes and school development overall. Looking back, we conclude that our work can be viewed as a 'transdisciplinary professional learning community' (cf. Stoll et al., 2006), since we have worked together as researchers, school leaders, principals and teachers. Within the reference group, our different experiences and professional knowledge have been useful in reflecting on the critical events and actions taken in the developmental work, and have given more strength and stability to the whole process, and in learning from each other's different professional points of views. We would also emphasize the involvement and participation of teachers in each school practice, forming their respective learning communities and strengthening their ESD work. From our experience, this is crucial in a holistic approach to ESD, or as a whole school approach, and in opposition to linear or top-down project-based developmental work (cf. Rieckmann, 2018).

14.6.2 The Development and Use of the Surveys, and the Support for ESD Dialogue

The development and use of the surveys, and the support for ESD dialogue have been most valuable. Despite the history and examples of good work with ESD in this municipality, the surveys provided both an overview and new knowledge beyond the established understandings of ESD in general. The comprehensive information in the teacher and principal surveys provided opportunities to analyse different dimensions of ESD in practice on a municipal level. We found educational dimensions (routines and structures) and actors (teachers in leisure-time centres, as well as principles and school leaders) that were somewhat invisible and forgotten in the previous ESD work (Manni & Knekta, 2022), which helped to direct support and change to those actors in need of improvement.

14.6.3 The Importance of Organizational Support

Another main conclusion from our work is the importance of the shift in focus from individual teachers to the organizational level to develop ESD as a whole school approach, engaging the entire 'education system' (Mathie & Wals, 2022). From a school organizational point of view, the results from the surveys, in combination with the analytical lens of the Scherp model (Scherp & Scherp, 2007), were crucial in transforming and communicating important aspects of the developmental work with all actors. We found that without the supporting structures from the very top of the organization, no lasting progress or transformation could happen. We dare to conclude this based on many years of individual engagement on ESD in this municipality without having reached a general acknowledgement of its importance. Not until ESD was properly included in the ordinary steering chain, in formal quality assessments, and in school leader agendas, was there a hope of successfully achieving the task.

14.6.4 The Holistic Understanding of ESD

This dimension of the school development model (Scherp & Scherp, 2007) has shown to be more important than we thought (even though previous research pointed in that direction [Mogren et al., 2019]). Scratching the surface, we have learnt that many teachers and school leaders have a shallow understanding of ESD, even less as a whole school approach with aims for the common good (UNESCO, 2021). Many are trapped in the traditional understandings of teaching and educational curricula (Johnston, 2009), where ESD is seen as something 'extra' and not 'within'; furthermore, current school organizations seem to support those traditions, complicating the necessary changes to reach a whole school approach on ESD (Holst et al., 2020).

We finally conclude that it has been most valuable to follow ongoing work over a long period of time, providing opportunities for sharing thoughts, systematically reviewing critical events and, with help of the school organization model, identify dimensions of importance for successful and inclusive ESD work.

References

Alvunger, D. (2022). *Exploring VET curriculum making through a review of Swedish research.* Conference presentation at NordYrk, Norrköping.

Bokova, I. (2015). *Rethinking education – Towards a global common good?* UNESCO.

Connelly, F. M., & Clandinin, D. J. (1990). Stories of experience and narrative inquiry. *Educational Researcher, 19*(5), 2–14. https://doi.org/10.3102/0013189X019005002

Dewey, J. (1966). *Democracy and education, an introduction to the philosophy of education* (1966th ed.). The Free Press.

Flyvbjerg, B. (2011). Case study. In N. K. Denzin & Y. S. Lincoln (Eds.), *The sage handbook of qualitative research* (4th ed., pp. 301–316). Sage.

Holst, J., Brock, A., Singer-Brodowski, M., & de Haan, G. (2020). Monitoring progress of change: Implementation of Education for Sustainable Development (ESD) within documents of the German education system. *Sustainability, 12*(10), 4306. https://doi.org/10.3390/su12104306

Jickling, B., & Wals, A. E. J. (2012). Debating education for sustainable development 20 years after Rio: A conversation between Bob Jickling and Arjen Wals. *Journal of Education for Sustainable Development, 6*(1), 49–57. https://doi.org/10.1177/097340821100600111

Johnston, J. (2009). Transformative environmental education: Stepping outside the curriculum box. *Canadian Journal of Environmental Education, 14*, 149–157.

Kolb, D. A. (1984). *Experiential learning. Experience as the source of learning and development.* Prentice Hall, Inc.

Lotz-Sisitka, H. (2017). Education and the common good. In B. Jickling & S. Sterling (Eds.), *Post-sustainability and environmental education* (pp. 63–76). Springer International Publishing. https://doi.org/10.1007/978-3-319-51322-5

Manni, A. (2018). Contingency and transformation. Teachers' and students' experiences of a climate council school project. *Nordidactica: Journal of Humanities and Social Science Education, 2018*(4), 1–20.

Manni, A., & Knekta, E. (2022). Fritidshemmet – en förbisedd potential i arbetet med lärande för hållbar utveckling? *Nordina: Nordic Studies in Science Education, 18*(1), 63–81.

Manni, A. (2023). Education "through" sustainable development in Swedish school-age educare – exploring how SAEC is responding to ESD in daily practices,

Education Inquiry. https://doi.org/10.1080/20004508.2023.2265634
Mathie, R., & Wals, A. (2022). *Whole school approaches to sustainability: Exemplary practices from around the world* (9464471514). Retrieved from: https://www.wur.nl/en/education-programmes/wageningen-pre-university/whole-school-approach.htm
Mogren, A., Gericke, N., & Scherp, H.-Å. (2019). Whole school approaches to education for sustainable development: A model that links to school improvement. *Environmental Education Research, 25*(4), 508–531.
Rieckmann, M. (2018). Learning to transform the world: Key competencies in education for sustainable development. In A. Leicht, J. Heiss, & W. J. Byun (Eds.), *Issues and trends in education for sustainable development* (pp. 39–59). UNESCO.
Sandell, K., Öhman, J., & Östman, L. (2005). *Education for sustainable development: Nature, school and democracy*. Studentlitteratur.
Scherp, H.-Å. and Scherp, G.-B. (2007). *Lärande och skolutveckling: ledarskap för demokrati och meningsskapande*. [Learning and school development: leadership for democracy and meaningmaking]. Estetisk-filosofiska fakulteten, Pedagogik, Karlstads universitet.
Sonday, A., Ramugondo, E., & Kathard, H. (2020). Case study and narrative inquiry as merged methodologies: A critical narrative perspective. *International Journal of Qualitative Methods, 19*, 160940692093788. https://doi.org/10.1177/1609406920937880
Sterling, S. (2004). Higher education, sustainability, and the role of systemic learning. In P. B. Corcoran & A. E. J. Wals (Eds.), *Higher education and the challenge of sustainability* (pp. 49–70). Springer.
Stoll, L., Bolam, R., McMahon, A., Wallace, M., & Thomas, S. (2006). Professional learning communities: A review of the literature. *Journal of Educational Change, 7*(4), 221–258. https://doi.org/10.1007/s10833-006-0001-8
Sund, P., & Gericke, N. (2020). Teaching contributions from secondary school subject areas to education for sustainable development – a comparative study of science, social science and language teachers. *Environmental Education Research, 26*(6), 772–794. https://doi.org/10.1080/13504622.2020.1754341
Swedish National Agency for Education. (2016). *Curriculum for the compulsory school, preschool class and school-age educare 2011 Lgr11: Revised 2016*. Swedish National Agency for Education.
Swedish National Agency for Education. (2018). *Curriculum for preschool, Lpfö18*. Swedish National Agency for Education. Retrieved from: https://www.skolverket.se/publikationsserier/styrdokument/2019/curriculum-for-the-preschool-lpfo-18?id=4049
Swedish National Agency for Education. (2019). Lärportalen: Hållbar Utveckling, åk 7–9. [Learning portal: Education for Sustainable Development, year 7–9]. https://larportalen.skolverket.se/#/modul/01-hallbar-utveckling/Grundskola/901-Hallbar-utveckling
Swedish National Agency for Education. (2022). Skolutveckling: Systematiskt kvalitetsarbete i skola och förskola. [School development: Systematic quality work in school and preschool]. https://www.skolverket.se/skolutveckling/leda-och-organisera-skolan/systematiskt-kvalitetsarbete/systematiskt-kvalitetsarbete-i-skola-och-forskola
Swedish Research Council. (2017). *Good research practise*. Swedish Research Council. ISBN 978-91-7307-354-7.
Uljens, M. (1997). *Didaktik -Teori, Reflektion och Praktik*. Studentlitteratur.
UN General Assembly. (2015). *Transforming our World: the 2030 Agenda for Sustainable Development*. https://sdgs.un.org/publications/transforming-our-world-2030-agenda-sustainable-development-17981
UNESCO. (2014). *UNESCO roadmap for implementing the global action programme on education for sustainable development*. UNESCO. http://unesdoc.unesco.org/images/0023/002305/230514e.pdf
UNESCO. (2021). *Reimagining our futures together. A new social contract for education*. UNESCO. Retrieved from: https://unesdoc.unesco.org/ark:/48223/pf0000379381
Webster, L., & Mertova, P. (2007). *Using narrative inquiry as a research method: An introduction to using critical event narrative analysis in research on learning and teaching*. Routledge.
Wickenberg, P. (2004). *Learning to change our world? Swedish research on education and sustainable development*. Studentlitteratur AB.
Yin, R. K. (1994). *Case study research, design and methods* (2nd ed.). Sage Publications.

15 Weaving Curriculum, Assessment, and Pedagogy: Global Citizenship Experience Lab School's Whole-School Approach to Sustainability and Global Citizenship Education

Stephanie Leite

Key Message

This case study suggests that weaving curriculum, assessment, and pedagogy together may act as a fulcrum around which other school dimensions can be reoriented toward whole-school transformation. The case demonstrates that transitioning to such an integrated, student-centered, and project-based learning model can be initially disorienting for both students and teachers. To overcome this challenge, the chapter argues that scaffolds are needed to support and prepare new school members for such a learning model.

15.1 Introduction

Education is widely recognized as playing an important role in sustainability transitions and climate change adaptation and mitigation (Kwauk, 2020; MECCE and NAAEE, 2022; Wals & Benavot, 2017). To leverage the potential of education in imagining and enacting more sustainable futures, mainstream approaches to schooling must expand beyond their current focus on test-driven cognitive learning to include social-emotional learning and action competencies that prepare learners to transform unsustainable behaviors, attitudes, and systems (MECCE and NAAEE, 2022; UNESCO, 2021).

Theoretical contributions to the field of education for sustainable development (ESD) and climate change education (CCE) recommend learning approaches that are project-based, experiential, localized, and grounded in real-world contexts (Demssie et al., 2020; Kwauk, 2020). Implementing these recommendations and reorienting schools toward a greater focus on sustainability requires more than making additions to curriculum; many ESD scholars advocate for a "whole system redesign" to rethink the underlying values and assumptions that influence school structures (Sterling, 2004; Wals & Benavot, 2017).

Schools that take such a whole-school approach (WSA) to sustainability education position sustainability not as a discrete disciplinary subject, but as a core purpose of schooling. WSAs explore sustainability in a relational way, embedding it in all aspects of the school's systems, from curriculum and pedagogy to school vision and leadership (Wals & Mathie, 2022). Support for WSAs as an effective and necessary way to implement environmental and sustainability education is backed by stakeholders at international, regional, and local levels (Hargreaves, 2008; UNESCO, 2016). Despite the wide theoretical support for WSAs, examples of schools that prac-

S. Leite (✉)
Department of Integrated Studies in Education, Faculty of Education, McGill University, Montreal, QC, Canada
e-mail: stephanie.leite@mail.mcgill.ca

A. E. J. Wals et al. (eds.), *Whole School Approaches to Sustainability*, Sustainable Development Goals Series, https://doi.org/10.1007/978-3-031-56172-6_15

tice such a model are still not mainstream. Aside from the challenges of implementing a WSA at the school level—such as rethinking school schedules, staff and teacher training, grading systems, school infrastructure, and teacher evaluation standards—transitioning schools to a WSA to sustainability is further complicated by the wider systems in which schools are embedded.

A whole-system redesign at the school level is therefore difficult because our educational systems have yet to undergo a similar redesign process; these ingrained systems in turn limit the extent of innovation that can take place at any individual school. For example, in their redesign process, schools will likely confront policies and practices that restrict innovation at: the district level (how funding is allocated and the types of professional development offered); state or provincial level (how curriculum standards are written and teacher certification requirements are designed); and national level (how grades are communicated and college admissions requirements are set). Any changes made at the school level must therefore consider the school's embeddedness within these systems, which may enable or constrain desired changes. The entangled relationships at different levels highlight the need to simultaneously engage stakeholders across the web to make any significant change (Ferreira & Ryan, 2013).

15.1.1 GCE Lab School in the Context of Sustainability Education in the United States

In the United States, there is no national mandate to teach about sustainability or climate change. The onus therefore falls on states, NGOs, schools, and individual teachers to integrate content into their programming (Feinstein & Carlton, 2013). Such efforts usually focus on curriculum reform, as it is the most feasible way of ensuring students build sustainability literacy. While an important first step, a sole focus on curricular reform is widely acknowledged as insufficient in preparing learners with the knowledge, skills, values, and agency needed to cultivate more sustainable and just futures (Kwauk, 2020; Sterling et al., 2018; Stevenson et al., 2017; UNESCO, 2018). However, when curriculum is designed around ESD themes and pedagogies, it necessarily acts as a fulcrum that connects other school structures and practices (Sterling, 2013).

One school that has served as a laboratory for imagining and testing alternatives to entrenched educational conventions is Global Citizenship Experience Lab School (GCELS), an independent upper secondary school in Chicago, Illinois, USA. A private nonprofit school, GCELS was founded in 2010 on the belief that, in order to "cultivate responsible global citizens in the 21st century, the traditional transmissive models of education" must be "rethought from the ground up" (Leite & Moring-D'Angier, 2022, p. 24). GCELS holds that students should graduate high school with the practical experience, empathy, and agency needed for careers that are both fulfilling and meaningful for fostering a sustainable world.

Built around an approach similar to the "head, hands, and heart" model (Sipos et al., 2008), GCELS's program aims to instill in students the knowledge, skills, and values of global citizenship. Over the 4-year program, students develop ESD competencies (see Redman & Wiek, 2021) including systems thinking, futures thinking, and inter- and intrapersonal competencies. There is also a special focus on building action competence, which entails a combination of: strengthening conceptual knowledge of action possibilities; developing a willingness to take action; and growing confidence that one has both the abilities to make change and that actions taken will make a positive contribution to sustainable development (Sass et al., 2022). As an independent school, GCELS is not mandated to follow the Illinois State Board of Education standards, nor does it have to follow the state assessment schedule, which gives the school more freedom to experiment with innovative and integrated approaches to curriculum, assessment, and pedagogy.

This chapter presents a case study of GCELS's approach to sustainability by examining artifacts from two Grade 9 courses as a way of exploring the interconnections between the school's curriculum, assessment, and pedagogy—as well as how connecting these three dimensions reverberates throughout the entire school model and beyond. In addition, this case study highlights the obstacles encountered when attempting to implement a WSA to sustainability and identifies some of the wider systemic changes that must be made to amplify innovation at the school level.

15.1.2 Theoretical Underpinnings

GCELS's learning model is influenced by the rich tradition of progressive education in Chicago, where the impact of educators like John Dewey and Jane Addams still resonates. The Progressive Era of education, focused on real-world problem-solving, social interaction, and personal reflection, has many pedagogical similarities with sustainability education and can be seen as a precursor to ESD (Armstrong, 2011). Progressive and ESD educators similarly embrace constructivist learning theory, where teachers act as facilitators of student-centered environments that allow learners to construct their own meaning and knowledge (ibid.).

The collaborative learning approach and instructional support embedded in GCELS's curriculum—to be discussed throughout this chapter—can be analyzed through the lens of Vygotsky's (1978) sociocultural learning theory. Central to Vygotsky's theory is his conceptualization of the zone of proximal development (ZPD), or the difference between what a learner can do *without* versus what a learner can do *with* assistance. Such assistance is provided through scaffolding, a concept introduced by Vygotsky (1978) and elaborated upon by Bruner (1986) for its instructional applications (De León, 2012). Scaffolding—or breaking learning into digestible chunks that assist with the learning process—might include a variety of strategies, including the use of tools that are appropriate to a learner's interests and knowledge level, or modeling behaviors and processes to support learners in digesting new concepts or developing new skills (Armstrong, 2015). Scaffolding is especially helpful in ESD, due to the complex and values-laden nature of the content (ibid.). While Bruner (1986) developed scaffolding for use with school-aged children, the technique is equally applicable to learners of any age (De León, 2012). At GCELS, scaffolding is built into both the student and teacher learning experiences. GCELS's process in identifying the need for such scaffolds and building them into their learning model may offer useful insights for other learning institutions transitioning from a WSA to sustainability.

15.1.3 Methodology

Considering the need for examples of how to transition schools to more student-centered and WSAs to sustainability, this case study builds on personal practitioner insights from 8 years of working at GCELS as a teacher, curriculum designer, and trainer between 2012 and 2020. While GCELS offers a full educational program for students in Grades 9–12, this chapter analyzes courses from the first term of Grade 9 and the program elements specifically designed to ease the transition of students and teachers into the GCELS learning community. Examining and analyzing publicly-available online student portfolios and course-related documents from the two Grade 9 courses featured was the main data source.

15.2 The GCELS Model

15.2.1 Curriculum

Curriculum is perhaps the most approachable starting point for integrating sustainability into school programming. Focusing on curriculum draws immediate attention to *what* is being taught, with less attention given to *how* it is taught, or *why*. At GCELS, an early commitment to the three keywords in the school name—global, citizenship, and experience—brought the *why* and *how* under scrutiny as part of the curriculum design process. A commitment to *global* awareness led to alignment with the UN Sustainable Development Goals; dedication to active *citizenship* manifested in thematic, inquiry- and project-based curriculum; and a commitment to practical *experience* gave rise to real-world experiential learning. In this way, curriculum became fundamental to the development of the school model, as structures, policies, and training had to be designed to support such learning approaches. In the following section, I discuss GCELS's learning model, which combines inquiry- and project-based learning through investigations that cycle from guiding questions to action projects.

Table 15.1 Unit 1 guiding questions for the *SDGs and You* course (2021 iteration)

Course GQ	What are the biggest challenges facing humanity, and how do we address them?
Unit 1 GQ	How can everyone have access to resources that meet basic human needs?
Internal investigation GQ	How do you know when basic needs are met?
External investigation GQ	How accessible are resources to satisfy *your* basic human needs?
Action project GQ	When do basic human *needs* become basic human *rights*?

15.2.1.1 Real-World, Inquiry- and Project-Based Learning

What are the biggest challenges facing humanity, and how do we address them? Students begin their first term at GCELS by grappling with this guiding question. The prompt is the first in a series of questions that guide students' initial integrated Humanities course, titled *SDGs and You: Mission 2030*. The *SDGs and You* course introduces students to the 17 UN Sustainable Development Goals (SDGs) and connects them to the city of Chicago, thereby locally contextualizing the global goals. Each step of the learning unit is initiated with a guiding question (GQ) to stimulate curiosity and critical thinking. The sequence of questions moves from exploring broad global issues to thinking about local and individual reflections and actions (see Table 15.1). These questions are posed by the teacher as model guiding questions, introducing students to the inquiry-based learning environment. As students progress through the 4-year program, they become more adept at generating their own guiding questions and designing corresponding investigations.

The *SDGs and You* course is structured as three cyclical investigations (or Units), each unfolding in three, often overlapping, stages: an internal investigation (I), which introduces core vocabulary, formulas, concepts, and academic skills needed to pursue the guiding question; an external investigation (E), which connects classroom concepts to real-life scenarios through city-wide field experiences; and an action project (A), whereby students synthesize and demonstrate learning from the unit through a hands-on, multimedia project. The action project balances cognitive, psychomotor, and affective learning domains (see Sipos et al., 2008) and builds action competence. All GCELS courses include these "IEA" phases, though delineations between the stages are not fixed.

Prompted by the guiding questions, students taking the *SDGs and You* course learn world geography, write comparative case studies, draw and analyze resource maps of Chicago neighborhoods, and interpret global statistics through the lenses of human rights and equity. Through these hands-on activities, they build core academic skills, including reading, writing, and social studies competencies that are aligned with the United States Common Core State Standards (CCSS) and the College, Career, and Civic Life (C3) Framework for Social Studies State Standards.

In addition to aligning with CCSS and C3 standards, GCELS interprets the SDGs as a set of global "standards" that drive a pragmatic definition of global citizenship education (Leite, 2021). While the SDG connections are evident in *SDGs and You,* every other core GCELS course also aligns with at least one SDG and indicator. If aligning curriculum with academic standards steers students toward literacy recommended at the national level, aligning with the SDGs contextualizes learning in both local and global settings.

During the first term of the school year, the *SDGs and You* Humanities course is paired with the *Water* STEAM (Science, Technology, Engineering, Arts, Mathematics) course, which takes a deep dive into SDG 6. Pursuing the question, *How is our understanding of water reflected in our interactions with it?,* students learn algebra, functions, and Earth and life sciences. Using data from sources ranging from local municipalities to the UN, the course explores the paradox of water as a resource that is at the same time abundant and rare. STEAM skills are developed within a sociological context, using timely real-world examples that are updated each year during an annual process of course revision. Offering the *SDGs and You* and *Water* courses during the same term allows teachers to collaborate and plan lessons that encourage students to draw connections between social and ecological sustainability.

GCELS's approach to project-based learning (PBL) covers the seven "gold standard" principles identified by the Buck Institute for Education, a leader in PBL since the 1980s. These principles include the following: a challenging problem or question; sustained inquiry; authenticity; student voice and choice; reflection; critique and revision; and a public product (Larmer et al., 2015). GCELS's curriculum departs from common interpretations of PBL—which foreground the *student voice and choice* principle—by including a recommended, pre-written course outline that can be used directly by teachers. Having a pre-written curriculum provides scaffolding for teachers who are new to facilitating PBL and may need the extra structure as they transition to a more student-centered classroom (Barron et al., 1998; Larmer et al., 2015). Additional scaffolds will be discussed in the following section on pedagogy.

15.2.2 Pedagogy

Sustainability education is as much about pedagogy as it is about curriculum (Evans & Ferreira, 2020). Teaching through real-world, inquiry- and project-based learning calls for "unlearning" ingrained habits and assumptions that both students and teachers bring to the school setting (Lozano et al., 2022). Some of these habits include siloed disciplinary thinking, expected student-teacher roles, and favored test-based measures of achievement.

By the time students reach Grade 9, they have had almost a decade of experiences that influence what they expect to do, feel, and learn in school. Likewise, teachers who join GCELS directly from other schools or teacher-training institutes have little practice in how to foster and facilitate an inquiry- and project-based environment. Despite broad support for the use of active, student-centered pedagogies in sustainability learning, "a lack of clear guidelines for 'how to' is off-putting" for many teachers (Evans & Ferreira, 2020, p. 29). An awareness of these reservations has influenced the onboarding and professional development processes at GCELS and led to the *Model the Learning* course, which asks teachers to complete course action projects as an example for the students, thereby shifting classroom power dynamics and opening a dialog between students and teacher on process and assessment criteria.

15.2.2.1 Modeling the Learning

When introducing a new action project, GCELS teachers share their own sample and ask students to evaluate it using the project rubric. An example of one such teacher project is shared in Fig. 15.1, which was posted on a GCELS teacher's public blog. This project asks learners to design an infographic for use on Chicago public transportation that raises awareness about water usage (see Table 15.2 for the project rubric). The infographic includes calculations on personal water usage and comparisons to national and international statistics. Students demonstrate a range of competencies in producing the infographic, including: math calculations and graphing; visual communication and graphic design; and global and intrapersonal awareness. All action projects are posted publicly on the student's blog, along with a personal reflection about the process and lessons learned.

Having teachers complete the action project and present it to students positions teachers as

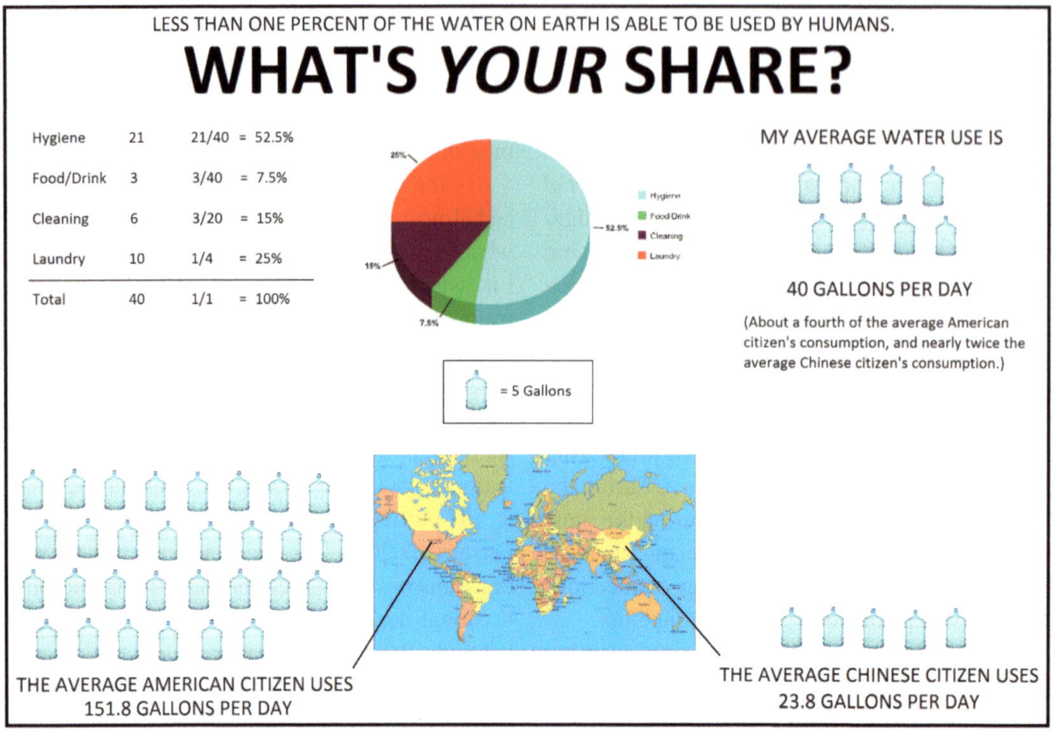

Fig. 15.1 Sample action project from the *Water* course by a STEAM teacher at GCELS (AMD, 2014)

students and students as teachers, promoting a power shift with various and far-reaching implications. First, it gives teachers credibility for having gone through the process students will go through. At the same time, teachers gain insights into the learning process that they would not otherwise have (for example, teachers may share what the experience was like, or how long it took them to complete each step). When teachers open their work up to critique from students, this builds transparency and understanding about the assessment criteria; students now understand the project from the perspective of teachers. Overall, this mutual shift helps transition the classroom to a more horizontal power dynamic; oftentimes, both teachers and students will offer improvements to a rubric based on their role shifting.

The added scaffolding in GCELS's curriculum is a response to several years of observing students and teachers struggle to adjust to the school's learning model. Aspiring to the "gold standard" of PBL requires significant time and energy from teachers as they not only design course content and structure but also undergo a process of letting go of the perceived needs to "control" their classroom and to be the sole content expert that disseminates information to students.

For these reasons, part of GCELS's onboarding process for new teachers is to complete a *Model the Learning* professional development course. *Model the Learning* is designed for teachers to experience the student perspective before teaching a new course. Teachers receive the written course curriculum and then actually take the course by reading embedded materials, completing formative assessments, conducting external investigations, and completing the unit action projects. While moving through the course, they write reflections on the flow of questions, swap out content, update case studies, plan field experiences, invite guest speakers, and revise the action project rubric and scaffolding based on their own learning experiences.

For teachers new to PBL, the *Model the Learning* process removes some of the unpredict-

Table 15.2 GCELS action project rubric from the *Water* course (2021 iteration)

Water Action Project #1 Guiding Question: How does water usage compare around the world?			
	Yes	Partially	No
Calculations Did you track at least 4 categories of water usage in your home? Did you correctly calculate the percentage of each type of water usage as part of the whole quantity? Did you present your data in an appropriate graph type of your choosing?			
Global Connections Did you compare your personal water consumption to the U.S. average, as well as another country of your choosing (in gallons, liters, and pounds)? Did you explain how you calculated the time it would take to carry your daily water usage from a nearby water source to your home or school (based on your field experience data)?			
Infographic Is your infographic neat and legible, with a creative title? Does your infographic include at least 2 relevant images (graphs, photos, etc.)? Do you communicate the main idea and motivation for this infographic—including a "call to action" for an interested reader?			
Blog Did you introduce the project in 300 words, summarizing and reflecting on everything from your Prep Worksheet? Did you emphasize your "call to action"—including ways readers can reduce their water consumption? Has your infographic and project been uploaded to your blog in accordance with the *blog posting rubric*?			

ability associated with action projects and affords them time to build their own reflexive practice. As teachers become more comfortable with their courses and the GCELS model in general, they rely less on the prescribed curriculum and become more confident in increasing student voice and choice. Having teachers complete *Model the Learning* requires a significant time commitment but making space for this experience early in their GCELS career alleviates some of the burnout that teachers experience as they adjust to being immersed in GCELS's learning environment. While only teachers undergo the *Model the Learning* process, everyone who works at GCELS participates in a professional development course called *Inquiry- and Project-Based Learning 101*. The course introduces the GCELS model and helps build a culture based on the understanding that learning takes place anywhere, with anyone.

15.2.3 Assessment

Assessments are notoriously challenging in learning environments designed around sustainability pedagogies. In contrast to *knowledge-based* learning, which focuses on what students know, ESD environments favor *competency-based* learning, which focuses on what students can do (Olema et al., 2021). While there is high agreement about what sustainability competencies entail (see Wiek & Redman, 2022), there is less consensus about how to assess these competencies in students (Cebrián et al., 2020; Wiek & Redman, 2022). Evaluating a student's action competence—which, as discussed previously, includes a combination of knowledge and skills, willingness to act, and confidence in one's ability to influence change (Sass et al., 2020)—poses a challenge for teachers.

For K-12 schools in the United States, which are tasked with preparing students for post-secondary education, assessment also involves the problem of how to communicate student performance in a way that colleges and universities understand. Schools that use PBL describe this challenge as having "a foot in both worlds," balancing the evaluation of student work with state and national test score reporting conventions (Boss, 2012, p. 48). This testing culture is embedded in all levels of the U.S. school system, and scholars argue that the "discourse of achievement and test-based accountability works against the goals" of sustainability education (Pizmony-Levy & Gan, 2021, p. 4).

The challenge of how to communicate student performance has influenced how grade transcripts are designed at GCELS. Transcript design in turn determines how students are evaluated at the classroom level. After several years of experimenting with alternative systems, GCELS currently uses a 100-point scale to evaluate action projects and communicate grades. Using the common language of grading scales and grade point averages (GPA) facilitates communication with colleges and universities, who already request additional explanations to interpret students' unfamiliar thematic course sequence. In the next section, I analyze GCELS's rubric design using an example from the *Water* course.

15.2.3.1 Rubric Design

Each summer before the new school year starts, GCELS teachers update their courses and participate in several weeks of professional development. In these annual cycles of curriculum revision, rubrics and assessments are the most scrutinized dimension of the GCELS learning model.

Rubrics that are used to evaluate action projects are based on a competency-based model. Competency-based assessments seek to capture a student's ability to, for example, transfer knowledge, apply skills to complex situations, solve problems, and make decisions in different contexts (Gallardo, 2020). Many competency-based rubrics are designed as charts, including an explanation of a desired competency, a rating scale of performance levels, and a description of achievement at each level. GCELS experimented with rubrics designed in this way, but they proved to be visually overwhelming for both teachers and students; the former did not find the rubrics any more helpful for grading, and the latter did not actively use them as a guide in developing their projects.

To make rubrics more student-friendly, GCELS's current approach draws from a single-point rubric design. Single-point rubrics are easier to read, leave more room for feedback, and facilitate student self-assessments (Fluckiger, 2010). At GCELS, rubrics are more than evaluative tools. From a sociocultural perspective, they are also pedagogical and dialogical, being used to introduce a project, ground the conversation around reasonable assessment criteria, and, ultimately, foster common understanding and accountability. GCELS structures rubrics as a series of questions intended to initiate dialog between teacher and student. A sample rubric from the first *Water* course action project is shown in Table 15.2.

Regarding action competence, the sample rubric in Table 15.2 asks students to demonstrate specific knowledge and skills, including mathematical calculations and graphing, comparisons, and data communication. Student reflections on the infographic action project, posted on their public blogs, provide anecdotal evidence of their willingness to act:

> I am enlightened by the fact that not everyone has the same privileges that I have, not everyone has the same education, the same quality of water (Student DAB, 2021).

> I have really enjoyed this unit regardless of how physically taxing it was to carry 7.5 gallons of water .3 miles, I'm thankful for the perspective it has offered me regarding water consumption (Student ER, 2021).

> This project made me realize just how important water is, and how not everyone is as lucky as I am to have easy access to it (Student 1126, 2022).

> I think that we can all do a better job at preserving water. Earth is a shared planet and we all as global citizens have to pitch in to take care of our planet (Student 1117, 2022).

Students submit the infographic action project approximately 4 weeks into their GCELS experience. At that early stage, it is difficult to measure any growth in confidence, but the 4-year program aims to build confidence in their ability to effect change and further strengthen their action competence.

In an effort to address some of the issues related to assessment, in 2019, GCELS joined the Mastery Transcript Consortium (MTC), a network of schools working to design, test, and mainstream a digital transcript that communicates student competencies without using grades or GPAs. Instead of a quantitative transcript that "reinforces outdated modes of education, constrains innovation, limits learning to single subjects, and impedes the pursuit of educational equity and excellence" (MTC, 2019), the Mastery Transcript offers an alternative. The MTC relieves some of the work of GCELS's school counselor, who spends hours each year explaining the GCELS model and translating the thematic courses to various college admissions officers. MTC works actively with K-12 schools, colleges, and universities to "promote sustainable change in higher education admissions" (ibid., 2021), but mainstreaming changes requires capacity building in both schools and higher ed. admissions officers, as well as a major cultural shift in how and why students are evaluated.

15.3 Conclusion

The artifacts examined in this chapter represent a snapshot in time of the GCELS learning model, and how it has sought to bridge some of the divisions that underpin mainstream schooling. With a focus on scaffolding, the chapter gives some practical examples of how schools taking a WSA to sustainability can face challenges posed by trying to imagine, test, and scale alternatives to the education system to which they are beholden. Approaching school redesign as a process of simultaneously rethinking curriculum, assessment, and pedagogy attempts to instill a more relational way of thinking, which is essential if we are to transcend and transform the social and ecological crises of today (Lange et al., 2021; Wals & Mathie, 2022).

By approaching education from an ecological perspective, connections between disciplines, people, and communities may be reestablished and strengthened. At GCELS, how to make these connections most effectively is an ongoing process of discovery. The school's learning model is at the same time constantly evolving and deeply rooted in commitments to *global, citizenship*, and *experience*. Ensuring there is transparency and self-reflection on this continuous process of change is one way in which schools can endeavor to prepare students for futures that are unpredictable, complex, and as of yet, unimagined.

15.4 Postscript

After this chapter was written, GCELS's board of directors abruptly made the decision to close the school. Citing low enrollment numbers, the board did not foresee a financial path for keeping the school open.

This chapter highlighted some of the challenges GCELS faced as it experimented with a whole-school approach to global citizenship and sustainability education. While these challenges proved insurmountable for the school, they served as provocations for innovative thought and practice. The fate of GCELS, thus, leaves several questions:

How to fund schools like GCELS? In the U.S. context, GCELS required, on the one hand, private independent status (to have greater freedom in curricular and programming design), but, on the other hand, a tuition-based model. How is this type of education made accessible to all students, then? In the absence of an endowment, schools must operate through a combination of tuition funds, partnerships with scholarship organizations, corporate backers, private donations, government grants, and continuous fundraising. These sources can be inconsistent, short-term, and may come with strings attached. In the case of GCELS, an over-reliance on a small base of private donors became untenable as the school moved past its first decade.

How to support schools to serve as laboratories for transformation? Vaughter (2016) advocates for schools to act as living laboratories that test and model learning for sustainability and climate education. Though there is no single model for Vaughter's ideal, "lab school" in North America generally means a partnership between a school and a university: University-backed research informs pedagogical and curricular decisions within the school and, in turn, the school provides a real-world setting to carry out academic research. Such partnerships ensure teachers have access to cutting-edge research and professional development. Without university backing, GCELS still called itself a laboratory school based on its dedication to innovation and iteration but needed support for longitudinal studies that could have documented the iterations of its learning model—including any failures and processes for addressing them.

How should schools like GCELS promote their learning model to prospective students? Despite the increasing intensity of climate change and related social issues, the primary measure for secondary school performance remains focused on academic measures. Real-world, project-based learning motivated by the SDGs is attractive for many families, but may not, alone, fulfill the desire for students to be well-prepared for university entrance exams and competitive actors in the global economy. How, then, to balance authentic, purpose-driven, student-led learning with test prep, ICT (information, communications, and technology) literacy, and other twenty-first-century learning competencies? In the case of GCELS, perhaps a different conjuncture and more institutional support were needed to serve a student body that was disillusioned by, yet beholden to, grade point average as a primary measure of achievement.

For schools working to influence systemic shifts in education, more investment is needed so they have the time and resources to imagine and test alternatives. GCELS's community believed in education as a powerful tool for transformative, systemic change toward sustainability. How much are we willing to invest in this belief?

Acknowledgments I would like to recognize the GCELS faculty, staff, students, and board members who worked tirelessly to design and redesign the GCELS model for learning since the school's inception. I would like to especially thank: the anonymous reviewers of this chapter for their constructive feedback; Carlos A. Pittella, Cabell King, Aaron Moring-D'Angier, Eleanor Nicholson, and Rosalie G. Mathie for their helpful comments on the development of this chapter; and Blane Harvey for his ongoing supervision and guidance. Finally, I am grateful to Vanier Canada Graduate Scholarships for their research support.

Bibliography

AMD. (2014, August 25). Public transit water usage advertisement. *Aarongce*. https://aarongce.wordpress.com/2014/08/25/public-transit-water-usage-advertisement/

Armstrong, C. M. (2011). Implementing education for sustainable development: The potential use of time-honored pedagogical practice from the progressive era of education. *Journal of Sustainability Education, 2*(1), 1–25.

Armstrong, C. (2015). In the zone: Vygotskian-inspired pedagogy for sustainability. *Journal of Classroom Interaction, 50*(2), 133–144.

Barron, B. J. S., Schwartz, D. L., Vye, N. J., Moore, A., Zech, L., & Bransford, J. D. (1998). Doing with understanding: Lessons from research on problem- and project-based learning. *The Journal of the Learning Sciences, 7*(3 & 4), 271–311.

Boss, S. (2012). The challenge of assessing project-based learning. *District Administration, 48*(9), 46–52.

Bruner, J. (1986). *Actual minds, possible worlds*. Harvard University Press.

Cebrián, G., Junyent, M., & Mulà, I. (2020). Competencies in education for sustainable development: Emerging teaching and research developments. *Sustainability, 12*(2), 579. https://doi.org/10.3390/su12020579

De León, L. (2012). Model of models: Preservice teachers in a Vygotskian scaffold. *The Educational Forum, 76*, 144–157.

Demssie, Y. N., Biemans, H. J. A., Wesselink, R., & Mulder, M. (2020). Combining indigenous knowledge and modern education to foster sustainability competencies: Towards a set of learning design principles. *Sustainability, 12*(6823), 1–20. https://doi.org/10.3390/su12176823

Evans, N. (S.), & Ferreira, J.-A. (2020). What does the research evidence base tell us about the use and impact of sustainability pedagogies in initial teacher education? *Environmental Education Research, 26*(1),

27–42. https://doi.org/10.1080/13504622.2019.1703908

Feinstein, N. W., & Carlton, G. (2013). Education for sustainability in the K-12 educational system of the United States. In *Schooling for sustainable development in Canada and the United States* (pp. 37–49). Springer. https://doi.org/10.1007/978-94-007-4273-4_3

Ferreira, J.-A., & Ryan, L. (2013). Working the system: A model for system-wide change in pre-service teacher education. *Australian Journal of Teacher Education, 37*(12), 29. https://doi.org/10.14221/ajte.2012v37n12.3

Fluckiger, J. (2010). Single point rubric: A tool for responsible student self-assessment. *Teacher Education Faculty Publications, 5*, 18–25.

Gallardo, K. (2020). Competency-based assessment and the use of performance-based evaluation rubrics in higher education: Challenges towards the next decade. *Problems of Education in the 21st Century, 78*(1), 61–79.

Hargreaves, L. G. (2008). The whole-school approach to education for sustainable development: From pilot projects to systemic change. *Policy & Practice: Education for Sustainable Development, 6*, 69–74.

Kwauk, C. (2020). *Roadblocks to quality education in a time of climate change* (p. 34) [Brief]. Brookings Institution. https://www.brookings.edu/wp-content/uploads/2020/02/Roadblocks-to-quality-education-in-a-time-of-climate-change-FINAL.pdf

Lange, E. A., Polanco O'Neil, J. K., & Ross, K. E. (2021). Educating during the great transformation. *Andragoška Spoznanja, 27*(1), 23–46. https://doi.org/10.4312/as/9692

Larmer, J., Mergendoller, J., & Suzie, B. (2015). *Setting the standard for project based learning: A proven approach to rigorous classroom instruction*. ACSD.

Leite, S. (2021). Using the SDGs for global citizenship education: Definitions, challenges, and opportunities. *Globalisation, Societies and Education, 20*(1), 1–13. https://doi.org/10.1080/14767724.2021.1882957

Leite, S., & Moring-D'Angier, A. (2022). USA: Chicago based global citizenship experience lab school. In R. G. Mathie & A. E. J. Wals (Eds.), *Whole school approaches to sustainability: Exemplary practices from around the world* (pp. 24–28). Education & Learning Sciences/Wageningen University. https://doi.org/10.18174/566782

Lozano, R., Bautista-Puig, N., & Barreiro-Gen, M. (2022). Developing a sustainability competences paradigm in higher education or a white elephant? *Sustainable Development*, 1–14. https://doi.org/10.1002/sd.2286

MECCE, & NAAEE. (2022). *Mapping the landscape of K–12 climate change education policy in the United States* (p. 54). Monitoring and Evaluating Climate Communication and Education Project and North American Association of Environmental Education. https://secureservercdn.net/45.40.155.193/38z.ea4.myftpupload.com/wp-content/uploads/2022/07/mapping_the_landscape.full_report.pdf

MTC. (2019, August 15). Mastery Transcript*: What we do*. Mastery Transcript Consortium. https://mastery.org/what-we-do/mastery-transcript/

Olema, D. K., Nabitula, A., Manyiraho, D., & Atibuni, D. Z. (2021). Analysis of the shift from knowledge based to competency based education among secondary school teachers in Uganda. *International Journal of Educational Research, 9*(1), 49–56.

Pizmony-Levy, O., & Gan, D. (2021). Introduction to special issue: Learning assessments for sustainability? Exploring the interaction between two global movements. *Education Policy Analysis Archives, 29*(August-December), 121–121. https://doi.org/10.14507/epaa.29.7171

Redman, A., & Wiek, A. (2021). Competencies for advancing transformations towards sustainability. *Frontiers in Education, 6*, 1–11. https://doi.org/10.3389/feduc.2021.785163

Sass, W., Boeve-de Pauw, J., Olsson, D., Gericke, N., De Maeyer, S., & Van Petegem, P. (2020). Redefining action competence: The case of sustainable development. *The Journal of Environmental Education, 51*, 292–305.

Sass, W., De Maeyer, S., Boeve-de Pauw, J., & Van Petegem, P. (2022). Honing action competence in sustainable development: What happens in classrooms matters. *Environment, Development and Sustainability, 1–22*. https://doi.org/10.1007/s10668-022-02195-9

Sipos, Y., Battisti, B., & Grimm, K. (2008). Achieving transformative sustainability learning: Engaging head, hands and heart. *International Journal of Sustainability in Higher Education, 9*(1), 68–86.

Sterling, S. (2004). Higher education, sustainability, and the role of systemic learning. In P. B. Corcoran & A. E. J. Wals (Eds.), *Higher education and the challenge of sustainability: Problematics, promise, and practice* (pp. 49–70). Kluwer Academic Publishers.

Sterling, S. (2013). The future fit framework: An introductory guide to teaching and learning for sustainability in HE (guide). *Journal of Education for Sustainable Development, 7*(1), 134–135. https://doi.org/10.1177/0973408213495614b

Sterling, S., Dawson, J., & Warwick, P. (2018). Transforming sustainability education at the creative edge of the mainstream: A case study of Schumacher College. *Journal of Transformative Education, 16*(4), 323–343. https://doi.org/10.1177/1541344618784375

Stevenson, R. B., Nicholls, J., & Whitehouse, H. (2017). What is climate change education? *Curriculum Perspectives, 37*(1), 67–71. https://doi.org/10.1007/s41297-017-0015-9

Student #1117. (2022). *Water, the resource we take for granted*. https://1117gcelabschool.blogspot.com/2022/10/water-resource-we-take-for-granted.html

Student #1126. (2022). *Save the* H_2O. https://1126gcelabschool.blogspot.com/2022/10/save-h2o.html

Student DAB. (2021). *Water isn't always clear*. https://dabgcelabschool.blogspot.com/2021/10/waterwe-have-it-easy.html

Student ER. (2021). *Drippy stats*. https://ergcelabschool.blogspot.com/2021/10/liquid-life-disparities.html

UNESCO. (2016). *Education for People and Planet* (Global Education Monitoring Report). https://unesdoc.unesco.org/ark:/48223/pf0000245752

UNESCO. (2018). *Issues and trends in education for sustainable development* (A. Leicht, J. Heiss, & W. J. Byun, Eds.). https://unesdoc.unesco.org/ark:/48223/pf0000261445

UNESCO. (2021). *Berlin declaration on education for sustainable development*. UNESCO. https://en.unesco.org/sites/default/files/esdfor2030-berlin-declaration-en.pdf

Vaughter, P. (2016). Climate change education: From critical thinking to critical action. *United Nations University Institute for the Advanced Study of Sustainability, 4*, 1–4.

Vygotsky, L. (1978). *Mind in society: The development of higher psychological processes*. Harvard University Press.

Wals, A. E. J., & Benavot, A. (2017). Can we meet the sustainability challenges? The role of education and lifelong learning. *European Journal of Education, 52*(4), 404–413. https://doi.org/10.1111/ejed.12250

Wals, A. E. J., & Mathie, R. G. (2022). Whole school responses to climate urgency and related sustainability challenges: A perspective from Northern Europe. In M. A. Peters & R. Heraud (Eds.), *Encyclopedia of educational innovation* (pp. 1–8). Springer Singapore. https://doi.org/10.1007/978-981-13-2262-4_263-1

Wiek, A., & Redman, A. (2022). What do key competencies in sustainability offer and how to use them. In P. Vare, N. Lausselet, & M. Rieckmann (Eds.), *Competences in education for sustainable development* (pp. 27–34). Springer International Publishing. https://doi.org/10.1007/978-3-030-91055-6_4

16 Students' Deliberation on 'Greening' the School's Energy Supply: A Case Discussed from a Whole School Approach Perspective

Øyvind Mathisen and Gerd Johansen

Key Message We propose that a culture of deliberative communication is one key aspect of strengthening a whole school approach (WSA) for sustainability. Inquiring the school energy technology regarding its sustainability may strengthen the students' energy awareness. Additionally, scrutinising the energy technology by dealing with concretes, such as technology, can support the students' oral deliberative practices. Through argumentative and respectful discussions, potentially leading to mutual consensus about the school's energy technology, can create school cultures that endorse WSA for sustainability.

What can a project like this contribute to making the whole school more sustainable? How can a technologically oriented 'green transition' provide students with opportunities to deliberate on technology?

16.1 Introduction

This chapter explores an in situ inquiry project in which students investigated and deliberated on how to make their school more environmentally friendly by using renewal energy sources in an effort to reduce energy consumption. We use this exploration as a starting point to discuss the project from a whole school approach (WSA) perspective.

Green transitions have become a political buzzword (Affolderbach & Schulz, 2018); however, what green transitions entail in an educational setting is less clear. We propose that students working with an inquiry project on local and renewable energy sources may exemplify such transitions and that the project can be a possibility for promoting the school's sustainability. Moreover, student participation and planning of school strategies through group work and discussions can strengthen students' perceptions of citizenship (Westheimer & Kahne, 2004). The utilisation of renewable energy technology can be used as a foundation for a potential green transition. Here, Levinson (2010) argues the importance of engaging students in deliberative discourses concerning technoscientific issues (Levinson, 2010). Hence, participation through a deliberative discourse regarding technology that can make the school more sustainable may enhance the goal of making students into active citizens. Moreover, Mogren (2019) claims that more research on the organisational characteristics of schools in which students engage with real-life environmental challenges is needed (Mogren, 2019).

Ø. Mathisen (✉) · G. Johansen
Faculty of Science and Technology, Norwegian University of Life Sciences, Ås, Norway
e-mail: oyvindm@afk.no; oyvind.mathisen@nmbu.no; gerd.johansen@nmbu.no

A. E. J. Wals et al. (eds.), *Whole School Approaches to Sustainability*, Sustainable Development Goals Series, https://doi.org/10.1007/978-3-031-56172-6_16

The students in this study engaged in an inquiry project on the school's renewable energy technology to allow for a deliberative discourse. The inquiry project was given as a commission by the county municipality. The local janitor service provided vital information that the students used to solve the commission, and the students handed in their proposals to the local leadership. We suggest that inviting several stakeholders can make students' inquiries more relevant and promote WSA because this project can create a multistakeholder discourse about the school's energy system. Discourses, such as deliberative communication (Englund, 2000, 2006, 2015, 2016; Englund et al., 2008), can strengthen oral practices among students, which is a central aspect of the school culture (Gan & Alkaher, 2021; Mathar, 2015; Mathie & Wals, 2022). School culture is one of several important features within WSA (Hunt & King, 2015; Mathar, 2015; Mathie & Wals, 2022). Hence, this inquiry project will be used as a point of departure to argue that certain characteristics within school culture, such as oral practices, can promote and support a potential WSA for sustainability.

16.2 Theoretical Perspectives

16.2.1 WSA That Supports Sustainability and Strengthens the Students' Citizenship

There are several ways of conceptualising a WSA. For instance, Mathie and Wals (2022) state that one can view a WSA more holistically from an education of sustainable development perspective, in which a WSA is 'meant to be used as a thinking tool to initiate and guide an ongoing multistakeholder dialogue about how sustainability can best unfold in a school setting' (Mathie & Wals, 2022, p. 6). From another perspective, a WSA can assess a single school regarding the current practices, difficulties, cooperation, support of interdisciplinary activities and assistance in creating new practices and actions on the route to enhancing more sustainable activities in the school (Mathar, 2015, p. 24). Furthermore, according to Hunt and King (2015), a WSA to citizenship education can be built on factors such as policy improvements, school cultures, teaching and learning and allowing students to voice their opinions (Hunt & King, 2015). Although these conceptualisations all bring valuable perspectives on a WSA, in this chapter, we predominantly focus on creating new practices (Mathar, 2013, 2015) and school culture (Hunt & King, 2015) to attain the aim of a WSA that supports schools for sustainability.

School culture consists of features such as rules, rituals, beliefs, values, traditions and norms (Deal & Peterson, 1990). These features influence how people perform and communicate. Research indicates that the successful implementation of education for sustainability (EFS) involves altering the school culture into a more sustainable everyday practice (Gan & Alkaher, 2021; Henderson & Tilbury, 2004). Thus, dialogues can play a central role in a school culture or ethos (Mathie & Wals, 2022); the culture can be developed by having the students participate in democratic discussions concerning sustainability issues, for instance, by examining the energy consumption at their school (Mathar, 2013, 2015). It is important to give students the opportunity to state their opinions and be listened to in accordance with what the students want to change in terms of the school's sustainability, though 'it will take time to cultivate another mindset in relation to education and learning' (Mathie & Wals, 2022, p. 57). Enabling these democratic discussions on sustainability issues and green transitions may strengthen the students' oral practices, which can be viewed as an important characteristic of school culture.

Through a WSA, EFS can change the school culture and practices, and this can, in turn, promote various student practices, in addition to stimulating student participation in decision-making (Gan & Alkaher, 2021). Therefore, we have chosen to see democratic discussions among the students as a practice and base of the cultivation of the students' mindset towards sustainability issues. Furthermore, our assumption is that such cultivation can take place through deliberative communication while working with renewable energy technology. In addition, the scrutinisation of a school's energy system corresponds well to a WSA that supports schools for sustainability (Mathar, 2013, 2015).

16.2.2 Participatory Citizenship and Deliberative Communication

Westheimer and Kahne (2004) state that *a participatory citizen* focuses on the participation and planning, for example, of school strategies and that this will require collective endeavours or efforts such as, for instance, how to participate during group discussions (Westheimer & Kahne, 2004). We suggest that this endeavour can be deliberative communication. The concept of deliberation is tightly connected to communication in a democracy (Englund, 2000, 2006; Englund et al., 2008). Deliberative democracy—or participatory democracy—'especially emphasizes responsibility and consequences, implying that socialisation into citizenship and the exercise of citizenship must be in focus' (Englund, 2006, p. 511). Ideally, the aim of deliberative communication is to reach a joint consensus or at least agree to disagree.

There are five distinct elements of deliberative communication in a school setting (Englund, 2006, 2015). Here, the participants:

(a) Are given space and time to present and challenge different views via argumentation.
(b) Listen to, tolerate and respect each other's argumentation.
(c) Reach a shared will formation, such as a joint consensus, or are aware of different views or at least agree to disagree.
(d) Do not accept everything as the 'truth', even if uttered by a teacher/authority and there are openings of challenging one's own beliefs and practices.
(e) Essentially fulfil elements a–d without teacher control. For instance, having students discuss challenges to solve or illuminate them from different perspectives.

These elements (a–c) make up the core parts of deliberative communication (Englund, 2006; Englund et al., 2008). Element (d) is about schools being an integrated part of the public sphere and, therefore, making up an arena where different views and values learned through socialisation will potentially be challenged by fellow students and teachers (Englund et al., 2008). Furthermore, deliberative communication may assist students in both shedding light and challenging those views that are nondemocratic/offensive or authoritarian because schools are a part of the public sphere. Element (e) is relevant because of group projects or plenary discussions, and element (e) is a communicative process, where meaning is established among equals without the presence of a teacher (Englund, 2006; Englund et al., 2008). Achieving these elements (a–e) may involve handling a specific theme to assess, and practising deliberation without teacher control may motivate the students 'to reconstruct and continue the theme initiated in institutionalised deliberative communication' (Englund et al., 2008, p. 43). Therefore, a school culture containing elements of deliberative communication related to themes such as renewable energy consumption can promote a WSA to a sustainable school.

16.2.3 The Use of Technology to Promote Deliberative Communication

A citizen may feel the responsibility of keeping up on technological development. Levinson (2010) argues that this will most likely occur through the process of deliberative discussions (Levinson, 2010, p. 82). According to Mitcham's *Thinking through Technology* (1994), one can conceptualise technology in four dimensions:

(a) *Objects* in technology can be tools, machines, other types of physical artefacts or technological processes (Custer, 1995). Technological objects have a 'social side', they have a purpose (Mitcham, 1994), and they are to be used for something by someone. Moreover, this applicability is also an important aspect of the design or redesign of technological objects (de Vries, 2016).
(b) *Knowledge* can be defined as a justified legitimate belief (Mitcham, 1994). Legitimate beliefs regarding the craftmanship and utilisation of an artefact can be validated by adjuring to theories, laws, rules, maxims or skills 'thus yielding different kinds of technology as knowledge' (Mitcham, 1994, p. 194). Declarative (factual), procedural ('know how'), conceptual ('know that') and metacognitive knowledge are different knowledge types that are codependent on each other and can develop symbiotically when handling technology (Barak, 2013).
(c) Activities in technology can include maintaining, operating, working, manufacturing, designing, inventing and crafting (Mitcham, 1994, p. 210). These *activities* rely on procedural and conceptual knowledge in the processes of creating technological objects, using and judging and assessing artefacts, hence taking place while students solve tasks and disclose their findings with fellow students (Pirttimaa et al., 2017).
(d) *Volition* is connected to choices, intentions, ambitions, motives and will (Mitcham, 1994, p. 247). Therefore, volition demands reflections and considerations regarding the design and use of technology. Moral and ethical considerations towards technology and society are at the centre of attention and can be linked to the perception of what is deemed 'good' (Ravanbakhsh & Taqavi, 2020).

These four dimensions are all present when dealing with technology. However, seen in light of sustainability, we draw attention to volition, especially those considerations that are connected to designing, using and assessing technology. Sandin (2013) states that the relationship between environment and technology is complicated. He further argues that there can be positions of admiration and deep scepticism towards the same technological object (Sandin, 2013). Stables (2015) takes the critique of technology a step further by claiming that environmental challenges are caused by technology—or rather that there were some side effects of the technology not deliberated upon (Stables, 2015).

We have used Englund's five elements (a–e) in combination with these four dimensions of technology (A–D) to investigate how they might interact.

16.3 Research Methodology

16.3.1 Research Design, Data Collection and Analysis

The findings reported here stem from a design-based research (DBR) project. One of the main objectives and advantages of DBR is to initiate changes and improvements in existing educational practices (Wang & Hannafin, 2005, p. 6). This particular inquiry project was initiated by the first author in close collaboration with teachers; it was part of a physics course in which students aged 17 took part (ISCED 3). Juuti and Lavonen (2006) state that one of the characteristics that make up DBR 'is to develop an artefact to help teachers and pupils to act … more intelligible' (Juuti & Lavonen, 2006, p. 59). In our case, these artefacts are mainly booklets concerning renewable energy technology. The booklets structured the students' inquiry of renewable energy technology. Our main data sources were video observations of two students collaborating over four lessons, in addition to a plenary discussion in which 12 students participated. Student dialogues from all lessons and the plenary discussions were fully transcribed.

The research design used a naturalistic approach to deliberative communication. This means that the students were not informed about the concept of deliberative communication or what it entailed before or during the inquiry project. As an analytical perspective, deliberative communication was incorporated after the empirical part of the research project has been conducted. Our analytical framework is based on Englund's view of deliberative communication (elements a–e) and Mitcham's dimensions of technology (a–d). The unit of analysis in a dialogue is student utterances. The data analysis is based on content analysis (Krippendorff, 2018) and involves both interpretations and measurements of the frequency of dialogue excerpts. We used the NVivo programme deductively, coding utterances by using the framework. The same excerpt could be attributed to technology dimensions and deliberative elements.

16.3.2 The Inquiry Project and Its Stakeholders

The task that was given to the students to delineate the inquiry project had many elements built into it. During the inquiries, the students were meant to take account of the economic, environmental, social and political aspects related to the design of a theoretical upscaled renewable energy system installed at their school.

The student inquiries in this study were based on inquiry teaching and learning methods/strategies (Bybee et al., 2006). Conducting an inquiry project requires some amount of scaffolding (Hmelo-Silver et al., 2007). The main scaffolding that helped the students with their inquiry was laboratory manuals, booklets with tasks and worksheets to guide the students in designing a renewable energy technology system. The scaffolding of the inquiry was to prevent students from stagnation during laboratory work and support their communication.

The municipality provided the students with the following mission:

> What kind of opportunities do renewable solar energy sources offer our local school regarding energy efficiency and sustainability, and what are the positive/negative aspects of these alternative energy sources? What can we/the students do, at our local Upper secondary school to promote and enhance energy efficiency in school building using solar technolody?

In return, the municipality received the students' solutions regarding the potential use of renewable energy technology within its building portfolio. The local janitor service was considerably involved in designing the booklets that the students worked on in each lesson. The booklets contained detailed information about the school's former and current energy consumption present through graphs, tables and diagrams. The janitor service also provided important information about the school's heat pump, which could not be neglected in an energy account. The cooperation with the janitor service made the student inquiry closer to reality and made it more likely that the results of the inquiry would be relevant for the school and municipality to implement.

The inquiry project consisted of four 4-hour lessons in which the students made inquiries concerning the technology to prepare for a well-informed plenary discussion.

- Lesson 1: (1a) A presentation of the mission from the municipality. Preliminary inquiry regarding renewable energy technology and current energy supply at the school. (1b) Experimentation and collection of data from a solar collector with the aim of calculating its efficiency to be used in a theoretical upscaling (see lesson 3).
- Lesson 2: Further scrutiny of renewable energy technology and the carrying out of an experiment concerning solar cells. Experimentation and collection of data from a solar cell with the aim of calculating its efficiency to be used in a theoretical upscaling (see lesson 3).
- Lesson 3: Student groups worked on various energy calculations. These calculations concerned the energy outputs from lessons 1 and 2, and the students investigated the school's energy consumption, including the school's heat pump. Based on this information, the students designed a theoretical large-scale energy system for the school.
- Lesson 4: Based on the inquiries the students had carried out during lessons 1–3, the groups argued for a solution to the mission. The teacher and researcher staged a plenary discussion that challenged the groups to make

arguments about the different technoscientific issues they had been working on. Three groups presented their suggestions in class; the local leadership was present.

16.4 Research Results

16.4.1 Occurrence of Deliberative Communication and Dimensions of Technology in the Inquiry Project

The NVivo results show that the student dialogues contained elements of deliberative communication (a–e) in all four lessons and in the plenary discussion. The four dimensions of technology (knowledge, object, activity and volition) were also represented in each lesson. Lesson (1a) was dominated by knowledge, lessons (1b) and 2 were dominated by activities, lessons 3 and 4 were dominated by objects, and the plenary discussion was dominated by both objects and (talk of) activities.

The results indicate that working with renewable energy technology can, to varying degrees, mediate the elements of deliberative communication both in group settings and plenary discussions. The plenary discussion contained the highest level of deliberative communication throughout the project, here as dominated by elements a (participants present argumentation) and c (participants reach a joint consensus or are aware of different views). Elements a and c occurred most frequently when the students deliberated on issues concerning the objects and actions related to technology.

16.4.2 Excerpts of Student Discussions That Contain Deliberative Features While Working on Technology

Example 1 (the Use of Energy)
In lesson 1a, the inquiry triggered the students to reflect on energy consumption within the school.

1. G1: Those loudspeakers are on, [looks in the direction of the loudspeakers] That uses electricity!
2. J1: But, it's not up to … thus we need the loudspeakers in order … [*interrupted*].
3. G1: Need a loudspeaker, but it does not have to be switched on all the time!
4. J1: No, but it is not up to the school to … thus, more up to the teacher,
5. G1: Indeed, we can have … [*G1 stops talking*].
6. J1: Yes, sustainable courses! [*laughing*].
7. G1: Yes, then we can train all the employees at the school, [*smiles*].
8. J1: Yes,
9. G1: The light is on over there, [*the science preparation room*] And no one is there!
10. J1: No, [*looks at G1 and nods affirmative*].
11. G1: Unnecessary. Unnecessary use of energy,
12. J1: Yes,

This excerpt can be interpreted as the students deliberating on issues regarding the energy consumption at their school in terms of objects (light sources), action (sustainability courses and turning off the light) and volition (unnecessary use of energy). In terms of deliberative communication, the students presented different scenarios (element a in lines 3 and 9); element b was present because the students listened to each other and followed up on each other's suggestions for improvements and measurements (lines 4, 6, 7 and 11), and the students agreed (element c in lines 8, 10 and 12), though quite hasty and without substantial counterarguments. Therefore, we argue that this excerpt shows that the two students agreed that the school should be more aware of its energy consumption related to the school's energy technology.

Example 2 (Renewable Energy Efficiency)
In lesson 4, during the plenary discussion, the students discussed different factors that could influence their choice to use renewable technology.

1. Teacher: If you want to do it [continue using solar cells after down payment] forever, then it would pay off! [*Teacher withdraws*].

2. E4: Yes, but it depends on how long the solar cells, like, really function, if they ...?
3. E3: ... about how long they [solar cells] will last.
4. E4: If there will be much better solar technology that has much better efficiency? That is also an issue to explore!
5. E2: Yes, most likely that will be the case!
6. E3/E4: Yes. [*Speak simultaneously*].
7. E3: But we do not know that!
8. E2: Anything else is idiotic.
9. E3: We do not know that.

In this excerpt, the students were found to engage in a discussion on economic issues concerning solar cells, their cost, efficiency and life expectancy. The discussion was initiated by the teacher, who withdrew and left the students to deliberate among themselves. The discussion can be interpreted as two different views of technology. Student E2 was far more optimistic about the technological development of solar cells, whereas student E3 seemed to be more negative—or perhaps more sceptical. In other words, their approach to technology as objects and knowledge differed in this case, and they, quite amicably, agreed to disagree.

Example 3 (Heat Pump)
In lesson 4, during the plenary discussion, the teacher presented the students with the dilemma of rejecting the already installed heat pump and only focusing on the use of solar power.

1. E1: I think that we can first consider the issue with sun ... or the heat pump's efficiency factor. Eh, because the heat pump has a very high efficiency factor. An efficiency factor of 2.9. That means that the energy we get from the heat pump is 2.9 times more than the energy amount we need to run the heat pump. So we are taking about a great energy output!
2. Teacher: In other words, not reject the heat pump?
3. E1: Yes.
4. Researcher: Do anyone here disagree with research group 1?
5. E3: Eh yes. No!
6. Researcher: But we might say that the heat pump uses electricity from the grid?
7. E3/E4: Yes. [*simultaneously*].
8. E4: We have thought about that. We thought that it would be unreasonable to reject a facility [heat pump system with several wells] that have already been constructed. A huge facility that the school has invested a lot of money. Eh, rather one ought to replace the electric energy that the heat pump requires to function with, for instance, solar energy.
9. E3: Yes.
10. Researcher: So, you would also not reject the heat pump?
11. E3/E4: No. [*Simultaneously*].

This excerpt can be interpreted as the students deliberating on issues regarding the heat pump at their school in terms of objects (heat pump) and volition (keeping the heat pump or rejecting it). The teacher and researcher scaffolded the plenary discussion to maintain deliberation among the students. The students presented different arguments (element a, lines 1 and 8) for keeping the heat pump, and there was a consensus for keeping it (element c, lines 3, 5, 9 and 11). Hence, they agreed on the role of the heat pump as an important part of the school's energy system.

16.5 Discussion and Conclusion

We have used Englund's definition of deliberative communication, here combined with Mitcham's understanding of how to view technology as our analytical framework, to identify elements of deliberative communication during the discussions of technology. Our research results indicate that there were a vast number of technology-related discussions not containing elements of deliberative communication. However, our empirical evidence suggests that the students engaged in deliberative communication while working with technology in group settings (lessons 1–4) but to a limited degree. The plenary discussion generated the highest degree of deliberative communication while discussing technology-related issues. However, in both the group settings and plenary discussion, technology as volition was almost totally absent.

The technological dimension of volition (the ethical and moral dilemmas) was not made explicit in the inquiry design. Research has indicated that participation does not automatically enhance students' analytical skills to criticise the causal effects of social challenges (Westheimer & Kahne, 2004, p. 264), such as technoscientific issues. For instance, one aspect related to social challenges is the question of who should have access to 'cheap' renewable energy technology, especially during an energy deficit. We believe that, if this challenge had also been implemented in our inquiry project, the dimension of volition during the entire inquiry project would have been strengthened; see also Stables (2015).

If a school chooses to emphasise deliberative communication, then this may influence the school culture in terms of the general practice of communication among students both in public and private domains (Englund, 2016). Tasks that stimulate deliberative communication can impact how students communicate in general, as well as their communication in class. In this research project, the students were not made aware of how to communicate in a deliberate way; even so, there were several instances where such communication occurred. If citizenship is seen as a vital part of WSA, then we would perhaps recommend a more explicit approach to deliberate communication. For instance, norms such as respect and tolerance are important both within deliberative communication and in a WSA culture (Englund, 2006; Raihani., 2011). Thus, deliberate communication can act as a way of concretising the norms of respect and tolerance, which also holds true when the topic is sustainable energy technologies. In the excerpts shown here, the students agree and disagree when they argue their case. Hence, when students are presented with the opportunity to investigate and deliberate in an inquiry project, it may strengthen their role as citizens.

Deliberative communication initiated by inquiry projects that focus on 'green transitions' may support a WSA by fostering schools for sustainability. This can be related to the scrutiny of the schools' energy system, which enables the students to critically investigate sustainable resources and renewable technologies (Mathar, 2015). For instance, the students' energy awareness concerning the school's ability to reduce its energy consumption can be activated and stimulated while deliberating on technology from a sustainable perspective (Example 1). Furthermore, student deliberation on maintaining existing renewable energy technology because of its energy efficiency and, therefore, its potential sustainability can also strengthen the students' energy awareness (example 3). In this project, this could have led to scrutiny of the school's energy consumption. Sustainable energy consumption could be confronted with practical obstacles and/or organisational constraints because of the integration of a student design (Mathar, 2015).

We suggest that the involvement of several stakeholders in the development of an inquiry project can strengthen the project because of the complexity of a school's energy system. The inquiry project can be strengthened by providing the objective (a mission) and essential information that acts as a starting point for the students' inquiry and deliberations. This promotes a student-made design of the school's energy technology that is 'closer' to reality. Also, the participation of several stakeholders can create the necessary dialogue between decision-making authorities, which can increase the probability of implementing changes at an organisational level (Henderson & Tilbury, 2004; Mathie & Wals, 2022). Additionally, an inquiry project with a multistakeholder dialogue can bring students closer to having decision-making powers (Gan & Alkaher, 2021). Students can investigate sustainable technology and discuss their findings with stakeholders inside and outside school to make the school more sustainable (Mathar, 2015), thus contributing to a green transition. According to Jasmi et al. (2019), green technologies may be utilised in issues concerning sustainability because they entail the use of an object to protect and conserve the ecosystem (Jasmi et al., 2019). Renewable energy technology may be regarded as a potential green technology that can possibly enable this 'green transition' at an organisational level. Nevertheless, renewable energy technology does not necessarily have to be a sustainable technology (Owusu & Asumadu-Sarkodie, 2016; Stables, 2015). An inquiry project related to energy technology can reveal the tensions between renewability and sustainability. These tensions may, for instance, amount to techno positive/negative opinions towards technology (Sandin, 2013). This can be confronted and challenged through deliberative communication (example 2), and we believe this can foster students' energy awareness when handling a potential 'green' transition.

The exploration and discussion presented in this chapter have led to the conjecture that deliberative communication can strengthen the students' citizenship. Participation is a fundamental democratic value because it gives everyone a voice, and in this case, deliberative communication is one way of cultivating the students' mindset and 'giving' them a voice. Even if there are indications in the presented material that the students use their voices to form opinions on the schools' energy system and green transition, our findings have suggested that further investigation is needed. There is a need to investigate how each dimension of technology influences the elements of deliberative communication and how this, in turn, influences the school culture, which is an important aspect within WSA. The application of deliberative communication combined with the conceptualisation of technology has turned out to be a fruitful approach and a great starting point for discussing the WSA supporting schools for sustainability.

References

Affolderbach, J., & Schulz, C. (2018). *Green building transitions*. Springer.

Barak, M. (2013). Teaching engineering and technology: Cognitive, knowledge and problem-solving taxonomies. *Journal of Engineering, Design and Technology, 11*(3), 316–333.

Bybee, R. W., Taylor, J. A., Gardner, A., Van Scotter, P., Powell, J. C., Westbrook, A., and Landes, N. (2006). The BSCS 5E instructional model: Origins and effectiveness. Retrieved from https://fremonths.org/ourpages/auto/2008/5/11/1210522036057/bscs5efullreport2006.pdf

Custer, R. L. (1995). Examining the dimensions of technology. *International Journal of Technology and Design Education, 5*(3), 219–244.

de Vries, M. J. (2016). *Teaching about technology: An introduction to the philosophy of Technology for non-philosophers* (2nd ed.). Springer International Publishing.

Deal, T. E., & Peterson, K. D. (1990). *The principal's role in shaping school culture: US Department of education*. Office of Educational Research and Improvement.

Englund, T. (2000). Rethinking democracy and education: Towards an education of deliberative citizens. *Journal of Curriculum Studies, 32*(2), 305–313.

Englund, T. (2006). Deliberative communication: A pragmatist proposal. *Journal of Curriculum Studies, 38*(5), 503–520.

Englund, T. (2015). Toward a deliberative curriculum? *Nordic Journal of Studies in Educational Policy, 2015*(1), 48–56.

Englund, T. (2016). On moral education through deliberative communication. *Journal of Curriculum Studies, 48*(1), 58–76.

Englund, T., Öhman, J., & Östman, L. (2008). Deliberative communication concerning sustainability and security: A Habermas-inspired approach. In S. Gough & A. Stables (Eds.), *Sustainability and security within liberal societies. Learning to live with the future* (pp. 29–48). Routledge.

Gan, D., & Alkaher, I. (2021). School staff perceptions on education for sustainability and sense of community as reflected in an elementary school culture in Israel. *Environmental Education Research, 27*(6), 821–847.

Henderson, K., & Tilbury, D. (2004). *Whole-school approaches to sustainability: An international review of sustainable school programs*. Australian Research Institute in Education for Sustainability: Australian Government.

Hmelo-Silver, C. E., Duncan, R. G., & Chinn, C. A. (2007). Scaffolding and achievement in problem-based and inquiry learning: A response to Kirschner, Sweller, and Clark (2006). *Educational Psychologist, 42*(2), 99–107. https://doi.org/10.1080/00461520701263368

Hunt, F., & King, R. P. (2015). *Supporting whole school approaches to global learning: Focusing learning and mapping impact*. DERC Research Paper no. 13 for the GLP. Retrieved from https://discovery.ucl.ac.uk/id/eprint/1473867/

Jasmi, N., Kamis, A., & Farahin, N. (2019). Importance of green technology, Education for Sustainable Development (ESD) and environmental education for students and society. *Journal of Engineering Research and Application, 9*(2), 56–59.

Juuti, K., & Lavonen, J. (2006). Design-based research in science education: One step towards methodology. *Nordic Studies in Science Education, 2*(2), 54–68.

Krippendorff, K. (2018). *Content analysis: An introduction to its methodology* (4th ed.). Sage publications.

Levinson, R. (2010). Science education and democratic participation: An uneasy congruence? *Studies in Science Education, 46*(1), 69–119.

Mathar, R. (2013). *The concept of whole school approach—a platform for school development with focus on sustainable development*. Resource Toolkit for Teacher Training. Retrieved from https://esd-expert.net/files/ESD-Expert/pdf/Concept-Paper-Mathar.pdf

Mathar, R. (2015). A whole school approach to sustainable development: Elements of education for sustainable development and students' competencies for sustainable development. In R. Jucker & R. Mathar (Eds.), *Schooling for sustainable development in Europe* (pp. 15–30). Springer.

Mathie, R., & Wals, A. (2022). *Whole school approaches to sustainability: Exemplary practices from around the world* (9464471514). Retrieved from https://library.wur.nl/WebQuery/wurpubs/fulltext/566782.

Mitcham, C. (1994). *Thinking through technology: The path between engineering and philosophy*. University of Chicago Press.

Mogren, A. (2019). *Guiding principles of transformative education for sustainable development in local school organisations*. (Doctorial dissertation). Karlstad Universitet, Retrieved from https://www.diva-portal.org/smash/record.jsf?dswid=-2823&pid=diva2%3A1368940

Owusu, P. A., & Asumadu-Sarkodie, S. (2016). A review of renewable energy sources, sustainability issues and climate change mitigation. *Cogent Engineering, 3*(1), 1–14. https://doi.org/10.1080/23311916.2016.1167990

Pirttimaa, M., Husu, J., & Metsärinne, M. (2017). Uncovering procedural knowledge in craft, design, and technology education: A case of hands-on activities in electronics. *International Journal of Technology and Design Education, 27*(2), 215–231.

Raihani. (2011). A whole-school approach: A proposal for education for tolerance in Indonesia. *Theory and Research in Education, 9*(1), 23–39. https://doi.org/10.1177/1477878510394806

Ravanbakhsh, R., & Taqavi, M. (2020). Muslim scholars and technological volition. *Technology in Society, 61*, 101263. https://doi.org/10.1016/j.techsoc.2020.101263

Sandin, P. (2013). Naturalness, artefacts, and value. In M. J. de Vries, S. O. Hansson, & A. W. M. Meijers (Eds.), *Norms in technology* (Vol. 9, pp. 207–221). Dordrecht Springer.

Stables, K. (2015). Environment, ethics and cultures: Design and technology education's contribution to sustainable global futures. In K. Stables & S. Keirl (Eds.), *Environment, ethics and cultures* (Vol. 5). Springer.

Wang, F., & Hannafin, M. J. (2005). Design-based research and technology-enhanced learning environments. *Educational Technology Research and Development, 53*(4), 5–23.

Westheimer, J., & Kahne, J. (2004). What kind of citizen? The politics of educating for democracy. *American Educational Research Journal, 41*(2), 237–269.

17 A Whole School Approach Towards Sustainability: Assessing the Implementation of a School Programme "*Recreos Residuos Cero*" (Zero Waste Break Time)

Anne-Marie Ballegeer, Álvaro Lozano Murciego, Enzo Rainiero Ferrari Lagos, Miriam Leirós, Carmen Gloder, and Camilo Ruiz

Key Message

Due to its holistic nature, assessing the implementation of Whole School Approaches is challenging. In this study, we combine quantitative and qualitative methods to evaluate the effectiveness of the implementation of a WSA approach towards waste management and sustainability. Our results highlight that the introduction of data-driven assessment tools to assess the effectiveness of WSA is a valuable contribution to the field.

A.-M. Ballegeer (✉)
Department of Mathematics and Science Education, Salamanca University, Salamanca, Spain
e-mail: amballegeer@usal.es

Á. L. Murciego
Department of Computer Science and Automatics, University of Salamanca, Salamanca, Spain

E. R. F. Lagos · C. Ruiz
Department of Mathematics and Science Education, Salamanca University, Salamanca, Spain

Department of Education, University of Bath, Bath, UK

M. Leirós
Teachers for future Spain, Vigo, Spain

C. Gloder
Teachers for future Spain, Madrid, Spain

17.1 Introduction

How we live, work, and act has a significant impact on the health of our planet (UNEP, 2021). Increased human activity and exploitation of resources are pushing our planet towards, and beyond, its boundaries (Galaz et al., 2012; Rockström et al., 2009). Pollution, land degradation, and climate change are only some of the issues that are posing a significant threat to our present and future well-being (UNEP, 2021). To ensure that human activity does not exceed the aforementioned boundaries, we need to live in a sustainable manner. There is an increasing number of references in the literature that underscore the urgent need to integrate sustainability in our education and training systems if we want to be successful in the transition towards a fairer and greener society (Molderez & Ceulemans, 2018; Wals, 2015; Wals & Benavot, 2017). In Spain, sustainability is an interdisciplinary topic that has become more visible during the latest national curriculum renewal (Real Decreto 157/2022, 2022; Real Decreto 217/2022, 2022) and the proposal of the Green Competence framework by the European Union (Bianchi et al., 2022). The Whole School Approach (WSA) is a holistic approach that integrates sustainability in all aspects of the learning environment (Buckler &

A. E. J. Wals et al. (eds.), *Whole School Approaches to Sustainability*, Sustainable Development Goals Series, https://doi.org/10.1007/978-3-031-56172-6_17

Creech, 2014; Hargreaves, 2018; Henderson & Tilbury, 2004; Rieckmann, 2017). The WSA approach is growing in popularity (Mathie & Wals, 2022 and references therein), and has made its way into policy discourse around education for sustainability (European Commission, 2022; UNECE, 2021; UNESCO, 2021). Despite the growing popularity, previous research alerts that the number of schools that can systematically follow a WSA is still scarce (Hargreaves, 2018; Holst, 2022; Wals, 2012). The holistic nature of the WSA makes its implementation and assessment more complex as education is integrated into the students' everyday activities, not only at school but also in their community (Gericke, 2022). Initial teacher education, lack of resources and expertise, and working as a team across the school have been identified as the key challenges in WSA implementation (Tilbury & Galvin., 2022). In this study, we use different instruments to evaluate the impact of a WSA approach towards waste management and sustainability in Spanish schools. A WSA approach focused on waste management is not new; the most relevant example is the Waste Wise School program implemented in Australia and reported on by Cutter-Mackenzie (2007, 2010) and Armstrong et al. (2004). In the current study, we have used a program of the collective "Teachers for Future" Spain that focuses on reducing waste during break time ("Recreos Residuos Cero" RR0 program). We discuss the alignment of the program with the theoretical framework of the WSA, examples of how different schools implement the program, and the benefits and difficulties of quantifying the results of the RR0 program.

17.2 Teachers for Future and the RR0 Program

In the year 2019, in different parts of the world, a growing concern on the current climate crisis gave rise to several social movements, such as "Fridays for Future," initiated by Greta Thunberg. Simultaneously, a network of teachers concerned with the climate crisis and environmental degradation created the organization "Teachers for Future". Schools and/or individual teachers have joined the organization voluntarily because they want to stimulate their students and peers to live and act in a sustainable manner. The organization Teachers for Future Spain has launched several programs to incorporate issues related to the environment in the school, for example "*Colecaminos*" focuses on more sustainable ways of transport to the schools, "*ecoadutiorias*" aims to improve the energy efficiency in the schools. These programs aim to involve all members of the school community and try to bring sustainability closer to the life of the students.

In this study, we will analyze one of the most important programs of Teachers for Future Spain: "*Recreos Residuos Cero*" (zero waste break time) or RR0, as we will call it (Teachers for Future Spain, n.d.). The RR0 programme aims to reduce the full and avoidable waste produced during break time in the school, addressing not only the need to reduce but also to understand the links between food and waste and how our eating habits can protect our health and the health of our planet. This programme uses an everyday activity, break time, to teach sustainability. Even though the programme initially focuses on reducing solid waste, one of the aims of the programme is to link sustainability with healthy eating habits. This programme is well accepted among students and parents as the link between nutrition and health is profoundly embedded in society. Health and nutrition are important to all parents and the RR0 programme motivates families to be more aware of what their children eat at school.

17.3 The RR0 Programme: A Whole School Approach to Sustainability

17.3.1 Theoretical Framework

RR0 was designed as a framework for teachers and schools to review and adapt their vision on solid-waste management at schools. In the framework used by Vare and Scott (2007), the RR0 programme could be classified as Education for Sustainable Development (ESD) level 1 learning as it provides guidance in shifting behaviour-related waste and food and provides a framework to discuss with students about how we live has an impact on the planet. The pedagogical environment created through the RR0 programme encourages collaboration and participation between schools, students, and families and provides a route to ESD level 2 learning: It builds capacity in the community to think critically about their impact on the planet and the convenience and difficulties to create sustainable communal practices. The RR0 programme is an opportunity for schools to put into practice the WSA. Waste management is key in the ecological footprint of schools and how students and staff look at waste can make an enormous difference. There are several theoretical models (and adaptations) for implementing the WSA approach published in the literature (Chopin et al., 2018; Henderson & Tilbury, 2004; Hunt & King, 2015; Mathar, 2015; Mathie & Wals, 2022; Scott, 2005; Shallcross & Robinson, 2008). We have adopted the WSA flower model (Mathie & Wals, 2022) (Fig. 17.1) as the theoretical model to implement the RR0 programme from a Whole School perspective. The flower model is an adaptation of the basic model of WSA proposed by Henderson and Tilbury (2004). The reason why we adopted the flower model is because its centrepiece is the school's vision, ethos, leadership, and coordination. Based on the experience of the teachers involved in the RR0 programme, school vision and coordination were put forward as crucial factors for a successful implementation of the RR0 programme. In the WSA flower model, the centrepiece is surrounded by 5 additional features that have been identified as key features for implementing the RR0 programme from a whole school perspective: the school curriculum, pedagogical environment, institutional practices, community connection, and professional development and capacity building. The school curriculum is a cornerstone of formal education and is important both to the teachers and parents. The national curriculum in Spain includes "education towards responsible consumption and sustainable development" as one of the six pedagogical principles (Real Decreto 157/2022, 2022; Real Decreto 217/2022, 2022). The RR0 programme can be used to complement the school curriculum. The programme offers an opportunity to connect items of the standardized curriculum (waste management, sustainability, and healthy eating) with an everyday activity. The RR0 programme draws the students' attention to interdisciplinary themes such as sustainable waste management and sustainable eating habits. The WSA approach aims to create a pedagogical environment that allows participation, communication, reflection, and discussion about things

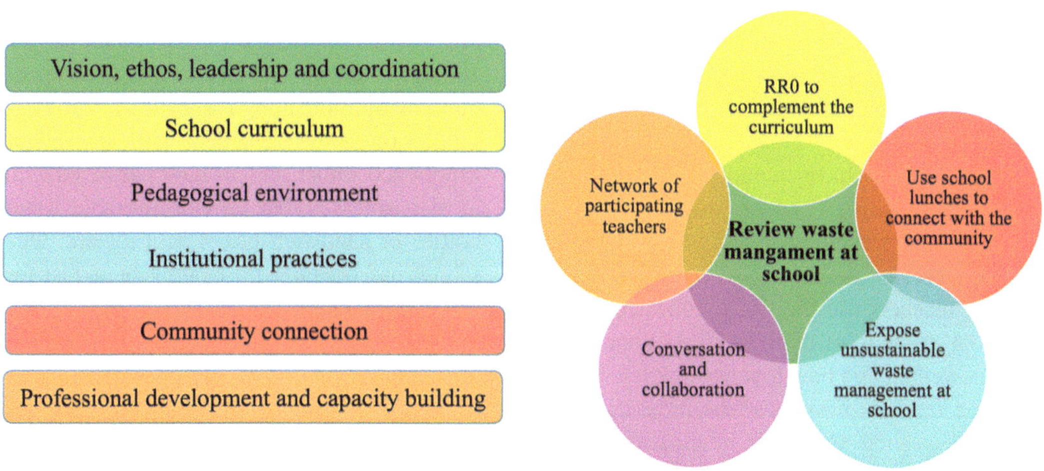

Fig. 17.1 Adaptation of the Whole School Approach flower model (Mathie & Wals, 2022) for the RR0 programme

that matter in the life of the students. The RR0 programme was created to open a conversation among the entire school community on how much waste is produced during break time and how this has a negative impact not only on the health of our planet but also on our own health.

17.3.2 The WSA Principles Put into Practice Through the RR0 Programme

The schools that have joined the programme have initiated an active collaboration between teachers, students, and parents to aim for a more sustainable attitude towards waste. Moreover, the RR0 programme takes sustainability outside the classroom and into everyday life. Based on the experiences of the participating teachers, we describe some examples of how the key principles of the WSA are put into action during the RR0 programme:

Vision, Ethos, Leadership, and Coordination

- Certain schools have incorporated the RR0 programme in the overall vision and mission of the school. The RR0 programme is announced on their website and when new students or staff members join the school, they are informed about the RR0 programme and are urged to use reusable water bottles and bring rubbish-free lunches. The school provides information and ideas on how to prepare healthy and sustainable lunches.

School Curriculum

- Some teachers have included the RR0 programme in their art classes: students organize exhibitions that visualize the waste generated during break time.
- Math teachers use the waste collected at the school grounds to teach students about volume, weight, and how to make graphs.
- Music teachers have used the programme to record a song on waste management and recycling.

Pedagogical Environment

- Many schools have introduced the figure of "*ecodelegados*" to implement the RR0 programme. The "*ecodelegados*" are students that are responsible for informing about the programme and they also collect data on how much and which type of waste is produced during break time. This strategy enables students to actively participate in the programme.

Institutional Practices

- Some schools have decided to remove the bins from the playground to encourage students to produce less waste. One of the schools drove this even further as they told their students that they had to take their waste back home.
- Some schools have installed a composting container on the school grounds to dispose of and recycle the organic waste produced during break time. Some teachers commented that they had trouble convincing the direction of the school as composting containers often are perceived as "dirty". Teachers that were able to install a composting unit confirm that this has been beneficial for waste management at school and that students from all educational levels show great interest in the process of composting.
- Many participating schools have organized different types of competitions on who is able to produce less waste during a certain period. Teachers confirm that this "healthy" competition is very motivating for the students. Most schools opt to award students with an outdoor experience. Other schools have partnered with local businesses that provide reusable water bottles or sandwich boxes as a reward. One school has organized a healthy breakfast for the winning class.

Community Connections

- One school has connected with local farmers. Students that have not produced any waste during the whole week get rewarded on Friday with fresh fruit donated by local farmers.

- Some schools have contacted the local health care center and invited health professionals to talk to students about the importance and benefits of healthy eating.
- Some schools have teamed up with parents to organize workshops where they make reusable sandwich bags for the kids.
- One school collaborates with a non-profit organization that makes reusable sandwich bags that are sold at the school. The money raised goes entirely to the non-profit organization. Students and parents are more inclined to buy these bags knowing that the money is used for a good cause.

Professional Development and Capacity Building

- Currently, the programme does not offer specific training for teachers, but the regular meetings connect all participating and interested teachers and staff, creating a network where experiences are shared and suggestions on improvement are made.

17.4 Evaluation of the RRO Programme

In this study, we propose two complementary strategies to evaluate the WSA principles. This approach aims to respond to the challenge of evaluating several aspects of a holistic approach simultaneously.

17.4.1 Web App

The first strategy is based on an app that was designed by our group to track the waste generated in different classrooms (Lozano Murciego et al., 2021), informing us on the efficiency of the programme in terms of waste reduction. The app divides the waste of each participating student into three categories: full waste, avoidable waste, and zero waste. Every day the teacher or "*ecodelegado*" collects data on the type of waste produced in the class. The app was launched in 2020, and this study includes data collected during the school terms 2020–21 and 2021–22. The data collected with the app are stored in a database that includes the name of the school, the name of the classroom, the age of the students, and the number of students in each category per day. The database has been used to determine the amount of waste generated and to verify if participation in the programme translates in a reduction of waste. The latter one is an indication of the overall performance of the programme; it allows us to demonstrate if the goal of the programme (to reduce waste) is met. Statistical analysis reveals a weak negative correlation ($r = -0.11$; $p < 0.05$; $N = 1058$) between the amount of waste (full waste + avoidable waste) and the date, indicating that there has been an overall reduction of waste in schools in 2021. In 2022, no significant correlation was found ($r = -0.04$; $p > 0.05$; $N = 323$).

In addition, the information about educational level and the age of the students allows correlation between age and the type and amount of waste generated. Our data reveal that the RR0 programme is more popular in primary education (63.6% of the students). Data show that participation of classrooms in the highest levels is low (22.5% for secondary education and 0.5% for baccalaureate) (Table 17.1).

Moreover, there is a significant correlation between the type of waste and the age. Our data show a positive correlation between the age of the students and the amount of full waste ($r = 0.29$; $p < 0.05$; $N = 2149$) and a negative correlation with the amount of avoidable and zero waste ($r = -0.20$ and $r = -0.21$, respectively; $p < 0.05$; $N = 2149$).

While this approach provides quantitative data on the amount of waste, it falls short of assessing other elements that are important for the programme such as the involvement of the community, institutional practices, school curriculum, and the attitude and behaviour of the students. To assess these additional elements, we used an additional instrument of assessment. We surveyed participating teachers on how they implement the

Table 17.1 Percentage of participating students according to the school level and age

School level	Ages	% of participating students
Kindergarten	0–5	13.3
Primary education	6–11	63.6
Secondary education	12–15	22.5
Baccalaureate	15–17	0.5

programme and the perceived benefits for the school and students.

17.4.2 Teachers Survey

To assess the implementation of the RR0 programme, a questionnaire was made available to participating teachers. The questionnaire was elaborated with the platform of Google Forms and was distributed among teachers by the organization Teachers for Future. To evaluate the implementation of the WSA principles, we have used 10 of the 25 items (Table 17.2). The questionnaire was answered by 81 teachers.

Data from our survey reveal that 94% of the teachers believe that schools and families should work together to achieve improvement in the eating habits of the students (item 9) and 73% of the teachers consider that the family is very important in determining the eating habits of the students (item 14). Moreover, 21% of the teachers think that there will be a reduction in food packaging at the homes of the participating students, while 47% think that there will be a general reduction of waste at home (items 17 and 25).

Three items of the questionnaire are related to the pedagogical environment created through the RR0 programme. About 97% of the teachers consider the learning experience associated with the programme beneficial for students (Item 10). Interestingly, 38% of teachers answered yes and 41% answered in most cases to the question of whether the RR0 offers a better approach to waste management compared to the current contents of the school curriculum (item 21). Teachers believe that through the programme, students learn about a number of interdisciplinary topics such as: reducing contamination, healthy eating, recycling, and responsible consumption (item 16).

Table 17.2 Listing of the items of the questionnaire and their relation with the WSA principles

WSA principle	Item
Community connection	Item 9: What is the role of the school and the students´ families in promoting healthy eating habits?
	Item 14: What factors influence the eating habits of your students (the school, friends, family, the media, social networks)?
	Item 17: Do you believe that participating in the programme will result in a waste reduction in the community outside school (at the homes of the students)?
	Item 25: Do you believe that there will be a general (not only food packaging) waste reduction at the homes of the students?
Pedagogical environment	Item 10: Do you think that the fact that the RR0 programme uses an everyday activity for learning is beneficial for students?
	Item 16: What topics do students learn through the programme?
	Item 21: Do you believe that the programme is a better approach to learn about waste management in comparison with what is included in the curriculum?
School curriculum	Item 22: How often do you use the programme as a complement to the contents of the curriculum?
	Item 23: Do you think that waste management should be more embedded in the curriculum?
	Item 24: What would you include in the curriculum to teach students about waste management (practical activities, transversal contents, complementary activities, or others)?

The last three items included in this study are related to the school curriculum. Our data reveal that 99% of the teachers use the RR0 programme to complement the curriculum (item 22). Moreover, 90% of the teachers surveyed consider that the school curriculum should include more items related to waste management and healthy eating (item 23). When asked what teachers would like to see included, 64% think it is very important to include more practical content, 44% would like to include more complementary activities, and 40% would like to include more transversal content (item 24).

17.5 Critical Discussion: Strengths and Challenges

A whole school approach is a holistic, systemic, and reflexive approach that involves all members from and around schools. Different key aspects of the education system need to be addressed simultaneously, creating difficulties for evaluation and assessment. We have proposed two complementary instruments as indicators of the results of the RR0 programme. The applied methodology allows us to assess the WSA principles proposed by Wals and Mathie (2020): The web app tracks the amount of waste that was created to evaluate institutional practices, and teachers were surveyed to evaluate the community connection, the pedagogical environments, and the school curriculum. In addition, we use the experiences of the participating schools and teachers to reflect upon the WSA principles applied during the RR0 programme. We have identified the strengths and challenges for each of the WSA principles.

17.5.1 Vision, Ethos, Leadership and Coordination

Strengths

Schools that have incorporated the RR0 programme into their school project and have involved the whole educative community have been able to change the habits of their staff and students. Some of the schools that have joined the RR0 programme report that the programme has provided a springboard to other sustainability initiatives. This is a positive outcome that also has been described by Cutter-Mackenzie et al. (2010) during the Australian Waste-Wise School project, another example of applying a Whole School approach towards waste management at school.

Challenges

The RR0 programme is offered to individual teachers and schools. There are many individual teachers that join the programme individually with their classes. These teachers comment that they are struggling to integrate the whole community and in many cases are not able to put into practice all aspects of the theoretically proposed WSA model. Nevertheless, it is fair to say that the programme could be used to create a bridge between the isolated effort of one teacher to a group of teachers that can steer the school in the right direction. As the school's leadership is so important for the WSA, this programme provides an opportunity to forge such leadership.

17.5.2 School Curriculum

Strengths

Teachers that participate in the RR0 programme use the programme to complement contents from their curriculum. The programme creates a space to address interdisciplinary challenges such as waste management and responsible consumption and allows teachers to connect items of the curriculum with an everyday activity outside the classroom. Teachers use this "new" space to put into practice alternative forms of learning. The programme puts into practice what has been described as essential for Education for Sustainable Development: cross-curricular teaching (Gericke, 2022; MCKeown & Hopkins, 2007; Sund et al., 2020).

Challenges

Teachers use the programme to teach items related to waste management and sustainability but consider that sustainability should be more embedded in the curriculum. The survey reveals that teachers would like to see more complementary activities and transversal content as part of the school curriculum.

17.5.3 Pedagogical Environment

Strengths

Learning about waste management and sustainability is transferred from the classroom to the school grounds and the homes of the students. This educational innovation is perceived by teachers as beneficial for learning. Moreover, teachers can teach interdisciplinary items such as recycling and responsible consumption through the activities of the programme. The teachers survey reveals that the RR0 programme facilitates interdisciplinary learning. Moreover, including the students in the management and implementation of the programme through the figure of "ecodelegado" has proven to be very effective. Students take ownership of the programme and receive a sense of empowerment.

Challenges

The data collected through the web app reveal that participation is lower in the higher educational levels. The quantitative data agree with the perception of the teachers; it remains difficult to engage the older students and teachers are looking for new strategies to motivate the older students.

17.5.4 Institutional Practices

Strengths

A strategy that, according to the participating schools, has had a positive impact is a healthy competition between classes where students are rewarded when they reduce the amount of waste. It gives the students a common purpose. Another approach that is well received by the school community and has a positive impact in reducing the waste produced on the playground is the installation of a composting container.

Challenges

Data from the web application is a quantitative measure of how many and which types of waste are produced by the students. Our data reveal a small reduction in waste in 2021 and no significant reduction in 2022. These results call for reflection as the goal of the programme is to reduce waste. Participating teachers comment that most schools start the programme with a short but intense campaign on waste reduction. This in most cases has positive results but after some time, students fall back into their old habits. Moreover, our data highlight something that is often discussed in meetings of the programme: it remains difficult to engage teachers and students from higher educational levels. The unequal participation forms a barrier to making a systemic change in the waste management of schools.

17.5.5 Community Connection

Strengths

In the WSA, the school is seen as part of a broader community that can enhance learning. Our survey confirms the importance of families in this learning experience. The RR0 creates a connection between schools, students, and families. This connection improves the relationship with the parents and the result of our survey suggests that the RR0 has the potential to promote sustainability in the homes of the students. This form of intergenerational learning was also observed as a result of previous studies (Armstrong et al., 2004; Cutter-Mackenzie, 2010).

Challenges

The participating teachers agree that families readily get involved in the programme, this is not the case for organizations or businesses surrounding the school. Establishing links with the community is one of the aspects of the WSA flower model that can, and should, be improved in the RR0 programme. As already mentioned, the holistic nature of the WSA challenges its implementation and based on empirical studies, combining the internal and external organization is problematic (Gericke, 2022).

17.5.6 Professional Development and Capacity Building

Strengths

The organization of Teachers for Future has different ways to support the participating schools and teachers. Their website offers numerous resources and ideas on how to implement the programme and involve the entire community. They have a blog where they publish experiences and share ideas. Moreover, they host regular meetings where teachers share their personal experiences and discuss ways to improve the implementation and success of the programme.

Challenges

Despite the resources and support offered by the organization Teachers for Future, some students have commented that they want their school to join the programme but are insecure about where to start. A more well-defined action plan for the programme would be beneficial for a more widespread implementation of the programme.

17.6 Conclusions

In conclusion, the RR0 programme is an opportunity to put into action the WSA approach and for many schools the programme is a first step towards a more sustainable school. The usage of the web app helps teachers to evaluate their progress and enables them to visualize the state of waste management at their school. The teachers survey exposes the perceived benefits and shortcomings of the programme and allows the direction to evaluate and redesign the structure and functioning of their school.

Acknowledgements The research leading to these results received funding from Spanish State Research Agency (AEI) 10.13039/501100011033 FEDER "Una manera de hacer Europa" under Grant Agreement No PID2020-114358RB-I00.

References

Armstrong, P., Sharpley, B., & Malcolm, S. (2004). The waste wise schools program: Evidence of educational, environmental, social and economic outcomes at the school and community level. *Australian Journal of Environmental Education, 20*(2), 1–11. https://doi.org/10.1017/S0814062600002159

Bianchi, G., Pisiotis, U., & Cabrera Giraldez, M. (2022). GreenComp The European sustainability competence framework (No. JRC128040). *Joint Research Centre* (Seville site).

Buckler, C., & Creech, H. (2014). *Shaping the future we want: UN decade of education for sustainable development (2005–2014)*. Final Report.

Chopin, N., Hargis, K., & McKenzie, M. (2018). Building climate-ready schools in Canada: Towards identifying good practices in climate change education. In *Sustainability and education policy network*. University of Saskatchewan.

Cutter-Mackenzie, A. (2007). *Waste wise schools research report* (Vol. 1). Monash University Publishing.

Cutter-Mackenzie, A. (2010). Australian waste wise schools program: Its past, present, and future. *The Journal of Environmental Education, 41*(3), 165–178. https://doi.org/10.1080/00958960903347471

Galaz, V., Biermann, F., Crona, B., Loorbach, D., Folke, C., Olsson, P., ... & Reischl, G. (2012). 'Planetary boundaries'—exploring the challenges for global environmental governance. *Current Opinion in Environmental Sustainability, 4*(1), 80–87.

Gericke, N. (2022). Implementation of education for sustainable development through a whole school approach. In G. Karaarslan-Semiz (Ed.), *Education for sustainable development in primary and secondary schools: Pedagogical and practical approaches for teachers* (pp. 153–166). Springer International Publishing. https://doi.org/10.1007/978-3-031-09112-4_11

Hargreaves, L. G. (2018). In J. Coriddi (Ed.), *The whole-school approach to education for sustainable development: From pilot projects to systemic change* (pp. 69–74). Policy and practice. A development education review.

Henderson, K., & Tilbury, D. (2004). *Whole-school approaches to sustainability: An international review of sustainable school programs*. Australian Research Institute in Education for Sustainability: Australian Government.

Holst, J. (2022). Towards coherence on sustainability in education: A systematic review of Whole Institution Approaches. *Sustainability Science*, 1–16.

Hunt, F., & King, R. P. (2015). *Supporting whole school approaches to global learning: Focusing learning and mapping impact.*

Lozano Murciego, A., Campos De La Torre, V., Ferrari Lagos, E., Delgado-MartÍn, L., & Ruiz, C. (2021). *Development of a Progressive Web App aimed to measuring the impact of the school program "Recreos Residuos Cero"* (pp. 72–77). Ninth International

Conference on Technological Ecosystems for Enhancing Multiculturality (TEEM'21) (TEEM'21). https://doi.org/10.1145/3486011.3486422

Mathar, R. (2015). A whole school approach to sustainable development: Elements of education for sustainable development and students' competencies for sustainable development. In *Schooling for sustainable development in Europe* (pp. 15–30). Springer.

Mathie, R. G., & Wals, A. E. J. (2022). *Whole school approaches to sustainability: Exemplary practices from around the world.* Wageningen University, Education and Learning Sciences.

MCKeown, R., & Hopkins, C. (2007). Moving beyond the EE and ESD disciplinary debate in formal education. *Journal of Education for Sustainable Development, 1*(1), 17–26. https://doi.org/10.1177/097340820700100107

Molderez, I., & Ceulemans, K. (2018). The power of art to foster systems thinking, one of the key competencies of education for sustainable development. *Journal of Cleaner Production, 186*, 758–770.

Real Decreto 157/2022. (2022). Real Decreto 157/2022, de 1 de marzo, por el que se establecen la ordenación y las enseñanzas mínimas de Educación Primaria. *Boletin Oficial Del Estado 52, 2 de Marzo de 2022.*

Real Decreto 217/2022. (2022). Real Decreto 217/2022, de 29 de marzo, por el que se establece la ordenación y las enseñanzas mínimas de la Educación Secundaria Obligatoria. *Boletín Oficial Del Estado, 76, 30 de Marzo 2022.* https://www.boe.es/buscar/pdf/2022/BOE-A-2022-4975-consolidado.pdf

Rieckmann, M. (2017). *Education for sustainable development goals: Learning objectives.* UNESCO publishing.

Rockström, J., Steffen, W., Noone, K., Persson, Å., Chapin, F. S., Lambin, E. F., ... & Foley, J. A. (2009). A safe operating space for humanity. *Nature, 461*(7263), 472–475.

Scott, B. (2005). Getting to the heart of the matter: Examining the efficacy of a whole-school approach to behaviour management. *Kairaranga, 6*(1), 29–34.

Shallcross, T., & Robinson, J. (2008). Sustainability education, whole school approaches, and communities of action. In A. Reid, B. B. Jensen, J. Nikel, & V. Simovska (Eds.), *Participation and learning: Perspectives on education and the environment, health and sustainability* (pp. 299–320). Springer. https://doi.org/10.1007/978-1-4020-6416-6_19

Sund, P., Gericke, N., & Bladh, G. (2020). Educational content in cross-curricular ESE teaching and a model to discern teacher's teaching traditions. *Journal of Education for Sustainable Development, 14*(1), 78–97. https://doi.org/10.1177/0973408220930706

Teachers for future Spain. (n.d.). *Recreos residuos cero.* Retrieved September 13, 2022, from https://teachersforfuturespain.org/que-hacemos/recreos-residuo-cero/

Tilbury, D., & Galvin. (2022). *European commission input paper: A whole school approach to learning for environmental sustainability.* Expert Briefing Paper in Support of the First Meeting of the EU Working Group Schools: Learning for Sustainability.

UNEP (2021). UNEP Annual Report 2021. Available at: https://www.unep.org/annualreport/2021/

Vare, P., & Scott, W. (2007). Learning for a change: Exploring the relationship between education and sustainable development. *Journal of Education for Sustainable Development, 1*(2), 191–198.

Wals, A. E. J. (2012). *Shaping the education of tomorrow: 2012 full-length report on the UN decade of education for sustainable development.* Unesco.

Wals, A. E. (2015). Beyond unreasonable doubt. *Education and learning for socio-ecological sustainability in the Anthropocene.*

Wals, A. E., & Benavot, A. (2017). Can we meet the sustainability challenges? The role of education and lifelong learning. *European Journal of Education, 52*(4), 404–413.

Wals, A. E. J., & Mathie, R. G. (2020). Whole school responses to climate urgency and related sustainability challenges. In M. A. Peters & R. Heraud (Eds.), *Encyclopedia of educational innovation* (pp. 1–8). Springer Singapore. https://doi.org/10.1007/978-981-13-2262-4_263-1

18 Practicing Climate Action in a K-12 School Using a Whole Institution Approach

Kristen Hargis

Key Message

A whole institution approach for climate change education and research can be a useful framework for shifting the focus from individual learners to collective practices. This is especially true when this approach is paired with a practice theory theoretical framework. Focusing on the elements of practice that currently exist at a school (and surrounding community) enables a strengths-based approach that works *with* not against established practices.

18.1 Introduction

Climate change is one of the most pressing sustainability issues, with social, economic, and environmental implications (such as potential loss of land and livelihood, displacement and migration, eco-anxiety, grief, and climate justice–related issues) (Eriksen et al., 2011; Yohe et al., 2007). Education has been identified as crucial to address climate change (UNESCO, 2017). The majority of climate change education (CCE) and research to date has focused on instilling *individual* scientific cognitive clarity instead of also learning how to take climate actions *together* (Cook, 2019; Brownlee et al., 2013; González-Gaudiano & Meira-Cartea, 2010; MECCE Project and NAAEE, 2022; Monroe et al., 2017). Understanding climate science, however, does not guarantee belief in or action on climate change (Kahan et al., 2012; Hornsey et al., 2016). It is becoming increasingly evident that the challenges associated with climate change necessitate a paradigmatically different type of education to bolster the agency and empowerment of citizens necessary to address climate change (Brownlee et al., 2013; UNESCO, 2010; Wibeck, 2014).

Approaching CCE and research from a practice perspective and a whole institution approach may present such an opportunity. Utilizing a practice lens means educators and scholars adjust conceptual focus away from the knowledge of individual learners to the practices they collectively "carry," (un)equally share, and mutually shape, wherein understandings, meanings, and purposes are irreducible to personal attributes (Reckwitz, 2002; Shove et al., 2012). This means emphasis is placed on communal educational activities, not individual learners (Nicolini, 2013). Such a shift in analytical focus enables a reconceptualization of learning as occurring "in the middle of everyday practical experience," wherein social milieus create the contextual conditions within which learning occurs (McKenzie & Bieler, 2016, p. 16, also see Lave & Wenger, 1991; Nicolini, 2013). This readjustment aligns with the broader turn within the social sciences

K. Hargis (✉)
College of Education, University of Saskatchewan, Saskatoon, Canada
e-mail: kristen.hargis@usask.ca

A. E. J. Wals et al. (eds.), *Whole School Approaches to Sustainability*, Sustainable Development Goals Series, https://doi.org/10.1007/978-3-031-56172-6_18

towards practice (Schatzki et al., 2001) and may hold potential for education and research to engender the large-scale transformation required by recent climate change. For the purposes of this study, a whole institution approach means that CCE is integrated within and across each of the domains of Overall Governance, Teaching and Learning, Community Partnerships, and Facilities and Operations. That is, within a whole institution approach, the focus of CCE is not only placed on individual learners, teachers, school staff, or administrators but also includes the activities of those in the entire school and local community.

This case study utilized Shove and colleagues' (2012) conceptualization of practices to critically describe climate action practices occurring at a Kindergarten to Grade 12 (K-12) school in Canada using a whole institution approach to CCE within and across whole institution domains. In the sections that follow, the version of practice theory utilized for this research is briefly overviewed, including why it was selected for the current research, before discussing what is known about quality CCE, including in relation to a whole institution approach. The methods used for this research and key findings are then summarized before ending with concluding thoughts for discussion.

18.2 Practice Theory

Lave and Wenger's (1991) conceptualization of learning as occurring within communities of practice is a common reference point for many educational practice theorists. While Lave and Wenger (1991) were not the first to acknowledge the value of experience within education, their conceptualization of learning as "legitimate peripheral *participation*" marked a significant paradigm shift wherein learning became conceptualized as a collective, participatory social process, as opposed to individual, cognitive acquisition (Hughes et al., 2007).

While a unified version of practice theory is nonexistent (Nicolini, 2013), there are some common components within its uptake in education. Most educational theorists of practice would agree in principle with Schatzki (2001) that practices are "embodied, materially mediated arrays of human activity (p. 11)." That is, practices are arrays of human activity in that they consist of the elements of human activity, whereby bodies and social activities are mutually constituted (embodied) in practice, a constitution that is, at minimum, mediated by material objects. This "constitutive entanglement of the social and the material" is often referred to as socio-material (Orlikowski, 2007, p. 1438). While educators have conceptualized learning as participatory outcomes of communities of practice for some time, socio-material approaches attend not only to "who" but also to "what" participates and how, as well as the resulting implications of those interactions (Fenwick et al., 2011).

The relationship between practice, change, and learning is debated (Hager, 2012), however, "understanding how practices change, as well as how they are stable and enduring, is a key issue in thinking through the relationship of practice to learning" (Hager, 2012, p. 10). Shove and colleagues (2012) account of the dynamics of practice was created due to an absence within practice theory literature of *how* practices change (Shove et al., 2012). Their account of the dynamics of practice also includes a focus on how practices appear, persevere, and disappear (Shove et al., 2012).

For Shove and colleagues (2012), practices are provisionally recognizable entities and performances that actively integrate the elements of which they are composed. These elements include at minimum: materials (such as objects, tools, technologies, and the body), competences (such as background knowledge, understanding, and skills), and meanings (such as ideas, emotions, aspirations, and symbolic meaning) (Shove et al., 2012). Practices exist when elements (*materials, competences,* and *meanings*) are linked, change upon the introduction or recombination of elements, and disintegrate when links between elements are broken (Shove et al., 2012). Shove and colleagues (2012) justify this reductive schematic due to its analytical capability to illustrate "the recursive relation between practice-as-performance and practice-as-entity"

(Shove et al., 2012, p. 15). That is, practices are product and process, noun, and verb.

For example, driving is dependent on owning a car (*material*), knowing how to drive (*competence*), and believing that driving to work is an acceptable form of transportation (*meaning*) (Shove et al., 2012). Historically, the practice of driving has changed upon the introduction of more reliable cars (*materials*), whereby drivers no longer needed to also be mechanics (*competence*) (Shove et al., 2012). This introduction of a new *material* element (that is, more reliable cars) enabled engagement with a more diverse range of practitioners and also changed the *meaning* of driving from one of adventure to one of practicality (Shove et al., 2012). Additionally, many elements required for driving existed before cars were invented and were only later linked. For example, the idea in England that one should drive on the left-hand side of the road (*meaning*) originated from the practice of horse riding in the 1700s where one might need to wield a sword with their right hand (Shove et al., 2012). Only later was this *meaning* linked to the practice of driving (Shove et al., 2012).

18.3 "Good" CCE

Research suggests "Good" CCE should employ a whole institution approach to support a culture of climate action (Bieler et al., 2018; Hargis & McKenzie, 2021; Hargis et al. 2021; UNESCO, 2016). A whole institution approach to CCE involves engagement within and across each of the domains of Overall Governance (such as policies and meetings), Teaching and Learning (such as curricula), Community Partnerships (such as field trips and guest speakers), and Facilities and Operations (such as solar panels; see Fig. 18.1). Within the Teaching and Learning domain, a whole institution approach also includes incorporating CCE in all subjects. If CCE is only included in science classes, for example, it sends the message that climate change solutions, causes, and effects are only scientific in nature versus also requiring social and political analysis and action (González-Gaudiano & Meira-Cartea, 2010; Hornsey et al., 2016). The success of a whole institution to CCE approach also depends on involving "students, teachers, principals, school staff at all levels, and the wider community – such as families and community members – in reflecting and acting on climate change" (CCUNESCO, 2020, p. 26). A whole institution lens to CCE also aligns well with practice theory as it shifts the focus from *individuals* to schools and local communities working *together* for climate action.

Research also suggests that "good" CCE should include a focus on cognitive, psychosocial, and action-oriented learning dimensions, all of which should include a justice orientation (see González-Gaudiano & Meira-Cartea, 2010; UNESCO, 2015, 2019). The cognitive learning dimension includes a focus on developing the knowledge and learning agility needed to understand climate change causes, impacts, and solutions (MECCE Project and NAAEE, 2022; UNESCO, 2015). As student knowledge of climate change grows, students may develop eco-grief or eco-anxiety (Doherty & Clayton, 2011; Norgaard, 2011; Randall, 2009). While small amounts of concern can lead to action, without the emotional resilience needed to process climate emotions, students may feel overwhelmed and hopeless (Dooley et al., 2021; Clayton et al., 2017). Finally, action-oriented CCE is also essential, as students may disengage with climate change if it is seen as distant and unsolvable (Amel et al., 2017; Monroe et al., 2017; Rowling, 2019). Climate justice should also be integrated across all holistic learning dimensions, including as those most affected by climate change have contributed the least to the problem (United Nations, 2019; Kanbur, 2015).

'Good' CCE should also incorporate Indigenous knowledges (Amsler & Jeannie, in press; Ferland, n.d.; Tanyanyiwa, 2019; Mbah et al., 2021; Viswanathan, 2020). Incorporating Indigenous knowledges in CCE (and education in general) is important because Indigenous knowledges "can help us to close some gaps in our knowledge about the environment that will enable us to counter the threats to the natural environment" (Nesterova, 2020, p. 1051).

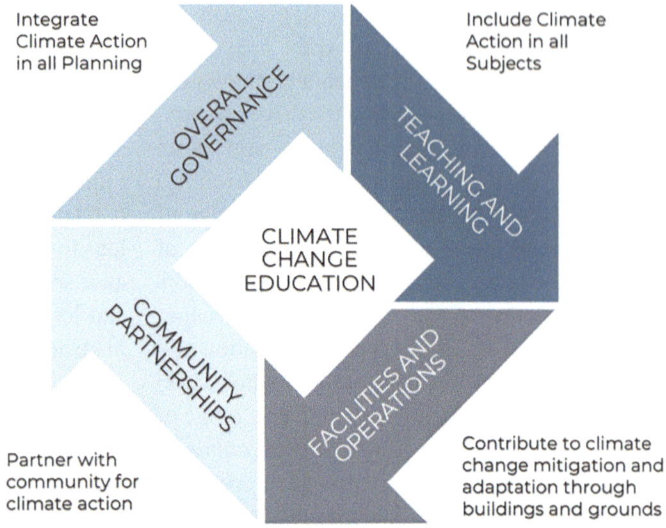

Fig. 18.1 This figure illustrates a whole institution approach to CCE. (Hargis & McKenzie, 2021; MECCE and NAAEE, 2022)

Additionally, in settler colonial societies, such as Canada, including Indigenous knowledges in CCE "can support the processes of transitional and historical justice to heal the damage inflicted on Indigenous peoples during colonialism and reconciliation and the building of new, just and equal relationships between Indigenous and non-Indigenous groups" (Nesterova, 2020, p. 1051).

18.4 Methods

In 2016, in response to recent international calls for CCE (General Assembly resolution 70/1, 2015; UNESCO and UNFCCC, 2016; UNFCCC, 2015), the United Nations Educational, Scientific and Cultural Organization (UNESCO) launched a Climate Change Pilot Project within their Associated Schools Network (ASPnet). From September 2017 to May 2018, ten primary and secondary education schools within Canada participated in this international project, which entailed a whole institution approach to CCE under the direction of the Canadian Commission for UNESCO (CCUNESCO). To implement the pilot project, schools were provided with a Getting Climate Ready (UNESCO, 2016) guide. The Sustainability and Education Policy Network (SEPN) conducted an evaluation of this project using interviews and a survey (see Chopin et al., 2018; Hargis et al., 2018), which highlighted a school exhibiting promising CCE for the current case study.

The site chosen for this study was a pre-Kindergarten to Grade 12 school, which also held designations as a UNESCO ASPnet school and an EcoSchool (Creswell & Poth, 2018). While SEPN's evaluation identified evidence of climate action practices already occurring, the current study further investigated those actions over 4 weeks.

During the 2018–2019 school year, the school used a whole institution approach to address the Sustainable Development Goals (SDGs), focusing on climate change during April 2019 as part of their Simply Living Simply program. The latter program was started several years previously as a whole institution approach to sustainability in the school. Data generation for the current case took place from late April to mid-May 2019. Participants included administrators, staff, teachers, students, and community members. Ethics approval was obtained through the Institutional Review Board at the University of Saskatchewan and at the school board level. All participants completed informed consent and/or informed assent forms. The total participant number was 96.

Date generation included a sensory walk of the school surrounds, observations of classroom lessons, field trips, and meetings, interviews with teachers and administrators (such as principals and vice-principals), and teacher and student focus groups. In the focus groups, teachers mapped where CCE practices were occurring at the school in relation to all four whole institution domains, and students drew pictures about the climate action practices currently happening at the school and actions they wished were happening at the school. Documents and photos from the site were also collected.

Analysis began with inductive hand coding and memo-writing to note emerging ideas (Creswell & Poth, 2018). Inductive codes included descriptive codes (which are codes summarizing the main topic discussed), In Vivo codes (which are codes using participants' own words), emotion codes (which are codes capturing emotion), and simultaneous codes (which are two or more codes applied to a single datum). Following Saldaña's (2016) coding recommendations, similar codes were grouped into emergent categories. The resulting themes were interrogated in relation to the chosen practice theory. This first cycle of coding preceded the development of a codebook, which included emergent themes as well as deductive codes related to practice theory and the whole institution domains (see Creswell & Poth, 2018). Once the codebook was finalized, all data were coded in the qualitative data analysis software NVivo 12.

Coded data supported theme development (Creswell & Poth, 2018). When developing themes, the method of zooming in and out (Nicolini, 2013) directed the researcher's analytical gaze to zoom in on practices within one domain before zooming out to see how practice elements were related across domains. This method was appropriate because "practices can only be studied *relationally,* and they can only be understood as part of a nexus of connections" (Nicolini, 2013, p. 229).

18.5 Findings

The findings below are presented by the whole institution domain (that is, Overall Governance, Teaching and Learning, Community Partnerships, and Facilities and Operations) in relation to selected themes that emerged from within and across each whole institution domain. Key practice elements are indicated parenthetically (that is, competences, materials, and meanings).

18.5.1 Overall Governance

Within the domain of Overall Governance, the importance of distributed leadership (between teachers, students, and administrators) for CCE practices was apparent. After noting few climate actions were mapped within the domain of Overall Governance during the teachers focus group, Teacher 2 mentioned, "[The map] reflects our governance. That [climate action is] across grades…to me this is school governance." Part of their governance structure also includes teachers "push[ing] each other on" to improve their CCE practices (Teacher 2). Teacher 10 noted, "It doesn't feel like a push. It feels like, Whohoo!" Teacher 10 continued, "it [that is, the push] fills you with life. It enlivens." While they all agreed that they support (*meaning*) each other to improve their CCE practices (such as sharing ideas and resources), several teachers noted the importance of having a key person (*material*) "to push the agenda gently along" (Teacher 2).

It was also understood (*competence*) that leadership for environmental education and CCE should be shared by teachers and students, and this style of leadership was supported by the presence of several key *materials*, particularly summer planning meetings, posters, and the staff room table. The central nature of the staff room table meant the teachers ate lunch together every day (see Fig. 18.2). During the teacher focus group, Teacher 2 asked one of the teachers to "put a bright star in the staff room. [It's where] all the intelligence and energy oozes out of us."

18.5.2 Teaching and Learning

Related to the theme of CCE content, two subthemes were found related to a cross-curricular approach and limited inclusion of Indigenous knowledges and climate justice.

The practice of CCE at the school is sustained, in part, by a provincial initiative for cross-curricular education (*material*), which happened

Fig. 18.2 Teacher's focus group map excerpt and picture of the staff room

15 years prior. Administrator 1 described how teachers can creatively (*meaning*) connect CCE to this provincial initiative by finding common threads across subjects:

> So, you could explore about [greenhouse] gas emissions in Math and use Math to explore that. And then you could write to public officials and do your Civics and Social Studies...So that although you're spending an hour on Math, you're actually spending three hours on climate education in kind of a sneaky way, not a sneaky way, but in a smart way.

The benefit of knowing how to include CCE within multiple subjects (*competence*) was described as not only a creative way to deal with barriers related to time (which was the most frequently mentioned barrier) but also enabled students to see that climate change had *meaning* in their lives beyond the classroom.

While there was some inclusion of Indigenous knowledges at the school, including in relation to climate change, and climate justice, this inclusion was rare. Reflecting on the overall inclusion of Indigenous knowledges (*competence*) at the school, Teacher 10 mentioned, "it would be better to have more Indigenous input on ways we could be making more of a connection with the Earth." Teacher 1 explained the history of Indigenous knowledges inclusion at the school, mentioning that since First Nations students no longer attended the school due to costs related to busing (*material*), there was even less focus on the inclusion of Indigenous knowledges than in years past. Teacher 6 described a children's book about residential schools (*material*) that she reads to her classes (see Fig. 18.3) but mentioned it was not related to climate change. That is, while Teacher 6 acknowledged the connection between residential schools and colonialism, she did not acknowledge the connection between colonial practices and climate change (*competence*). In addition, while some students were familiar with climate justice (*competence*), this familiarity was not the norm.

18.5.3 Community Partnerships

There was also an understanding (*competence*) at the school of the importance of bringing everyone (at the school, in the community, and at the school board) along on the journey towards increased climate action. Teacher 2 mentioned the importance of "bring[ing] all of your staff

Fig. 18.3 Picture of the book read in Teacher 6's classes

along" by inviting them (principals, office managers, custodians, cafeteria staff) to CCE activities at the school (such as assemblies). In discussing how the cafeteria staff have joined the school's journey towards increased CCE, she mentioned, the cafeteria staff have "started making suggestions, 'What if we do this differently?' They've switched to reusable containers [*material*], and they got rid of straws [*material*] at our behest." Teacher 2 also mentioned the custodians have joined in by allowing them to do waste audits of recycling, garbage, and compost. She mentioned "without the cooperation of your custodial staff, [waste audits] could be a very contentious thing [*meaning*]."

In relation to the broader community, there was also a focus on how to communicate about climate change. The school is in a politically conservative area, so during the first few years of the Simply Living Simply program, the teachers and staff never used the word "climate change" to keep the investigation of the topic "safe." Slowly, several staff and students came to see the connection between the program and climate change themselves. The teachers also used context-specific framings when discussing climate change (such as in relation to protecting jobs) that resonated with the community. That is, the campus and local community were 'brought along' through the inclusive language used to discuss climate change.

18.5.4 Facilities and Operations

When participants were asked what it meant to take climate action at the school, their responses usually referred to taking Facilities and Operations-related actions (*meaning*). Similarly, students usually mentioned Facilities and Operations-related activity when describing what the school is currently doing for climate action (such as recycling, composting, and gardening). Ideas for future climate action practices from students, teachers, and parents were also often related to Facilities and Operations activity.

The effect *materials* can have on facilities and operations practices related to CCE was evident during the research visit. For instance, there was evidence that the presence (or not) of particular *materials* significantly affected drinking practices related to CCE at the school. The school wanted to be a community where everyone uses reusable water bottles (*meaning*); however, they found that the water fountains needed to be refrigerated for students to use them. After refrigerated water fountains were installed, the staff asked the community for reusable water bottle donations (*material*) so they could give them to every student. It was important (*meaning*) to the staff that the onus not be put on the parents to buy water bottles.

Having facilities and operations-related *materials* all around the school also made it easier to use those objects as teaching moments related to climate change since the teachers knew how those materials were related to climate change (*competence*). Discussing these materials, Teacher 1 mentioned that she was able to, "grab [something] and say, 'Oh you want to learn about this. Let's talk about this.'" She also said that because they were able to have these types of conversations frequently, it meant they did not have to explain the basics of climate change as often because students were already familiar with the topic. This familiarity allowed for more meaningful conversations about climate change (*meaning*).

18.6 Discussion and Conclusion

This research has implications for how to implement a whole institution approach, in general, as well as the potential of using a whole institution approach for CCE (also see Hargis & McKenzie, 2021; Hargis et al., 2021; Mathie & Wals, 2022; Wals & Mathie, 2022). This research also indicates the potential of practice theory for researching CCE.

18.6.1 How to Implement a Whole Institution Approach

The whole institution approach at the school was put in place largely due to the school's Simply Living Simply program, as well as their designations as an EcoSchool and a UNESCO ASPnet school, which have targets, programs, and resources aimed at all domains of a whole institution approach. Schools looking to implement a whole institution approach to CCE may want to create their own program (such as the Simply Living Simply program) and/or join other certification programs that have environmental and climate action targets and resources across one or more domains (such as EcoSchools Canada and UNESCO ASPnet schools).

A whole institution approach requires working with individuals from across all domains of school activity. While working with so many different people could have been a barrier to climate action, the school's focus on bringing everyone along, wherein they invited all staff to CCE assemblies, framed messages about climate change according to their audience, shared resources, and supported each other, helped to ensure uptake of CCE. By framing climate change messages in relation to the community's priorities, the school was able to avoid any major pushback from the community related to CCE. Context-specific framings related to climate change have also been found to have a positive effect on climate change engagement and the shifting of climate change views (Callison, 2014; Goldberg et al., 2021; Li & Su, 2018).

18.6.2 Overall Governance

In relation to Overall Governance, the importance of distributed leadership is a key implication for future CCE practice and research. According to Harris (2003), "distributed leadership is characterized as a form of collective leadership in which teachers develop expertise by working collaboratively" (p. 11). While distributed leadership can be implemented 'naively' if practitioners do not know how to lead, do not want to lead, or are prevented from leading (McKenzie & Locke, 2014), when done effectively, distributed leadership is "more than the sum of the component parts or practices" rather it is "a system of practice comprised of a collection of interacting components: leaders, followers, and situation" (Spillane, 2005, p. 15). Past research has emphasized the importance of champions in relation to environmental education (Wood et al., 2014); however, CCE had endured at the school not so much because of individual champions but because of collective and distributed efforts to take environmental action among all practitioners at the school. Distributed leadership also aligns well with a whole institution approach, which seeks collaboration across all domains of school activity, and practice theory, which shifts the focus from individuals to collaborative practices.

18.6.3 Teaching and Learning

Several implications for practices within the domain of Teaching and Learning also emerged from this research. The most commonly mentioned barrier at the school for CCE was time. Teachers often overcame this barrier by integrating CCE across subjects. Cross-curricular approaches to environmental and sustainability education and CCE are variously supported in the literature. Proponents of cross-curricular approaches for CCE have highlighted their potential to illustrate that multiple disciplines have a role to play in addressing climate change (Field et al., 2019; Hargis & McKenzie, 2021; Hargis et al., 2021). Critics of cross-curricular approaches have highlighted the large amounts of time and support needed to embed a topic with which teachers are not overly familiar with across subjects (Dyment & Hill, 2015; Hill & Dyment, 2016; Nicholls & Thorne, 2018). While not dismissing prior findings regarding how lack of time and support may impede the successful implementation of a cross-curricular approach to CCE, the present case illustrates that the support teachers gave each other at the school enabled them to overcome barriers related to time and to ensure that CCE surfaced across all subjects.

Finally, while the school has connected many topics to climate change, there are several key topics that have not yet been connected to climate change as strongly as they could be (such as Indigenous knowledges and climate justice). Implications for future practice, policy, and research include the importance of acknowledging that Indigenous knowledges are important for Indigenous and non-Indigenous students (Antoine et al., 2018; Kapyrka & Dockstator, 2012; Vizina, 2018), that Indigenous land-based education is CCE (Wilson in UNESCO, 2021), and that colonialism is directly related to climate change (Funes, 2022; Sultana, 2022). There were also limited discussions of climate justice at the school. The topic of climate justice, however, is very relevant for this community due to its location in an economically depressed area that is heavily reliant on farming and tourism, both of which could be negatively affected by climate change. As the impacts of climate change increasingly affect local environmental conditions and the human and nonhuman populations who reside there disproportionately, locally relevant discussions of climate justice will become an increasingly important part of CCE.

18.6.4 Community Partnerships

When discussing climate actions at the school, all participants referred to practices in the Community Partnerships domain. Despite living in a remote location, the school had many community partners who were essential to the climate actions occurring at the school. Prior research has also found that schools in rural locations are more likely to have community partners for environmental and sustainability education (Regier, 2019). No matter their location, schools should seek out community partners to support their climate action practices (Hargis & McKenzie, 2021).

Through the school's focus on engaging with community partners, they created their own mini-network. The potential of networking for climate action is also illustrated in prior work (see Hargis et al., 2021). Due to this collaborative and community-driven approach to CCE, the school faced minimal backlash, despite its location in a politically conservative area. Implications for future practice and research include the potential for a whole institution approach that actively involves and works with the school and local community to overcome prior differences of opinion and to create a culture of climate action *with* not in opposition to existing practices.

18.6.5 Facilities and Operations

When practice elements were connected across domains at the school, one of the domains was usually Facilities and Operations. For instance, physical *materials* within the room, such as light timers and recycling and compost bins, which were functioning within the Facilities and Operations domain were also often brought into the domain of Teaching and Learning to discuss how those practices were connected to climate change mitigation, thereby strengthening and reinforcing both practices. Schools making the transition to climate-friendly Facilities and Operations practices should consider ways to bring those practices into the classroom, which will enable the school to become a living lab and can be used to educate the school and local community about (through) climate action (see Hargis & McKenzie, 2021; MECCE Project and NAAEE, 2022; UNESCO, 2016).

Most of the ideas for future climate action practices at the school mentioned by participants were related to Facilities and Operations. While necessary and important, there is also potential for new practices in other domains (Overall Governance, Teaching and Learning, and Community Partnerships). Policies can support expanded imaginaries of potential actions by suggesting actions schools can take across all domains of school activity (see Hargis & McKenzie, 2021; MECCE Project and NAAEE, 2022).

18.6.6 Potential of a Whole Institution Approach and Practice Theory to Support CCE

Using a whole institution framework for the current research served to direct the researcher's gaze across the entire school and out into the local community. Pairing this approach with practice theory allowed the researcher to focus on key elements of practice instead of the individuals carrying those practices. By focusing on the elements that make up practices, and how those elements and their connections may change across time/space, the researcher was better able to see what "ingredients" for practice are available, as well as possible connection points for future practices. A whole institution approach to CCE can also enable a culture of climate action practices (Hargis & McKenzie, 2021). When climate action practices exist within and across whole institution domains, the practice of CCE is strengthened. A whole institution approach to CCE ensures that all individuals at the school and the local community are involved in climate action and correspond well with theories of practice, which focus on collaboration, not individuals.

Theories of practice harbor the potential to envisage social change regarding the major issues of our time, such as climate change through their focus on everyday practices (Buegger, 2014; Shove, 2010; Shove & Spurling, 2013). The methodological shift from following individuals to "the elements of practice" they carry (Shove et al., 2012, p. 22) enables a departure from traditional knowledge, attitude, and behavior approaches within CCE and research (Brownlee et al., 2013). This shift is necessary because "more knowledge does not necessarily equate with changed actions; but rather…it is practice itself that enables change" (McKenzie & Bieler, 2016, p. 123). By better attending to practical experience, educational approaches can "build the potential for cultural change as a response to the pressing critical issues of our times" (McKenzie & Bieler, 2016, p. 9).

Acknowledgments Words cannot express my gratitude to my supervisor, Dr. Marcia McKenzie, whose guidance, suggestions, feedback, and support were integral to the completion of this research. I would also like to express my deepest appreciation to the members of my advisory committee for their continual support and direction: Drs. Dianne Miller, Paul Orlowski, James Robson, Chelsea Willness, and Alex Wilson. I'm also extremely grateful to my research participants, who volunteered their time and welcomed me into their community. Finally, this work would not have been possible without generous funding and support from the Sustainability and Education Policy Network.

References

Amel, E., Manning, C., Scott, B., & Koger, S. (2017). Beyond the roots of human inaction: Fostering collective effort toward ecosystem conservation. *Science, 356*, 275–279.

Amsler, S., & Jeannie, K. (in press). Challenging complacency in K–12 climate change education in Canada: Decolonial and Indigenous perspectives for designing curricula beyond sustainable development. In

M. Lam, C. Skyhar, & A. Farrell (Eds.), *Teaching in the anthropocene*. Canadian Scholars' Press.

Antoine, A., Mason, R., Palahicky, S., & Rodriguez de France, C. (2018). *Pulling together: A guide for curriculum developers*. BCcampus. Retrieved from https://opentextbc.ca/indigenizationcurriculumdevelopers/

Bieler, A., Haluza-Delay, R., Dale, A., & McKenzie, M. (2018). A national overview of climate change education policy: Policy coherence between subnational climate and education policies in Canada (K-12). *Journal of Education for Sustainable Development, 11*(2), 63–85.

Brownlee, M. T. J., Powell, R. B., & Hallo, J. C. (2013). A review of the foundational processes that influence beliefs in climate change: Opportunities for environmental education research. *Environmental Education Research, 19*(1), 1–20. https://doi.org/10.1080/13504622.2012.683389

Buegger, C. (2014). Pathways to practice: Praxiography and international politics. *European Political Science Review, 6*(3), 383–406.

Callison, C. (2014). *How climate change comes to matter: The communal life of facts*. Duke University Press.

CCUNESCO. (2020). *Teachers' toolkit: UNESCO schools network in Canada*. CCUNESCO.

Chopin, N., Hargis, K., & McKenzie, M. (2018). *Building climate-ready schools in Canada: Towards identifying good practices in climate change education*. Sustainability and Education Policy Network, University of Saskatchewan, Saskatoon, Canada. Retrieved from https://sepn.ca/wp-content/uploads/2018/12/CCUNESCO-ASPnet-Pilot-Project-Data-Driven-Report-NoHLs-2018-12-05-1.pdf

Clayton, S., Manning, C. M., Krygsman, K., & Speiser, M. (2017). *Mental health and our changing climate: Impacts, implications, and guidance*. American Psychological Association and ecoAmerica.

Cook, J. W. (Ed.). (2019). *Sustainability, human well-being, and the future of education*. Palgrave Macmillan.

Creswell, J. W., & Poth, C. N. (2018). *Qualitative inquiry and research design: Choosing among five approaches* (4th ed.). SAGE Publications.

Doherty, T. J., & Clayton, S. (2011). The psychological impacts of global climate change. *American Psychologist, 66*(4), 265–276. https://doi.org/10.1037/a0023141

Dooley, L., Sheats, J., Hamilton, O., Chapman, D., & Karlin, B. (2021). *Climate change and youth mental health: Psychological impacts, resilience resources and future directions*. https://seechangeinstitute.com/sci_project/climatechangeandyouthmentalhealth/

Dyment, J. E., & Hill, A. (2015). You mean I have to teach sustainability too? Initial teacher education students' perspectives on sustainability cross-curriculum priority. *Australian Journal of Teacher of Education, 40*(3), 21–35.

Eriksen, S., Aldunce, P., Bahinipati, C. H., Martins, R. D., Molefe, J. I., Nhemachena, C., O'Brien, K., Olorunfemi, F., Park, J., Sygna, L., & Ulsrud, K. (2011). When not every response to climate change is a good one: Identifying principles for sustainable adaptation. *Climate and Development, 3*(1), 7–20.

Fenwick, T., Edwards, R., & Sawchuk, P. (2011). *Emerging approaches to educational research: Tracing the sociomaterial*. Routledge.

Ferland, N. (n.d.). *This is indigenous land: An indigenous land-based approach to climate change education*. Global Environmental Education Partnership.

Field, E., Schwartzberg, P., & Berger, P. (2019). *Formal report for learning for a sustainable future*. York University Printing Services.

Funes, Y. (2022, April 4). Yes, colonialism caused climate change, IPCC reports. *Atmos*. https://atmos.earth/ipcc-report-colonialism-climate-change/#:~:text=The%20Intergovernmental%20Panel%20on%20Climate,driver%20of%20the%20climate%20crisis

General Assembly resolution 70/1. (2015, September 25). *Transforming our world: The 2030 agenda for sustainable development*, A/RES/70/1. Retrieved from http://www.un.org/ga/search/view_doc.asp?symbol=A/RES/70/1&Lang=E

Goldberg, M. H., Gustafson, A., Rosenthal, S. A., & Leiserowitz, A. (2021). Shifting Republican views on climate change through targeted advertising. *Nature Climate Change, 11*, 573–577.

González-Gaudiano, E., & Meira-Cartea, P. (2010). Climate change education and communication: A critical perspective on obstacles and resistances. In F. Kagawa & D. Selby (Eds.), *Education and climate change: Living and learning in interesting times* (pp. 13–34). Routledge.

Hager, P. (2012). Theories of practice and their connection to learning: A continuum of more or less inclusive accounts. In P. Hager, A. Lee, & A. Reich (Eds.), *Practice, learning and change: Practice theory perspectives on professional learning* (pp. 17–32). Springer.

Hargis, K., & McKenzie, M. (2021). *Responding to climate change education: A primer for K-12 education*. Sustainability and Education Policy Network. https://sepn.ca/resources/report-responding-to-climate-change-education-a-primer-for-k-12-education/

Hargis, K., Chopin, N., & McKenzie, M. (2018). *Ten Canadian schools' stories of climate action*. Sustainability and Education Policy Network, University of Saskatchewan, Saskatoon, Canada. Retrieved from https://sepn.ca/wp-content/uploads/2018/12/CCUNESCO-ASPnet-Pilot-School-Stories-EN-2018-12-04-1.pdf

Hargis, K., McKenzie, M., & Levert-Chiasson, I. (2021). Whole-institution approaches to climate change education: Towards preparing school systems to be climate-ready. In R. Iyengar & C. Kwauk (Eds.), *Curriculum and learning for climate action: Towards an SDG 4.7 roadmap for systems change* (UNESCO-IBE book series) (pp. 43–66). Brill Publishers.

Harris, A. (2003). Distributed leadership in schools: Leading or misleading? *Management in Education, 16*(5), 10–13. (SAGE Publications).

Hill, A., & Dyment, J. E. (2016). Hopes and prospects for the sustainability cross-curriculum priority: Provocations from a state-wide case study. *Australian Journal of Environmental Education, 32*(3), 225–242.

Hornsey, M. J., Harris, E. A., Bain, P. G., & Fielding, K. S. (2016). Meta-analyses of the determinants and outcomes of belief in climate change. *Nature Climate Change, 6*, 622–626. https://doi.org/10.1038/NCLIMATE2943

Hughes, J., Jewson, N., & Unwin, L. (2007). Introduction. Communities of practice: A contested concept in flux. In J. Hughes, N. Jewson, & L. Unwin (Eds.), *Communities of practice: Critical perspectives* (pp. 1–16). Routledge.

Kahan, D. M., Peters, E., Wittlin, M., Slovic, P., Ouellette, L. L., Braman, D., & Mandel, G. (2012). The polarizing impact of science literacy and numeracy on perceived climate change risks. *Nature Climate Change, 2*, 732–735.

Kanbur, R. (2015). *Education for climate justice: The many faces of climate justice: An essay series on the principles of climate justice*. Mary Robinson Foundation, Trinity College. http://www.mrfcj.org/pdf/faces-of-climate-justice/Education-for-ClimateJustice.pdf

Kapyrka, J., & Dockstator, M. (2012). Indigenous knowledge and Western knowledges in environmental education: Acknowledging the tensions for the benefits of a "two-worlds" approach. *Canadian Journal of Environmental Education, 17*, 97–112.

Lave, J., & Wenger, E. (1991). *Situated learning: Legitimate peripheral participation*. Cambridge University Press.

Li, N., & Su, L. Y.-F. (2018). Message framing and climate change communication: A meta-analytical review. *Journal of Applied Communications, 102*(3), 1–14.

Mathie, R. G., & Wals, A. E. J. (2022). *Whole school approaches to sustainability: Exemplary practices from around the world*. Education and Learning Sciences/Wageningen University.

Mbah, M., Ajaps, S., & Molthan-Hill, P. (2021). A systematic review of the deployment of Indigenous knowledge systems toward climate change adaptation in developing world contexts: Implications for climate change education. *Sustainability, 13*, 4811.

McKenzie, M., & Bieler, A. (2016). *Critical education and sociomaterial practice: Narration, place, and the social*. Peter Lang.

McKenzie, K. B., & Locke, L. A. (2014). Distributed leadership: A good theory but what if leaders won't, don't know how, or can't lead? *Journal of School Leadership, 24*, 164–188.

MECCE Project and NAAEE. (2022). *Mapping the landscape of K-12 climate change education policy in the United States*. MECCE Project and NAAEE.

Monroe, M. C., Plate, R. R., Oxarart, A., Bowers, A., & Chaves, W. A. (2017). Identifying effective climate change education strategies: A systematic review of the research. *Environmental Education Research*, 1–22. https://doi.org/10.1080/13504622.2017.1360842

Nesterova, Y. (2020). Rethinking environmental education with the help of Indigenous ways of knowing and traditional ecological knowledge. *Journal of Philosophy of Education, 54*(4), 1047–1052.

Nicholls, J., & Thorne, M. (2018). Queensland teachers' relationship with the sustainability cross-curriculum priority. *Australian Journal of Environmental Education, 33*(3), 189–200.

Nicolini, D. (2013). *Practice theory, work, and organization: An introduction*. Oxford University Press.

Norgaard, K. M. (2011). *Living in denial: Climate change, emotions, and everyday life*. MIT Press.

Orlikowski, W. J. (2007). Sociomaterial practices: Exploring technology at work. *Organization Studies, 28*(9), 1435–1448.

Randall, R. (2009). Loss and climate change: The cost of parallel narratives. *Ecopsychology, 1*(3), 118–129. https://doi.org/10.1089/eco.2009.0034

Reckwitz, A. (2002). Toward a theory of social practices: A development in culturalist theorizing. *European Journal of Social Theory, 5*(2), 243–263.

Regier, R. L. (2019). *Teachers' experiences in engagement with partners in environmental and sustainability education* [Master's thesis, University of Saskatchewan]. HARVEST.

Rowling, M. (2019, September 24). UN climate summit exposes struggle to ditch fossil-fuel 'status quo.' *Reuters*. https://www.reuters.com/article/us-climate-changesummit-policy/u-n-climate-summit-exposes-struggle-to-ditch-fossil-fuel-statusquo-idUSKBN1W924R

Saldaña, J. (2016). *The coding manual for qualitative researchers*. SAGE.

Schatzki, T. R., Knorr Cetina, K., & Savigny, E. V. (Eds.). (2001). *The practice turn in contemporary theory*. Routledge.

Shove, E. (2010). Beyond the ABC: Climate change policy and theories of social change. *Environment and Planning A, 42*, 1273–1285.

Shove, E., & Spurling, N. (Eds.). (2013). *Sustainable practices: Social theory and climate change*. Routledge.

Shove, E., Pantzar, M., & Watson, M. (2012). *The dynamics of social practice: Everyday life and how it changes*. Sage.

Spillane, J. P. (2005). Distributed leadership. *The Educational Forum, 69*(2), 143–150.

Sultana, F. (2022). The unbearable heaviness of climate coloniality. *Political Geography, 99*, 1–2638.

Tanyanyiwa, V. I. (2019). Indigenous knowledge systems and the teaching of climate change in Zimbabwean secondary schools. *SAGE Open, 9*(4), 1–11.

UNESCO. (2010). *The UNESCO climate change initiative: Climate change education for sustainable development*. UNESCO.

UNESCO. (2015). *Global citizenship education: Topics and learning objectives*. UNESCO. https://unesdoc.unesco.org/ark:/48223/pf0000232993

UNESCO. (2016). *Getting climate ready: A guide for schools on climate action*. UNESCO.
UNESCO. (2017). *Changing minds, not the climate: The role of education*. https://www.gcedclearinghouse.org/sites/default/files/resources/190248eng.pdf
UNESCO. (2019). *Educational content up close: Examining the learning dimensions of education for sustainable development and global citizenship education*. UNESCO.
UNESCO. (2021, June 21). *Land as teacher: Understanding Indigenous land-based education*. https://en.ccunesco.ca/idealab/indigenous-land-based-education
UNESCO and UNFCCC. (2016). *Action for climate empowerment: Guidelines for accelerating solutions through education, training and public awareness*. UNESCO.
UNFCCC. (2015). *Adoption of the Paris agreement. 21st conference of the parties*. United Nations. Retrieved December 1, 2017 from https://unfccc.int/resource/docs/2015/cop21/eng/l09.pdf
United Nations. (2019). *Climate justice*. https://www.un.org/sustainabledevelopment/blog/2019/05/climate-justice/
Viswanathan, L. (2020). *Indigenous land-based learning: A way to take action on climate change*. Indigenous Climate Hub. https://indigenousclimatehub.ca/2020/09/indigenous-land-based-learning-a-way-to-take-action-on-climate-change/
Vizina, Y. N. (2018). *Indigenous knowledges and sustainability in post-secondary education* [Doctoral dissertation, University of Saskatchewan]. HARVEST.
Wals, A. E. J., & Mathie, R. G. (2022). Whole school responses to climate urgency and related sustainability challenges. In M. A. Peters & R. Heraud (Eds.), *Encyclopedia of educational innovation* (pp. 1–8). Springer Nature.
Wibeck, V. (2014). Enhancing learning, communication and public engagement about climate change – Some lessons from recent literature. *Environmental Education Research, 20*(3), 387–411. https://doi.org/10.1080/13504622.2013.812720
Wood, B. E., Cornforth, S., Beals, F., Taylor, M., & Tallon, R. (2014). Sustainability champions? Academic identities and sustainability curricula in higher education. *International Journal of Sustainability in Higher Education, 17*(3), 342–360.
Yohe, G. W., Lasco, R. D., Ahmad, Q. K., Arnell, N. W., Cohen, S. J., Hope, C., Janetos, A. C., & Perez, R. T. (2007). Perspectives on climate change and sustainability. In M. L. Parry, O. F. Canziani, J. P. Palutikof, P. J. van der Linden, & C. E. Hanson (Eds.), *Climate change 2007: Impacts, adaptation and vulnerability. Contribution of working group II to the fourth assessment report of the intergovernmental panel on climate change* (pp. 811–841). Cambridge University Press.

Utilizing Collaborative Self-Study to Explore Pedagogies for Sustainability

19

Christina Phillips and Patrick Howard

Key Message
This case study has significance in the field of critical and reflexive self-study for teacher education and may provide ideas about practical approaches to sustainability from a whole school perspective. This work may contribute to the disruption of a status quo regarding sustainable practices in institutions, disrupting racism through the inclusion of Indigenous perspectives, and serve as a counter-narrative to the exploitation of our environments.

19.1 Utilizing Collaborative Self-Study to Explore Pedagogies for Sustainability

19.1.1 Context Setting

Defining sustainability, and how best to implement sustainable approaches, remains contested in academe and in elementary, secondary, and post-secondary schooling systems. Whole school approaches (WSA) (Wals & Mathie, 2022) to sustainability where various facets or currents of environmental education (for example, Sauvé, 2005) are integrated with subject matter, leadership practices, and everyday classroom routines remain elusive in many contexts as they serve as counter-narratives to the status quo promoting over-consumption and exploitation of environments and people.

In this chapter, we present a case study exploring how a Canadian post-secondary institution, Cape Breton University (CBU), has enacted integrative (for example, Indigenous perspectives), whole school approaches to sustainability in creative and immersive ways. We have utilized reflexive, collaborative, critical self-study to examine program objectives in teacher education and how sustainability goals have been interpreted, translated, and implemented at the course level in pre-service teacher education (that is, science methods courses) and at the graduate level (that is, an applied research project course for the Master of Education in Sustainability, Creativity, and Innovation).

Cape Breton University (CBU) is a small university located in Sydney, Nova Scotia, Canada. The university's commitment to sustainability initiatives makes it an emerging leader on a variety of fronts among post-secondary institutions in the province of Nova Scotia and, nationally, in Canada. CBU has invested in renewable technologies in new building construction, facility infrastructure programmes, and in meeting its present and future energy demands. Most notable was the

C. Phillips (✉)
Ontario Institute for Studies in Education, Toronto, ON, Canada
e-mail: c.phillips@utoronto.ca

P. Howard
Cape Breton University, Sydney, NS, Canada

A. E. J. Wals et al. (eds.), *Whole School Approaches to Sustainability*, Sustainable Development Goals Series, https://doi.org/10.1007/978-3-031-56172-6_19

Fig. 19.1 The Cape Breton University Campus located in Sydney, Nova Scotia, Canada

construction of a wind farm to move the campus toward being carbon neutral (Aschaiek, 2016). Cape Breton University is also home to the Verschuren Centre for Sustainability in Energy and the Environment (see Fig. 19.1).

The Verschuren Centre, established in 2009, is a registered not-for-profit research, development, and demonstration facility providing contracted services to industry and community in the areas of bioprocessing, carbon transformation, energy storage, and sustainable resource use (Government of Canada, 2022). The Centre was established to further the commitment to local and global sustainability.

19.1.2 L'nu Communities and Connections

Cape Breton Island is home to five L'nu communities which represent the largest population concentration of Mi'kmaw people in Mi'kma'ki—the ancestral home and un-surrendered territory of the Mi'kmaw people. L'nu: or L'nuk is the term the Mi'kmaq use to describe themselves as Indigenous people. It means "the people" (see: http://bit.ly/3IeXH8K). The territory includes all of what is now Nova Scotia and Prince Edward Island, the Gaspé Peninsula of Quebec, the north shore of New Brunswick and inland to the Saint John River watershed, eastern Maine, and part of Newfoundland, including the islands in the Gulf of Saint Lawrence as well as St. Pierre and Miquelon (Mi'kmaq Resource Centre, n.d.). In 1998, CBU established what is now called Unama'ki College to broaden Mi'kmaq and Aboriginal courses and programming at the university. That year also saw the opening of the Mi'kmaq Resource Centre as a repository of academic and research documents related to Mi'kmaw history, culture, and language. New investments were announced recently to support a research centre for Mi'kmaw environmental justice and climate change studies (Connors, 2020).

19.1.2.1 Etuaptmumk (Two-Eyed Seeing)

The groundbreaking concept of Two-eyed Seeing (Etuaptmumk) was developed at Cape Breton University. This is an Integrative Science program based on the collaborative efforts of a Western scientist and Mi'kmaw scholars (Institute for Integrative Science and Health, n.d.) where issues pertaining to science and the environment are viewed through the 'two eyes' of both Western and Mi'kmaw world views.

19.1.3 Education for Sustainability and Cape Breton University

Cape Breton University has endeavoured to make education for sustainability a core focus in its teacher education programming. A decade ago, the Council of Ministers of Education, Canada, in partnership with the International Institute for Sustainable Development and Learning for a Sustainable Future, surveyed Canadian teacher education institutions to "gain a better understanding of how they are incorporating education for sustainable development into their pre-service programs, research, and other activities" (CMEC, 2012. p. 1). At the time of the 2012 Canadian Council of Ministers of Education survey, Cape Breton University was in the early stages of reorienting its teacher education programming for sustainability. The CBU reorientation was based largely on theoretical and conceptual frameworks related to the values-driven and normative undertaking that is education for sustainable development; in other words, faculty and staff engaged in the hard work of "decid(ing) which themes to emphasize within their curriculums, programs, practices, and policies to ensure that teacher-education programs fit the environmental, social, and economic conditions and goals of their communities, regions, and nations" (Hopkins & McKeown, 2005, p. 15).

A decade later, the understanding of education for sustainability within the CBU Department of Education and across the institution continues to mature and develop. The concept of education for sustainability has blossomed in new directions to inform program development in both pre-service teacher education and in professional learning for in-service teachers. Education for sustainability influences approaches to equity, diversity, and inclusion; issues of social justice; student and teacher mental health and well-being; the responses to truth and reconciliation; and the efforts to support the reclamation and revitalization of the Mi'kmaw language. In recent years, the Department of Education at CBU continues to find ways to deliver teacher education that are "transformative, which empowers lifelong learners with the knowledge, values and competencies to not only make informed decisions but also to bring about the individual and collective change required to positively impact our societies..." (UNESCO, 2022, p. 3).

19.2 Literature Review

19.2.1 Pedagogies for Sustainability Education in Higher Education

A great deal has been written about the long history to transform K-12 education through progressive, experiential educational approaches. Dewey (1938/1997) showed that complexity and change meant that education could not be reduced to formulaic preparation for life; it needed to be *life* itself, messiness and unpredictability included. Progressive educators looked for ways to realize Dewey's vision through valuing diversity, building on the interests of students, organizing learning in larger more holistic units, connecting school to the surrounding community, and developing citizenship (Howard, 2020; Waks, 2013).

Transformative teaching and learning that incorporates values of sustainability and progressive pedagogies is still considered an emerging field. Studies conducted reveal "that the concept of education for sustainable development has not been sufficiently integrated into the concept of transformation in higher education institutions" (Filho et al., 2018, 286). Systematic literature reviews have found that higher education institutions have direct and significant impacts on sustainable development through research that impacts society, the environment, and the economy. Higher education institutions must do more to embed sustainability into their systems, and major gaps in the literature provide ample space for future research in this rapidly evolving field of inquiry. "[M]ore research with a holistic perspective... a whole institution approach would also help identify impact areas and stakeholder groups that are currently underrepresented in the literature".

Research into pedagogical approaches for sustainability in higher education institutions has increased in the past two decades. Frameworks gave been offered to help educators in creating and updating courses to create a more "complete, holistic, and systemic sustainability education for future leaders, decision makers, educators, and change agents". However, more research is needed on pedagogical approaches and the potential to develop sustainability competences. Such forms of pedagogy and learning are only beginning to emerge in higher education and it has been argued that to respond to the climate crisis and the wicked problems of sustainability "there is a need for more exploratory, transgressive forms of learning in our institutions…[that] will require an *integration* of sustainability-oriented higher education teaching research and community engagement processes into possibilities for learning...". Recent research has taken up the role of HEIS in contributing to the United Nations Sustainability Goals (SDGs) and conclusions that call for sustainability principles needing to be at the heart of higher education institutions' strategic planning and incorporated into the organizational culture are significant findings (Agbedahin, 2019; Zaleniene & Pereira, 2021).

19.2.2 School-Oriented Models of Sustainability

Many important frameworks to support environmental and sustainability education have been developed and to varying degrees, implemented in different schooling contexts (that is, typically in elementary and secondary schools; not higher education). For example, frameworks such as those represented here are a sampling of the excellent efforts to reorient education for peace, social justice, citizenship, well-being, and ecological literacy:

- Eco and green school frameworks (Metzger, 2015),
- Living schools (O'Brien & Howard, 2020),
- Whole school approaches to sustainable development (Wals & Mathie, 2022),
- Education for sustainable development educator competences (UNECE, 2011),
- United Nations and UNESCO global education for sustainable development programs (UNESCO-UNEP, 1977).
- (UNESCO, 2005, 2006, 2016) including the Sustainability Development Goal (SDG).
- Quality Education for all (UNESCO, 2021a, b).

These lenses all provide clarity about key skills and competencies, for both teachers and students, around an educational vision for future-oriented teaching and learning with an overarching goal of realizing a global shift to reimagined education systems to create a "safer, kinder, and flourishing world and planet" (Sterling, 2017, p. 42).

In higher education research, a gap has been identified relative to frameworks that would allow for a more comprehensive understanding of the approaches and impacts of sustainability education on teaching and learning, policy, individual behaviour, student values and beliefs, faculty responsiveness and institutional culture are currently underexplored and merit further attention. Qualitative approaches may be a tool to help understand how these concepts might be functioning in higher education contexts. Findler et al. (2019) report that the use of qualitative data such as narratives might help to assess impacts and change in higher education that would otherwise be difficult to measure using quantitative data.

The work of Lucie Sauvé (2005) is a helpful extension of these frameworks, ideas, and notions regarding concrete approaches to sustainability and various facets of environmental education, and by extension sustainability education. In her explorations, she characterizes many common forms of environmental education envisioned through the metaphor of 'currents' such as: naturalist; holistic; conservationist/resourcist; bioregionalist; problem-solving; praxic; systemic; socially critical; scientific; feminist; ethnographic; value-centred; eco-educational; and sustainable development/sustainability currents may serve as a useful guiding framework to critically analyse programming and course delivery at the higher education level. Of note, is the absence of a comprehensive current to devote exclusively to Indigenous world views on sustainability and environmental education (that is, although this seems to be a dimension of the 'ethnographic current').

The case study reported here, addresses a gap in the literature (that is, how higher education is enacting sustainability education) and explores how, and to what extent Cape Breton University has enacted integrative, whole school approaches to sustainability as per descriptions and the theoretical framing of Wals and Mathie (2022) in teacher education programming via collaborative and reflexive self-study methods.

19.3 Methods

This reflexive, small-scale qualitative case study explored the concept of immersive and inclusive sustainability in higher education programs at Cape Breton University (that is, in both pre-service teacher education and graduate-level education programs) via collaborative self-study methods. We have drawn from the ideas of Merriam (1988) to characterize our work where she described a case study as "...an intensive, holistic description and analysis of a single instance, phenomenon, or social unit" (p. 21). The case study explores holistically the extent to which whole school approaches (Wals & Mathie, 2022) to sustainability education at Cape Breton University have been enacted in our programming in the Department of Education within the School of Education and Health. In alignment with case study approaches, our case has clearly demarcated institutional boundaries (that is, both physical and virtual) (Merriam, 1988; Miles & Huberman, 1994).

19.3.1 Self-Study as a Research Method in Teacher Education

Self-study is a relatively new field of qualitative inquiry that emerged in the 1990s (Loughran, 2018). Education researchers and teachers alike have recognized the power of this methodology to deconstruct classroom practices and educational observations via critical reflexivity and situated inquiry.

This approach is commonly referred to as S-STEP or the Self-study of Teacher Education Practices in teacher researcher circles. Loughran (2018) describes this methodology as an emergent field based on the recognition of the importance of "reflective practice, action research and practitioner inquiry..." (p. 1). Loughran also connects the self-identification of problematic aspects of practice with the seminal work of Dewey (1933) where a problem often forms the foundation for reflective practices. The work of Samaras and Freese (2006) may also be beneficial when exploring scholarly definitions of self-study where they characterize the method as "... situated inquiry, process or paradoxical: individual and collective, personal and interpersonal and private and public..." (pp. 40–53).

Our work, as described throughout the following segments, aligns well with these descriptions of self-study methods and intersects with the fusion of individual and collective practices. Loughran (2018) reminds us that the 'self' and change are intimately linked and also that self-study is a method tied with a research collective that focuses on both the mentoring and development of both individual and community. This method is not focused on the quantity of voices, but rather quality engagement with practices, reflexivity, and the commitment to developing a better practice for self and community.

19.3.1.1 Data Sources

Data sources that informed the case included:

- Departmental documents (for example, course syllabi).
- Journal entries and responses (collected and stored online in a collaborative document

accessible to both professors engaged in this work) to reflexive questions about sustainability and institutional practices regarding sustainability based on the work of Wals and Mathie (2022).

19.3.1.2 Reflexive Engagement and Data Analysis

Each professor involved in the study explored prompting questions on whole school approaches to sustainability based on a framework presented in Wals and Mathie (2022). These questions guided critical and self-reflexive analysis of the courses taught and the challenges and successes regarding the implementation of broader university objectives promoted by the school. The prompts used to guide critical self-reflection regarding whole school approaches to sustainability included:

- What does sustainability mean to me? How does this manifest in my day-to-day life? How do I enact sustainability in my work? Classes? In other contexts?
- -To what extent has our institution experimented with sustainability? Creating sustainability? Examples?
- What new learning processes contribute to sustainable practices? Are there learning environments associated with CBU that enhance learning about sustainability?
- How do we as educators manifest and immerse our students in sustainability via course design, content, assessment, and other ways? What are some examples?
- To what extent are all staff engaged in PD that supports sustainability? What kinds of PD? Organizations?
- To what extent do we engage in connections between the school and community to enhance sustainable practices? How do we do this?

These prompts were posted on a private, collaborative, and interactive platform (that is, Microsoft™ OneDrive), housed by our institution so that each researcher could reflect upon these questions and add their perspectives to the digital platform. Kitchen (2022) draws from the work of Loughran and Brubaker (2015) to remind us of the importance of the use of collaboration and critical friends in self-study work to facilitate understandings of data interpretation and to promote trustworthiness in analysis work. The outcomes of our journaling, discussions, and collaborations using these prompts (that is, based on WSA) supported out deconstruction and analysis of our institutional WSA used at Cape Breton University. These prompts helped us to determine successes and challenge areas.

Additionally, Sauvé's (2005) environmental education currents framework was used to analyse course syllabi and our journal entries (that is, responses to the above prompts) for various facets of sustainability, connections with WSA, and environmental education (please see Appendix A for a more fulsome description of Sauvé's currents). Documents such as course syllabi and journal entries were assigned codes using the 'review and add comment' function embedded in Word™ documents. Sauvé's (2005) currents served as a framework for an a priori coding approach (Saldaña, 2012) to help unpack focal areas that seemed connected with facets of environmental education.

19.4 Findings

In consideration of WSAs to sustainability and sustainable practices, we have engaged with the work of Wals and Mathie (2022) to critically and reflexively engage in self-study to deconstruct, unpack, and critically reflect upon our educational practices, departmental engagement, and institution-commitments, as in alignment with the call to oscillate between the 'self' and the community when using a methodology such as collaborative self-study. We also found it beneficial to engage in an a priori coding approach to analyse documents and artefacts that are bounded within our institutional case study exploring how and perhaps why we are engaging with sustainable practices.

19.4.1 A Portrait of Sustainability

Our findings paint a portrait of sustainable practices at Cape Breton University grounded in our collaborative self-study and document coding processes. Our findings are organized in this section by sub-headings to present emergent themes followed by descriptions and examples of how these themes were constructed from various examples and quotations from our data. The following graphic (see Fig. 19.2) may provide a useful overview of some key emergent findings from this work; a sustainability 'DNA fingerprint' for Cape Breton University.

19.4.2 Incorporation and Prioritization of L'Nu Ways

19.4.2.1 Individual/Departmental Level

Education Courses and Program. Our post-secondary education programs and courses emphasize commitments to sustainability and Indigeneity in different ways. Our Bachelor of Education students can opt to receive additional training and learnings by pursuing a focus in Indigeneity, sustainability education, or French. Pre-service teachers can opt for an education for sustainability concentration and take Teaching and Learning for a Sustainable Future and Sustainable Well-being and Deep Learning. Mi'kmaw language preservation and revitalization was deemed to be a priority area when L'nu communities were consulted as part of the pre-service education program planning. Mi'kmaw language courses and the preparation of teachers for the Mi'kmaw language immersion schools provides students with additional opportunities to support Indigenous language revitalization and growth.

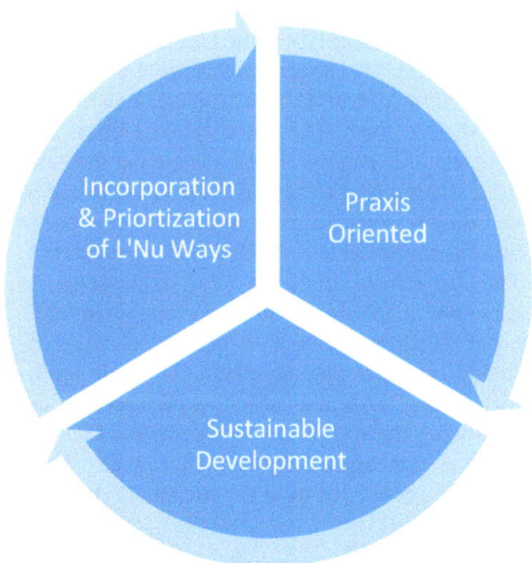

Fig. 19.2 Graphic depicting sustainability 'fingerprint' of key facets of individual, departmental, and collective institutional focus

Our pre-service teacher candidates engage in a number of courses focused on social justice, inclusion, and connections to specialized subject areas where currents of sustainability are expected to flow across all offerings. A commonality across all course syllabi is the inclusion of our Acknowledgment of Territory,

> Cape Breton University's Department of Education recognizes that Cape Breton Island is in Mi'kma'ki, the ancestral and unceded territory of the Mi'kmaq People. This territory is covered by the "Treaties of Peace and Friendship" which Mi'kmaq and Wolastoqiyik (Maliseet) people first signed with the British Crown in 1725. The treaties did not deal with surrender of lands and resources but in fact recognized Mi'kmaq and Wolastoqiyik (Maliseet) title and established the rules for what was to be an ongoing relationship between nations.

Our courses engage students in reflection and connection with Indigeneity, drawing from the work done at Cape Breton University and local Indigenous communities. For example, our science education courses explored Indigeneity and Indigenous forward perspectives in many ways. Christina provides commentary here from her reflections:

> In my science methods courses, I choose materials and resources that my students can read and process in connection with the grade level they intend to teach and work with such as Two-Eyed Seeing; ReKindling Traditions; Deeper toolkit; Learning for a Sustainable Future. These sources have many practical suggestions and ideas for educators to implement sustainable practices in courses. (CP, July 18, 2022)

Indigenous perspectives, such as Two-Eyed Seeing, that draw from reciprocal relationships built between Cape Breton University and local Mi'kmaw communities, were emphasized using in-class viewing of videos and explorations of lesson plans and ideas that could be used in science education (for example, ReKindling Traditions Project see: https://education.usask.ca/ccstu/welcome.html). This theme is perhaps not surprising given the strong ties to the Indigenous community and the larger institutional prioritization of this connection to be described in our following section.

Fig. 19.3 An example of Mi'kmaw artwork displayed on CBU campus

19.4.3 Institutional Level

19.4.3.1 Physical Environments

The institution attempts to embody and enact sustainability in many ways such as through the construction and use of environmentally sustainable buildings and structures (for example, wind turbines, living wall, light-emitting diode, street lighting, and solar roof arrays) as well as cultural, aesthetic considerations including the prominent display of Mi'kmaw artwork (see Fig. 19.3). Cape Breton University is home to Unama'ki College (as described in our introduction/context setting segment). In addition to the work done in individual departments and courses, CBU has an entire college devoted to Indigenous thought and scholarship housed at our institution.

These connections to the more than human world and land promote a physical connection to environments where our students can tangibly learn about sustainability and about Indigenous perspectives from Indigenous communities.

19.4.3.2 Whole School Approaches Via Strategic Academic Plans

There is a deep commitment across the university to strengthen relationships with our L'Nu communities. This has taken on renewed urgency as a response to the Truth and Reconciliation Report (please see: https://www.rcaanc-cirnac.gc.ca/eng/1450124405592/1529106060525). Patrick reflects here about this important institutional commitment:

> CBU developed an in-depth self-directed online professional development opportunity that all employees in both academic and service units complete – it is called "Learning About the L'nu Way." This learning experience provides staff and faculty of CBU an opportunity for professional development that addresses some of the calls to action from the Truth and Reconciliation Commission Report of Canada, as they relate to post-secondary education and institutions of higher learning. "Learning About the L'nu Way" is a multifaceted learning experience that provides foundational knowledge of who the L'nu are, their history, and their culture. The learning experience also includes a Mi'kmaw cultural component that sees participants attend a cultural event or workshop in or hosted by a local L'nu community. Some of the cultural components that CBU employees participated in ranged from attending a powwow or mawiomi, to taking part in workshops using traditional beading techniques making earrings or dream catchers, among others. Difficult topics such as the Residential School experience and the continuing impacts of the 60s Scoop and the contribution to intergenerational trauma are also a part of the impactful learning experience. (PH reflections, Jan. 2023)

It is important to note the monetary considerations tied with these examples. There has been a significant budget allocated to the promotion and integration of L'Nu worldviews through workshops, outreach, and of course, Unama'ki College which is housed at CBU.

19.4.4 Sustainable Development and Socially Critical Sustainable Practices

Sustainable development and socially critical sustainable practices were apparent in many of our reflections and analysis. We are drawing from the work of Sauvé (2005) here to describe the emergent currents below at both the individual/departmental and institutional levels.

19.4.4.1 Individual/Departmental Level

Socially critical pedagogies. An emergent theme from explorations of course materials and critical reflections was the emphasis placed on culturally responsive teaching in science and engagement with communities. This was explored in a number of ways such as through the Draw-A-Scientist Test; deconstructions and analysis of media depictions of science and scientists (for example, Big Bang Theory US television series) and students conducting self-analysis with the support of digital tools such as the Harvard Implicit Bias Test (see: https://implicit.harvard.edu/implicit/takeatest.html). These pedagogies also intersect with a focus on the promotion and inclusion of L'nu ways; however, this broader characterization also considers connections with other socially critical activities such as the role of gender in science and problematic stereotypes of scientists that may perpetuate the notion that science is not for all.

19.4.4.2 Innovation and Science s'mores

Science, technology, engineering, and mathematics (STEM) and its associated fields (for example, STEAM) have become a topic of intensive study and speculation in both academic and practitioner circles. Blendings of STEM and environmental connections could be yet another interesting facet of study and debate. Engineering thought and the introduction of aspects of the design process and prototype building are more and more frequently being introduced in science courses. An activity that may blend these ideas with aspects of sustainable development is the design and testing of solar ovens made from cardboard, paper, aluminium foil and plastic wrap (for example, see https://www.homesciencetools.com/article/how-to-build-a-solar-oven-project/). The use of such ideas can intersect with engineering concepts of design and also facets of sustainable development such as reducing waste or substituting more environmentally hazardous materials for less harmful materials. Christina describes this activity here from her reflections about engaging her pre-service teachers in this activity in her science methods course:

> We also build solar ovens (see Fig. 19.4) to make 's'mores', a traditional Canadian treat made of cookies, marshmallows and chocolate. The challenge is to design and build a small solar oven that can provide enough heat to melt the marshmallow and chocolate. We then debrief and discuss the science of solar power and connections to sustainability.

Fig. 19.4 Construction of solar ovens to make s'mores in Christina's pre-service science methods course

19.4.5 Institution Level

19.4.5.1 A Focus on Sustainable Development

Sustainability and sustainable development are also about building sustainable, resilient communities and providing access to good work and healthy living conditions. Patrick shares his reflections about the historical context out of which CBU was established and how this has formed the basis of the spirit of sustainable development that grounds much of what permeates the campus and programming here,

> Cape Breton University was created out of a community effort to establish a post-secondary institution to continue to educate the young and contribute to the development of the island communities. It was a symbol of a maturing independence to forge a unique institution separate from the satellite campuses of larger, established institutions that previously provided post-secondary education to the people of the island. Building on a proud tradition of adult education designed to lift people out of poverty and lives of dangerous exploitative work in the extractive industries of coal mining, logging, fishing and the boom bust cycles of steel making, Cape Breton was a beneficiary of the Antigonish Movement led by two Roman Catholic priests in the early 20th century who pioneered a blend of adult education, microfinancing, cooperatives, and community economic development for the purpose of providing a sustainable future for people and to increase the wellbeing of communities trapped in intergenerational poverty and exploitative labour conditions. (PH Reflections, January, 2023)

As a newcomer to the Cape Breton University community, Christina shares perspectives about facets of sustainable development that have manifested in physical structures and institutional practices. She comments on the beautiful living wall and the numerous walking trails across the campus. Christina reflects here about additional campus structures such as windmills and the practice of hiring local musicians to support the local economy,

> The windmills are incredible and I think a great example of how CBU has embraced moving toward sources of renewable energy in a highly visible way. The other example that really struck me was how many local musicians are hired to play music in the institution hallways to celebrate different events and to welcome students, staff and faculty on special occasions. (CP Reflections, July, 2022)

19.4.6 Praxis-Oriented Approaches

19.4.6.1 Individual/Department Level

As noted above, Cape Breton has a rich history of action-oriented approaches to community betterment. As such, it is probably not surprising that this praxis orientation permeates course offerings and departmental approaches in education and in the larger institution. At the level of department, we offer a project-based master's level course for in-service teachers and administrators, part of which is described here:

> Students will assess the current strengths and areas for development regarding sustainability education in their school or organization. Based on the assessment students will design, implement, evaluate and report on a project that is completed within the time frame of the course. (EDUC 6800 Course Syllabus, July, 2022)

A clear connection to a praxis orientation (Sauvé, 2005) is evidenced here with the focus placed on action. Our students who choose to take this course focus on a sustainability education challenge at their school or organization and devise a plan to address the issue which they report on at the end of the course. We have had our students complete fantastic sustainability projects in this work including building outdoor classrooms, educational materials for teaching others about sustainability education, and health and wellbeing resource development and implementation at the school level.

19.4.6.2 Institution Level

CBU is committed to community development and sustainable development across its departments. Patrick describes in his reflections below the offering of the Bachelor of Community Studies when the institution was first established here,

> A one-of-a-kind program at the time (it continues to this day) it was a progressive program for its interdisciplinary nature, and its reliance of

> authentic, experiential, project-based community learning driven by student interests and the identification of challenges and big question to which students proposed real-world action-oriented projects to address the challenges. (PH Reflections, 2023)

Another program that aligns with a focus on sustainable development across the institution is the establishment of the Centre for Sound Communities at CBU demonstrated the evolution of community-focused scholarship into the twenty-first century. The CSC is The Centre for Sound Communities is an arts-led social innovation lab at Cape Breton University that is internationally recognized and involved in carrying out research through artistic practices (mainly dance, music, theatre, and digital media). It provides training for students, faculty, and community partners and addresses systemic inequities through a focus on research that serves the needs of under-represented and under-resourced populations. See http://soundcommunities.org/

Finally, the institution's commitment to the Indigenous L'nu communities and the collective responses to the Truth and Reconciliation Report could fall under the earlier theme of L'nu inclusion or as a tie with action-oriented praxis (Sauvé, 2005). It is important to note that although we present these themes as discrete entities, that in their living embodiment, are flowing and interchangeable, much like Sauvé's (2005) overlapping currents of environmental education that we have drawn from in this work.

19.5 Conclusions and Next Steps in Whole School Approaches to Sustainability

Our reflexive practices have indicated successes at Cape Breton University in promoting L'nu (Indigenous) perspectives; a focus on sustainability and sustainable development and a praxis/action orientated curriculum. Areas of further study and need include increased capacity building for all staff and community members at Cape Breton University and increasing the connections between more traditional institutional siloes on campus. It was unclear from our critical analysis work the extent to which all faculty and staff engage and learn in a more fulsome and participatory way about sustainable practices (that is, excluding the PD available to the entire institutional community about the L'Nu Way).

The use of both WSA (Wals & Mathie, 2022) and Sauvé's (2005) currents have been useful methodological frameworks to guide our reflexive work. We suggest, based on the outcomes of this exploration that each framework might benefit from the addition of a 'current' or 'petal' that engages in Indigenous or other inclusive world views regarding sustainability.

This case study has significance in the field of critical and reflexive self-study, for teacher education and for higher education for sustainability education. in general. Gaps in the research have been identified relative to holistic assessment approaches and frameworks that allow for a more comprehensive understanding of the approaches and impacts of sustainability education on teaching and learning and institutional culture. Qualitative and narrative approaches like the one used in this research may provide insights into how to understand sustainability education from a whole school perspective. This work may contribute to the disruption of a status quo regarding sustainable practices in higher education institutions; supporting anti-racist practices through the inclusion of Indigenous perspectives on sustainability and serve as a counter-narrative to the exploitation of our environments and the people and communities who co-exist in these spaces.

References

Agbedahin, A. (2019). Sustainable development, education for sustainable development, and the 2030 agenda for sustainable development: Emergence, efficacy, eminence and future. *Sustainable Development, 22*, 669–680. https://doi.org/10.1002/sd.1931

Aschaiek, S. (2016). *Cape Breton U goes carbon neutral with its own windfarm.* University Affairs. https://www.universityaffairs.ca/news/news-article/cape-breton-u-goescarbonneutralwith-its-own-wind-farm/

CMEC. (2012). *Education for Sustainable development in Canadian faculties of education.* www.cmec.ca/

publications/lists/publications/.../esd_dean_reporten.pdf*CMEC*

Conners, H. (2020). *CBU to honour Donald Marshall Jr. with new research centre: Marshall Institute will focus on environmental justice, Indigenous approaches to climate change*. https://www.cbc.ca/news/canada/nova-scotia/cape-breton-university-marshallinstitute1.5779103

Dewey, J. (1933). *How we think*. Heath and Company.

Dewey, J. (1938/1997). *Education and experience*. Simon and Schuster.

Filho, W., Raath, S., Lazzarini, B., Vargas, V., de Souza, L., Anholon, R., Quelhas, O., Haddad, R., Klavins, M., & Orlovic, V. (2018). The role of transformation in learning and education for sustainability. *Journal of Cleaner Production, 199*, 286–295. https://doi.org/10.1016/j.jclepro.2018.07.017

Findler, F., Schonherr, N., Lozano, R., Reider, D., & Martinuzzi, A. (2019). The impacts of higher education institutions on sustainable development: A review and conceptualization. *International Journal of Sustainability in Higher Education, 20*, 23–38. https://doi.org/10.1108/IJSHE-07-2017-0114

Government of Canada. (2022). *Verschuren Centre expansion will help cleantech entrepreneurs find solutions to environmental challenges*. https://www.canada.ca/en/atlanticcanadaopportunities/news/2022/06/verschuren-centre-expansion-will-help-cleantech-entrepreneurs-find-solutions-to-environmental-challenges.html

Hopkins, C., & McKeown, R. (2005). *Guidelines and recommendations for reorienting teacher education to address sustainability*. https://unesdoc.unesco.org/ark:/48223/pf0000143370.

Howard, P. (2020). Living schools and 21st century education: Connecting what and how with why. In C. O'Brien & P. Howard (Eds.), *Living schools: Transforming education* (pp. 13–24). EWSB Press. https://www.eswb-press.org/uploads/1/2/8/9/12899389/living_schools__2020_.pdf

Institute of Integrative Science and Health. (n.d.). *Two-eyed Seeing*. http://www.integrativescience.ca/Principles/TwoEyedSeeing/

Kitchen, J. (2022). Critically inquiring as community through self-study communities of practice. In B. Butler & S. Bullock (Eds.), *Learning through collaboration in self study: Communities of practice, critical friendships, and collaborative self-study* (pp. 221–233). Springer.

Loughran, J. (2018). Learning about self-study of teacher education practices. In J. K. Ritter, M. Lunenberg, K. Pithouse-Morgan, A. P. Samaras, & E. Vanassche (Eds.), *Teaching, learning, and enacting of self-study methodology: Unraveling a complex interplay* (pp. 1–7). (Self-Study of Teaching and Teacher Education Practices; Vol. 19). Springer. https://doi.org/10.1007/978-981-10-8105-7_1

Loughran, J., & Brubaker, N. (2015). Working with a critical friend: A self-study of executive coaching. *Studying Teacher Education, 11*(3), 255–271.

Lozano, R., Merrill, M., Sammalisto, K., Cuelmans, K., & Lozano, F. (2017). Connecting competences and pedagogical approaches for sustainable development in higher education: A literature review and framework proposal. *Sustainability, 9*(10), 1889. https://doi.org/10.3390/su9101889

Merriam, S. B. (1988). *Case study research in education: A qualitative approach*. Jossey-Bass.

Metzger, A. B. (2015). Green school frameworks. In *Marketing the green school: Form, function, and the future* (pp. 1–14). IGI Global.

Mi'kmaq Resource Centre. (n.d.) *The Mi'kmaq*. https://www.cbu.ca/indigenous-affairs/mikmaqresource-centre/the-mikmaq/#:~:text=The%20tribal%20territory%20included%20all,Lawrence%20as%20well%20as%20St.

Miles, M. B., & Huberman, A. M. (1994). *Qualitative data analysis: An expanded sourcebook*. Sage.

O'Brien, C., & Howard, P. (Eds.). (2020). *Living schools: Transforming education*. EWSB Press. https://www.eswb-press.org/uploads/1/2/8/9/12899389/living_schools__2020_.pdf

Saldaña, J. (2012). *The coding manual for qualitative researchers*. Sage.

Samaras, A., & Freese, A. (2006). *Self-study of teaching practices*. Peter Lang.

Sauvé, L. (2005). Currents in environmental education: Mapping a complex and evolving pedagogical field. *Canadian Journal of Environmental Education (CJEE), 10*(1), 11–37. https://files.eric.ed.gov/fulltext/EJ881772.pdf

Sterling, S. (2017). Assuming the future: Repurposing education in a volatile age. In B. Jickling & S. Sterling (Eds.), *Post sustainability and environmental education: Remaking education for the future* (pp. 31–45). Palgrave.

UNECE. (2011). *Learning for the future: Competences in Education for sustainable development*. https://unece.org/fileadmin/DAM/env/esd/ESD_Publications/Competences_Publication.pd

UNESCO. (2005). *The DESD at a glance*. UNESCO. https://unesdoc.unesco.org/ark:/48223/pf0000141629

UNESCO. (2006). *Framework for the UNDESD International Implementation Scheme* (Vol. 6, p. 2012). UNESCO. https://unesdoc.unesco.org/ark:/48223/pf0000148650

UNESCO. (2016). *Education for people and planet: Creating sustainable futures for all: Global education monitoring report*. UNESCO. https://en.unesco.org/gem-report/report/2016/education-people-and-planet-creating-sustainable-futures-all

UNESCO. (2021a). *Reimagining our futures together—A new social contract for education*. UNESCO. https://unesdoc.unesco.org/ark:/48223/pf0000379707

UNESCO. (2021b). *Berlin declaration on education for sustainable development*. UNESCO. https://en.unesco.org/sites/default/files/esdfor2030-berlin-declaration-en.pdf.

UNESCO. (2022). *The concept of sustainability and its contribution towards quality transformative education: thematic paper*. https://unesdoc.unesco.org/ark:/48223/pf0000381528

UNESCO-UNEP. (1977). The Tbilisi declaration. In *Intergovernmental Conference on Environmental Education* (pp. 14–26). UNESCO-UNEP. https://unesdoc.unesco.org/ark:/48223/pf0000032763

Waks, L. (2013). John Dewey and the challenge of progressive education. *International Journal of Progressive Education, 9*(1), 73–83. . https://dergipark.org.tr/tr/download/article-file/258513

Wals, A. E. J., & Mathie, R. G. (2022). *Whole school responses to climate urgency and related sustainability challenges: A perspective from northern Europe*. Springer. https://doi.org/10.1007/978-981-13-2262

Zaleniene, I., & Pereira, P. (2021). Higher education and sustainability: A global perspective. *Geography and Sustainability, 2*, 99–106. https://repository.mruni.eu/bitstream/handle/007/17549/1-s2.0-S2666683921000195-main.pdf?sequence=1&isAllowed=y

20 Critical Eco-Reflexive Approaches: A Case Study of "Teaching for Sustainability" Towards SDG 4.7 Transition Via a Whole School Approach Perspective in Higher Education

Birgitta Nordén

Key Message
Considering the art, science, and philosophy of teaching, professional teachers have autonomy in decoding innovative ideas into practices. In teaching, teachers perform a key role in applying didactic modelling. Teachers' professional knowledge integratively connects the contemporary with future world-centred visions for schooling in the Anthropocene. Regarding self-determination, participation, and solidarity, Bildung is an essential part of sustainability didactics. Bildung entails a dynamic worldview that values the independence of mind and spirit grounded in ecological and social interdependence.

20.1 Introduction

Teachers hold a key role in decoding innovative ideas into practices and accomplishing applied didactic modelling. The case study discussed in this chapter consistently displays the challenges of accumulating fragmented facts without relevance, and it proposes a holistic approach to HESD-based teacher education embedded in integrative worldview perspectives, pluralism, and sustainability awareness.

This chapter first provides a general introduction to the Whole School Approach (WSA) in relation to sustainability and introduces the context of the case study and its theoretical underpinnings. The subsequent section contains an in-depth presentation of the case study, followed by a short explanation of how the case was selected, analysed, and reported. This is followed by a presentation of the results and a discussion of the study outcomes. The closing section offers some lessons learned from a critical eco-reflexive analysis of the case study and suggestions to strengthen the WSA from a Higher Education for Sustainable Development (HESD) perspective.

20.1.1 The Whole School Approach to Sustainability

UNESCO describes the Whole School Approach (WSA) as a key thinking tool for Education for Sustainable Development (ESD) "to enable learners to live what they learn and learn what they live" (UNESCO, 2020). In the context of sustainability, the WSA can be traced back to the 1990s, when educational reforms started to engage more in holistic integrated sustainability agendas highlighting how environmental issues

B. Nordén (✉)
Department of Science, Mathematics and Society, Malmö University, Malmö, Sweden
e-mail: birgitta.norden@mau.se

A. E. J. Wals et al. (eds.), *Whole School Approaches to Sustainability*, Sustainable Development Goals Series, https://doi.org/10.1007/978-3-031-56172-6_20

interconnect with social and political issues (Henderson & Tilbury, 2006). According to Mathie and Wals (2022), the WSA "provides a framework for re-orienting and redesigning education considering emerging global sustainability challenges. It invites a holistic, systemic, co-creative and reflexive effort by all participants included in education to meaningfully involve students in complex sustainability challenges." By "holistic", they refer to the effort to explore and address sustainability issues from multiple perspectives in a relational and integrated way. "Systemic" refers to concurrently considering main aspects of the education system (curriculum planning, educational development, visions, leadership, school–community associations, and teaching practices). "Co-creative" refers to including multiple voices and social actors in developing the WSA within a given context. Finally, "reflexive" refers to the continuous learning, monitoring, evaluating, and re-adjusting that are necessary since the world is in "constant flux" (Mathie & Wals, 2022).

While established governmental and non-governmental certification programmes award schools with certificates if they implement ESD into their daily practice, they often depart from an accountability perspective. Several studies (Pauw et al., 2015; Pauw & Van Petegem, 2018) compare schools in certification programmes with schools not participating—in terms of ESD outcomes—and often show negative effects, particularly when estimating the effects on student perception of the occurrence of holism and pluralism in their classrooms. It is difficult to generalise the results regarding the effectiveness of ESD, particularly in the eco-schools in Sweden, which have more potential for effectiveness and show promise in this context. However, Olsson and Gericke (2016) found such schools were underperforming in educating students on sustainability and environmental issues. This finding is remarkable, considering the amount of effort spent on having an ESD school profile and applying a WSA as a thinking tool for educational innovation in general and as a more comprehensive approach that considers the engagement and commitment of the students from both a short- and a long-termed perspective.

20.1.2 Teaching for Sustainability in a Swedish Context: Introducing the Case Study

Sweden has a tradition of collaborative pedagogies, including features such as theme-based team teaching (Avery & Nordén, 2021). However, in the past decade, the country has turned towards more conventional standards and a curriculum-driven system, which affects the scope left for teacher collaboration and transdisciplinary approaches in ESD (Nordén, 2018). While policy and curricula play important roles, practical considerations related to teachers' professional competence are informed by how teaching *subject knowledge* is structured today. Thus, this case study focuses on the *unconventional* WSA-inspired course Teaching for Sustainability (TfS) for international student teachers engaged in practice teaching within higher education at Malmö University (MAU). Their initial training in subject teaching focusing on *mode, resources, content, and assessment* has consequences for their ability to teach sustainable development goals (SDGs) across subjects. One strategy of teaching these students in the course has been to work with transdisciplinary teaching in teacher teams (Yueh & Barker, 2011). This, in turn, requires increased attention to teacher collaboration, the role of school leadership, and issues of WSA for educational development.

This chapter discusses professional education and development based on the concept of *Bildung*, which describes the self-formation of independent yet socially reflective, ethically aware, and democratic-minded individuals (Sörlin, 2021). Earlier research in this area has examined the learning, content, and activities required to develop appropriate critical knowledge capabilities (Bowden, 2004; Steiner & Posch, 2006; Nordén, 2018), drawing on experiences of teaching ESD in a variety of non-traditional upper secondary and pre-school context settings (Nordén, 2016; Nordén & Avery,

2021). A (more "humanistic") critical-reflexive approach to sustainability education that discusses and problematises its content and practices creates a broader cultural milieu where teacher education/training can develop. With a didactic foundation in eco-reflexive approaches, rarely used in teacher training and professional development programmes, the study presented in this chapter seeks to aid educators in their sustainability-oriented educational planning and analysis, but from a more critical perspective. The multifaceted aspects of anchoring holistic, systemic, and sustainable perspectives meaningfully in education are emphasised in the course Teaching for Sustainability (TfS); thus, due to the need for more knowledge and examples of WSA in practice, this course was chosen as the case study. The study considers which unconventional (or non-traditional) changes from the educative concept of a WSA would most effectively support transitions towards sustainability when putting these ambitions into practice.

20.1.3 Re-designing Methodologies for Transformative Teaching Towards Eco-reflexive Bildung

Given the nature of modern societies and the global challenges that we face, sustainability education has to be reconceptualised to open diverse opportunities for students to critically and reflexively question and engage with the world. This reconceptualisation demands developing and adopting curricula and instructional approaches that foster eco-reflexive sustainability thinking and participatory action research as students explore complex systems and phenomena of relevance to them and their communities. This chapter elaborates on this educational perspective by incorporating contemporary ideas on eco-reflexive Bildung processes and sustainability education (Sjöström & Talanquer, 2018). Although not commonly practiced in the sustainability education literature (Sjöström et al., 2017), these processes can be used to conceptualise a humanistic (Nussbaum, 2010) and critical-reflexive approach to sustainability education development in an "eco-version". Specifically, the thinking, saying, learning, knowing, doing, and practice of sustainability education should take into consideration moral-philosophical-existential-political alternatives (Kemmis et al., 2014).

Rarely conducted in teacher training and professional development programmes, this case study focuses on achievable co-determining indicators of progress and reflexive methodologies for transformative teaching towards sustainability as an essential component of learning-based change and innovation. It aims to examine how international teacher education students in heterogenous groups experience their learning process in the context of a student-led *non-traditional inquiry-based* learning assessment. With a critical eco-reflexive approach assisted by participatory action research (Kemmis et al., 2014), the students analyse and describe their interpretation of critical consciousness, critical literacy (content), and critical agency (mode) involved in sustainability education. Accordingly, they also engage in reflecting upon the standards of their upcoming profession. The study findings may indicate whether the current teacher education ought to be revised didactically. The research question is as follows:

> From the perspective of Higher Education for Sustainable Development (HESD), what should be included in a revised WSA didactic that is informed by a Bildung-oriented eco-reflexive approach?

20.1.4 Theoretical Underpinnings for Teaching Towards Sustainability at Malmö University

Malmö University (MAU) is in the challenging process of re-designing adult learning in higher education through Challenge Based Learning (CBL). With emphasis on contributing to a global learning society and fulfilling the UN sustainable development goals (SDGs)—particularly the connection between quality education for all (SDG 4); sustainable cities and communities (SDG 11); peace, justice, and strong institutions

(SDG 16); and partnerships to achieve the goals (SDG 17)—MAU is devoted to the dialogues on the role of higher education in the knowledge transformation of society. As a young and socially inclusive university, MAU appeals to people who have traditionally not had access to higher education and cultivates their previous experience and knowledge as valuable learning resources. In line with WSA, "this engagement sets the arena for inclusive education and the co-creation of knowledge, originating from the values and experiences everyone brings to the learning environment" (Christersson et al., 2022). MAU prioritises aligning theories, methods, and practices to provide learning experiences and a common foundation for knowledge formation, independent of disciplines, grounded in a holistic HESD based on CBL (Leijon et al., 2021).

According to Goldberg et al. (2022), well-being and engagement should be used as daily guidelines for teaching—that is, teaching should not be based only on the learning outcomes measures. Thus, in line with experience-based learning, the indicators of the learning process should be well-being and engagement. This requires teachers to focus on the students interests and strengths. Qualitative studies investigating the role of well-being and engagement on academic achievement show the potential of a process-oriented teaching approach (Goldberg et al., 2022).

In an attempt to map CBL in higher education, this case study might contribute to research that explores CBL and student voices, praxis, and effect on learning. CBL aims to generate readiness to handle unknown future challenges holistically on an HESD base (Christersson et al., 2022). Thus, both *critical thinking* and *critical doing* need to be integrated when approaching global and local challenges. A broader approach, including theories and methods not explicitly labelled as CBL, could provide a deeper understanding of a growing field. Comparing other school forms from a WSA lens might also be valuable. Furthermore, based on MAU experiences, Christersson et al. (2022) claim higher education institutions could "contribute to the sustainable development of a global learning society by adopting the CBL-domains" (p. 11). The social mission of research methods in higher education is pivotal, underpinning the CBL-domain *Diversity and Inclusion*, particularly in emerging innovative and collaborative learning and social innovation (cf. Avery & Nordén, 2021). Also relevant are the fast-changing technology and a labour market demanding active, creative citizens ready to re-skill and re-learn. This requires the CBL-domain of *Collaboration and Co-Creation* that enables students, staff, and societal stakeholders "to tackle complex societal challenges through inter- or transdisciplinary perspectives on learning, where praxis and theory are interwoven" (p. 11). Moreover, through adopting the CBL-domain *Change Agents and Contextual Challenges*, higher education institutions can contribute to societal transformations (Christersson et al., 2022). Working with CBL is challenging as it provokes the roles and the relationships between institutions, students, staff, and stakeholders. Transformative CBL processes require an open-minded and inclusive co-creation of knowledge by bridging research and education in collaboration. They also require institutional and individual courage to allow and support creating innovative didactics for teaching and learning that would shape a sustainable learning society.

20.2 Method

This case study sheds light on how international exchange students (n = 37) are undertaking the course Teaching for Sustainability (TfS). The course is annually offered on site at Malmö University. The syllabus for the course was tested, modified, and revised by the course leader (the author of this paper) for approval of revision number 3 (Malmö University, 2022).

20.2.1 TfS Objectives and Assessment by 2022

The aim of the course TfS is to develop the teacher students' competence and knowledge on planetary boundaries and in sustainability education. The student should develop deep approaches to learning towards sustainability and knowledge formation on challenges regarding the UN Sustainable Development Goals in global–local contexts, as well as teaching skills in relation to the student's learning process. The course's learning outcome assesses capability to describe the Anthropocene, analyse humans' effect upon their environment, and actively discuss sustainability issues and challenges to sustainable development in their home countries and worldwide. The course involves different assessment forms, including workshops and a home assignment. Further, the students reflect and share their insights on the value of professional transformative teaching and individual learning skills by participating in a joint platform to progress knowledge formation among themselves explicitly.

20.2.2 Study Design and Methodology

The study's method design is inspired by *effective WSA* to sustainability as characterised by Tilbury and Galvin (2022b), summarised in the following five points (Tilbury & Galvin, 2022a, emphasis added):

(a) RELEVANT—to school's mission; national educational priorities; *community identity*; as well as, environmental priorities of the region.
(b) RESOURCED—with expertise and support in sustainability and learning for sustainability; physical resources and technologies *to make the transition*; and, medium-term finance to execute plans.
(c) REFLECTIVE—skilled in critical reflection and evaluation at all levels; developed *critical thinking competences in its staff* and students; striving to become *a learning organisation*.
(d) RESPONSIVE [cf. Participatory action research]—embraced a flexible structure and adapted to local and cultural settings; developed *learner capabilities that helped recognise complexity* as well as the changing nature of sustainability challenges and rejected a one size fits all approach to sustainability.
(e) REFORMATIVE—appreciated that the agenda is not simply one of adding on environmental or SDG themes to the curriculum but that of *reframing the entire educational experience*.

This case study analyses and describes how the concept of WSA can be used as a general thinking tool for educational innovation addressing an overarching umbrella of "sustainability" within international teacher training programmes in higher education. With this re-balancing transition (Wals, 2021), the specific HESD course TfS at MAU seeks to re-orient education to be responsive, relevant, responsible, and re-imaginative, considering pressing local and global encounters (cf. Li, 2021).

The method holistically applies a Whole Institution Approach (WIA), which is inspired by constructive alignment (Trigwell & Prosser, 2014), illustrated by Fig. 20.1. The study relies on an interactive process viewing the curriculum as a dynamic, emergent, and collaborative process of learning for both student and teacher, and the beliefs or conceptions of the teacher are also considered a crucial tool for understanding successful curriculum design towards WSA and WIA. As course leader and teacher, I felt that my students were well prepared for the subject when attending the TfS classes because the teaching applied already from the start of the TfS course provided the prerequisite knowledge by adopting constructive alignment (see Trigwell & Prosser, 2014).

The different steps conducted in this case study could be compared with the didactic modelling cycle (Fig. 20.2), with four steps based on a humanistic Bildung-oriented tradition as interpreted by Sjöström (2022). Didactic models involve theoretical tools that teachers can use in planning, implementing, and analysing teaching (namely, didactics). They could differ but share that they support teachers in reflecting on teach-

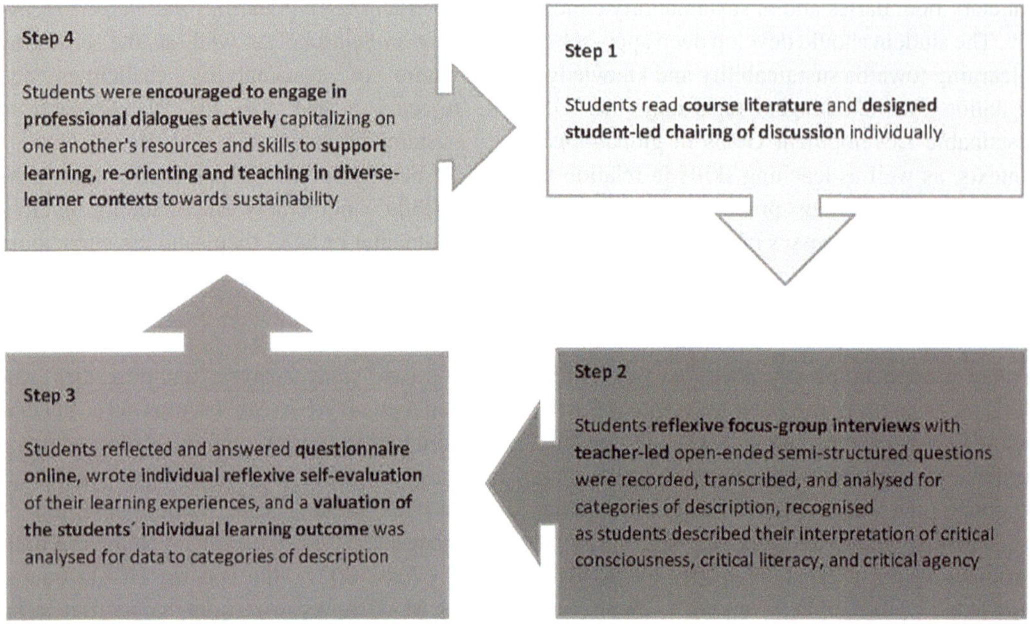

Fig. 20.1 Visualisation of the study's methodology and procedural steps

Fig. 20.2 Didactic modelling cycle with four steps. (Sjöström, 2022, p. 255)

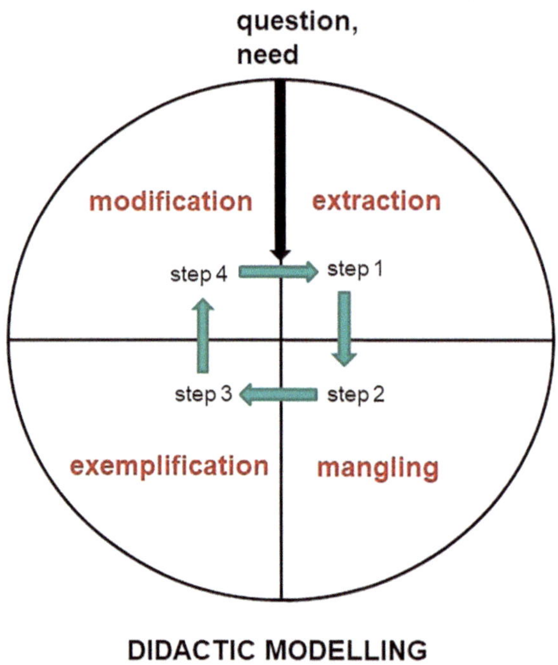

ing practices, grounded on the didactic questions of why, how, and what to apply in teaching arrangements for specific learning situations. Didactic models are useful in constituting a basis for teachers' professional judgment, and they can be seen as bridging the gap between theoretical perspectives and teaching practices (Fig. 20.2). They often have equally theoretical–philosophical as empirical–analytical foundations. Synonyms to didactic models are frameworks, compasses, instruments, and tools (Sjöström, 2022).

In this study, where I explored inclusive pedagogy, I have devised multiple tools to engage the participating student teachers in reflective teaching and critical dialogue. Didactic modelling can be considered as one of these tools that could be appropriate and helpful towards WIA. The steps of the study were as follows:

Step 1 consisted of an oral literature examination seminar on the course book. Therefore, it initially required the students to individually read the course book *The World We'll Leave Behind: Grasping the Sustainability Challenge* (Scott & Vare, 2018) and prepare three questions each on sustainability challenges to chair in the upcoming student-led group discussions. (See Step 1, extraction, in Fig. 20.2.)

In Step 2, the students participated in reflexive focus-group interviews with semi-structured questions raised by the lecturer. The interview questions covered students' previous experiences of various assessment forms in general during their earlier higher education studies; their ways of studying the course book; and their own approaches to deciding and transforming content knowledge to adequate questions, taking command in the learning discussion, and sharing expectations and reflections on the learning processes. The student teachers were divided into eight separate groups (consisting of three to five participants each) to conduct focus-group discussions *for idea exchange and experience sharing* among themselves as well as for the purpose of collecting empirical research data for current case study.

The focus-group discussions with the international student teachers (n = 18 in the year 2022; n = 19 in the year 2021) were recorded, transcribed, and analysed. Thus, prominent characteristics were distinguished in the data in line with the qualities indicated by the *effective WSA* points above, and categories of descriptions were recognised. This revealed how the students analysed and described their interpretation of critical consciousness, critical literacy (content), and critical agency (mode). (See Step 2, mangling, in Fig. 20.2.)

In Step 3, the students anonymously and individually (n = 37) answered an extended online questionnaire by writing their individual reflexive self-evaluation of the learning outcome. This added further data for analysis in the study. (See Step 3, exemplification, in Fig. 20.2.)

Finally, in Step 4, I encouraged the student teachers to engage in some form of professional dialogue with one another (for example, actively take part in one another's presentations and the final papers for the remaining examination parts of the TfS course) to discuss what they could do to support everyone's learning and what they believe about re-orienting and teaching for sustainability in diverse-learner contexts. (See Step 4, modification, in Fig. 20.2.) In addition, reflections at the conclusion of the case study improved the data collection.

20.2.3 Data Analysis

Audio files were transcribed with Amberscript™ (Version 2022 Global B.V., Amsterdam, The Netherlands), which is a GDPR-compliant and ISO27001 and ISO9001-certified tool that Malmö University has chosen for converting audio to text in a safe and anonymous manner. I reviewed all transcripts to ensure accuracy. Any information related to a specific person was replaced by a non-identifiable descriptor. Transcripts were explored for themes through qualitative analysis. This form of thematic analysis emphasises the use of hierarchical coding but balances a highly structured process of analysing textual data with the flexibility to adjust to the needs of a study (Brooks et al., 2015).

A coding template was developed for both the validation in focus-group discussion (transcripts from the interviews in Step 2) and the written post-validation questionnaire (in Step 3). To define an initial coding template, I used open coding to create codes based on a subset of the transcripts. The created codes were revised and refined based on subsequent transcripts during an iterative reflective process. All quotes were carefully read and analysed and then slightly edited for readability.

The analysis of the empirical material can be described as abductive analysis (Peirce, 1934) and involved identifying descriptions in the material in relation to the research aim. Alternating between theory-loaded empiricism and empirically loaded theory revealed qualitative patterns. Concretely, the analysis followed the interpretation paths of close listening and reading to identify distinctive descriptions, to put critically problematised distinctive categories of descriptions in relation to earlier research and concepts (Marton & Booth, 1997; Trigwell & Prosser, 2014). A cohesive analysis was performed considering the didactical modelling, resulting in a conceptualising focus (Sjöström, 2022). Empirical and theory-based interpretation paths were characteristically intertwined. Quotations were selected for their clear exemplification of the categories of descriptions in the data gathered.

20.3 Results

Drawing on the empirical evidence, the case study provides a better and a more comprehensive understanding of learning in sustainability transitions. Teaching and learning in transition contexts can be perceived as a non-linear, iterative process of meaning-making of experiences in communicative interaction; this process is in a reciprocal relationship with the critical social, (bio-)physical, and institutional context, and by gaining knowledge and insights, it helps generate new forms of WIA. Furthermore, it can be seen as a process where opportunities for collaborative action are experimented with and developed by ideas and put to practice in various networks.

20.3.1 Why Use the WSA Concept?

The conducted steps in the didactic modelling cycle showed the impact of WSA on academic achievement. Analysis of the reflexive focus-group interviews emphasised aspects regarding why we should use WSA. A conscious desire to change the world through a WSA within sustainability studies showed implications for changing the teaching of and with student teachers.

Consequently, these expectations need to be met and contextualized to strengthen the student teachers' ability to extend and develop their teaching of the SDGs across subjects. Nourishing their request for continued training in transdisciplinary teaching and subject didactics may form a novel solid foundation. To further facilitate the way ahead, encouraging whole institution approaches, also student interactions should be considered to power transgressive learning (Wals, 2021). The course component areas, identified as needing some transitional change pedagogically, are as follows—mode, resources, content, student interactions, and assessment:

Challenging the MODE is critically reflecting on the relationships between roles, resources, and content. Influenced by a WSA, TfS focuses not on learning tasks but on better compositions of experiences to facilitate learning. This demands a flexible, holistic, relational, and educative teaching mode. One question, though, is how educators could play a more effective role in student groups and whether—or to what extent—the students and teachers/educators share responsibilities. Is the educator really needed for learning to occur?

Challenging the RESOURCES is using materials, tools, and approaches purposefully to enhance learning. What would happen if each student group could curate their own resources? This case study presents how the students shared concepts but learnt different content.

Challenging the CONTENT is filtering and evaluating the skills, practices, and information with which the students engage.

Challenging STUDENT INTERACTIONS. The students were given opportunities to be the experts, and they took on the expert roles while collaborating with students from other countries. The students developed a support network out of campus when preparing for the forum discussion (Examination 1) in class. Thus, in a way, they took command and "led the teaching".

Challenging the ASSESSMENT concerns rebalancing the activities and artefacts that describe, support, and reflect learning. A relevant question is, how does the assessment (namely, the unconventional examination format) help students learn about the standards of their field or future profession? Moreover, how does the educator, efficiently and effectively, make clear to the students what they need to know prior to the assessment?

20.3.2 What Is Useful in WSA and How?

To illuminate what could be of value to include in a WSA to HESD didactics informed by a Bildung-oriented eco-reflexive approach and how, the analysis focused on the third step (exemplification) in the didactic modelling cycle. The textual analysis of the individual reflexive self-evaluation in the questionnaire answers showed what to consider when using the WSA concept and how. Three categories emerged from analysing the students' questionnaire answers: powerful knowings, understanding, and reimagining ethics. They are presented below through quotes (emphasis added).

Quotes Exemplifying POWERFUL KNOWINGS

> Thanks to this type of examination i have been able to talk and reflect about sustainable development with my classmates sharing different opinions and reaching common conclusions. I really liked learning about the different aspects that affect SD such as circular economy, gender equality or the influence of educations, and I think that this experience has *enriched my knowledge about this topic.*

> I started with very low knowledge but was able to gain many opinions and different insight of other cultures. I had to do a lot of research on my own because the lectures were sometimes too professional and overwhelming for me, but as I digged deeper and gained more knowledge I started to like the topics.

> The only thing I would change could be to show the students *how* can they *really show* that they have *acquired the knowledge.*

> From an academic prospective, I added new informations to the knowledge I had about climate change, about what is climate change and *how educations is a powerful tool to solve many wrong behaviours* that caused it. Studying sustainable development from an educational point of view, gave me access to new practical tools that I'm sure I will use in future projects, especially ides for activities and teaching methods that I would like to explore.

> I feel more secure to talk about envrionmental issues and the complicated problems of climate change. I always felt like not knowing enough or having wrong information and often said nothing because I would have felt embarassed to talk about possible "fake news": I often hesitate to contribute to specific topics but this literature seminar and discussion really makes me feel more confident.

Many students experienced that they went from a more general view in the whole topic to seeing the whole complexity and interconnection within the different topics. The variety of learning assessments was enriching, and many students felt they could acquire *knowledge* by applying them. Some deepened their knowledge through discussions found useful, while engaging with other students´ opinion and connecting these to their own or disagree. Talking to one another helped most of the students "to build sustainable knowledge". Many of the students realized that the examination forms enabled them to show different acquired abilities and understanding.

Quotes Exemplifying UNDERSTANDING

> I learnt to consider environmental challenges and solution as part of an educational agenda in which I am both a learner and a teacher. Learning about climate change issues from the prospective of someone who needs to share informations in a easily accessible way, changed how I interact with learning material. Discussing about social implications of the climate crisis in an *engaging way gave me more understanding of interconnected dynamics* that I haven't considered before.

> I read the book and was able to connect the course classes with what I was reading *I got a feeling of deep understanding of the subject.* I believe without the learning activities the teaching process would not have been the same…

> *I found myself looking for more content that could help me better understanding the topic and shape my thoughts from a different point of view.* Being responsible for my own performance in a collective environment, as the examination, helped me being responsible about my education.

It is not that my idea of sustainability changed, it is that I didn't have one. This course made me open my eyes. As a future teacher, one of my aims is to teach in sustainable terms. Implementing concepts such as "circular economy", or "living within the limits".

I think my idea of sustainability and ESD has expanded beyond belief. I came from an education where sustainability wasn't really talked about nearly as much as it is in Sweden, so I was really grateful to get to learn about these concepts in such a deep way. Doing so in a more heterogeneous context (with people from different countries) was also really rewarding, since I could hear about the initiatives in different places. The examination was also really interesting and I plan to implement something like this in my career as a teacher since it was a really rewarding experience.

I sometimes wondered if I—as an individual—am even able to change the world. *I now know: I surely do!* Even though climate change is a global disaster and we have to globally fight against it, I know, that I can rethink my routines and habitats to be able to influence this world positively. Also as a future teacher I now know, that I will have great influence on my future pupils. It won't be easy, but with the knowledge I gained I now know that I can...//... I will go on to listen to speeches and podcasts and to watch documentaries because I think they need to be heard and seen.

The students thought the learning process was perfect, with an actual book that talks about issues of the subject and from daily life. The book was easy to read and understand. After the creation of the question, some students made lots of their own ideas, connections, and other subquestions that made them reflect more deeply on the book. Finally, to prepare their own questions was really productive because they learnt from other perspectives, people, and points of view. The activity was not stressful.

Quotes Exemplifying REIMAGINING ETHICS

Both the lessons and the book assigned gave me new insights and knowledge about the topic. The book assigned was compact and filled with a big amount of content and data, without focusing too much on science but *providing a bigger understanding of how different topics interact with each other.* The examination dynamic, consisting in making questions and answering in a comfortable and relaxed environment, resemble how a discussion in a public informal setting works, and I appreciated the freedom of choosing how to present my question and deciding if I wanted some specific tools to help me and my classmates to perform better (ex. a classmate gave us a sheet with informations for understanding better the concept of system thinking, and gave us opportunities to develop our thoughts on that content).

... appreciate the fact that the course tries to use innovative methods to teach us about sustainability. Even though it is a very interesting topic, learning about it can sometimes be boring if the teaching method is not the right one.

In the group activities, you can help another and *everyone has different knowledge which grows to a big one.*

The book did meet my expectations, but sometimes it *feels like a tale.* I think that ending hunger could be done with one person's money, but the world works by interests, economy... In some kind of way I think that the same people that is truly responsible for polluting are the ones making the rules for sustainability.

Very (!) good (learning) experience, something new for me—Best and favourite way of learning (interaction with others, talking, discussing)—Benefits from interaction with others, it is nice to reflect on different opinions—also after the exam—Sustainable and effective way of learning, I'll for sure remember more of it than through a written exam—I was reading with more concentration and a more intense way of attention than I would read normally—Time frame was fine for me, really short breaks (at least 2 min) are important—Nice to get responsibility to co-create the exam through our own questions.

It was great for me to experience learning to be fun after a long time. I was a bit tired of the monotonous way of learning at my home university and it feels really refreshing to work this way. Everything felt well-thought-through and connected—a red thread was clearly visible for me! I can really integrate this new knowledge into my life and into the facts I already knew. It felt like this kind of learning made much more sense to me and that reading the course literature was really important and interesting as a preparation for the exam—it's not so often the case that a course literature is really "taken seriously" and such a big part of the class/lecture/seminar.

The learning situation was encouraging, since the students experienced autonomy and got the opportunity to take responsibility for their learning outcome. The students experienced the freedom to make out thoughts, opinions, and reflections. They felt deliberated, free to share, and had the opportunity to set a sustainable mentality that they thought could be put in practice every day. So, from a social aspect, many students learned a lot through critical discussions in learning situations with other people. They valued the various forms of learning situations (lecture, seminar, workshop, study visits, group discussion), trying to evolve out of university in the practical learning settings (e.g. museums) to have a broader view on the topic.

Reading the course book helped the students absorb the concept of sustainability, but the assessment was a new learning experience. Many students were anxious, unacquainted with predicting and being accountable for the assessing discussion. After the assessment, students continued to reflect and talk about the questions raised. They said it generated a deep, long-lasting learning opportunity compared to a written examination. They described their learning processes as ongoing, even after the closing seminar.

20.4 Discussion

This section offers an advanced critical analysis and reflection and calls attention to the effectiveness of the Whole School Approach to ESD in the course Teaching for Sustainability.

From a Bildung perspective, subjects within HESD (and its ambition towards WSA-responsive education) need to be open to reflectivity and debate (Herranen et al., 2021). To achieve this, Herranen et al. (2021) underscore three kinds of knowledge: (1) *content knowledge*, that is, ontological knowledge; (2) epistemological knowledge, which is basically *understanding*; and (3) ethical knowledge, which allows critically seeing ESD in a social context and considering ethical aspects. These three kinds were recognised in the course and consistent with the assessment of the participating student teachers. They were to a great deal found to be achieved in various examination situations, as stated by the course evaluation of the learning outcome of the TfS.

20.4.1 Double Unlocking for Powerful Knowings

First, the course's *content knowledge* was new and much appreciated, since many students were unfamiliar with the UN SDGs and how they could be applied in education. From a Bildung perspective, discussions about the connection between content and students can be elevated based on Klafki's concept of "double unlocking" (Sjöström & Eilks, 2020, p. 56), which denotes unlocking both the knowledge and the student. During teaching and learning, students can be offered cases that they can use to verify the subject matter; further, the students can participate in discussions about the content and how to study it. Thus, powerful (HESD) knowledge and capabilities can be applied as specific (HESD) knowings (Bladh, 2020). From an eco-reflexive Bildung perspective, such HESD knowings are framed by a socio-eco-critical awareness. They are examples of "powerful knowings" (Carlgren, 2020).

20.4.2 Intended Deep, Responsive, and Eco-Reflexive Study Strategy

Second, some student teachers experienced the knowledge formation process as troublesome, and they encountered having to find their own *understanding* of what—and how much—was expected of them, since they were not used to the type of examination form used in the course. It became a challenging transition for individual students to step up and take the floor. Some students expressed that they thought communication and feedback should be clearer; since they were not used to taking responsibility for selecting holistic approaches, they felt insecure about having to decide how to prioritise areas for knowledge formation. This gave an opportunity for some critical reflexivity on the shared learning experience on the part of both students and lecturers. These lessons learned influenced how the learning environment was set up next time. Some factors that were experienced as helpful to the students (e.g. concerning the performance) were knowledge of the procedure, guidance, and supervision (Step 4, modification). The support of peers and of the teacher will be developed further since these were, in particular, seen as beneficial to mentally process the procedure.

Qualitative variation in approaches to teaching is related to variations in students' approaches to learning. When teaching has the intention to develop or change students' conceptions and to question students' understanding rather than present information, the students are more likely to adopt deeper approaches to learning. In particular, when the teachers focus on transmitting the course concepts that might be available in a good book, their students adopt more of a surface approach (Trigwell & Prosser, 2014). On the contrary, this was not the case in this case study, due to the conducted constructive alignment that was employed considering the object of study (the intended learning object), the teaching approach, and the assessment focus. According to Trigwell and Prosser (2014), students' perceptions of the assessment requirements are related to variation in the quality of their approach to learning. When tests are perceived to measure understanding rather than reproduction, students describe a deeper intended study strategy, which was the case for the student teachers in the TfS course.

20.4.3 Didactic Modelling as a Framework for Reformative Transitioning Towards WIA

For many student teachers, *critical pedagogy* is equal to *ethical considerations*, which they may find relevant and can *reimagine*. Nonetheless, the relationship between knowledge, values, and action is complex. Consequently, in addition to providing knowledge of environmental issues, education should draw attention to values, emotions, and social norms. Education should be framed with democratic values. Furthermore, teaching always needs to expose students to well-informed alternative reasoning. According to Alexander (2018), pedagogy worthy of the designation 'critical' must not only offer exposure to different viewpoints but also initiate certain ethical perspectives.

Students must be able to transform epistemological knowledge and get skills to question mainstream discourses. As necessary meta-skills, self-efficacy, and creativity strengthened by responsive critical thinking come into question, for example, in challenging learning situations (cf. Herranen et al., 2021). A holistic view is necessary in sustainability education. In practice, it means including philosophical, ethical, and socio-political perspectives into HESD, focusing on problematisation to facilitate understanding of uncertainties.

The WSA-influenced didactic in this case study frames HESD through a wider and more socio-philosophical perspective. The reason is to emphasise critical notions, assess the understanding of knowledge in the course, and evaluate student teachers' possibilities to apply the knowledge in practice. That purpose inevitability demands describing and analysing teachers' and students' views and roles and highlighting eco-reflexivity. In contrast to the fact-based tradition, which focuses on scientific factual knowledge, or the normative tradition, which focuses on changing people's way of living in different directions, the pluralistic teaching tradition stresses the uncertainty of knowledge; it also views environmental issues as moral and political problems (Sund et al., 2020).

Shedding more light on content issues, Herranen et al. (2021) remark that didactical teaching and Bildung require normativeness, so the teachers—and in this case, student teachers—are challenged to be aware of the inevitable normativeness present in interacting with any subject matter. Testing the WSA influences in teacher practices empirically demonstrated its potential, usefulness, and limitations. Since teachers are in the position of decoding innovative ideas into practices in their teaching, they perform a key role in the applied phases of didactic modelling. This case study correspondingly highlights the challenges of accumulating isolated facts without relevance (fragmentation) and proposes a holistic approach to HESD-based teacher education embedded in integrative worldview perspectives, pluralism, and sustainability awareness.

20.5 Conclusions

Along with assessing learning outcomes, teaching should focus on developing learning needs—not only within the course but also in a wider perspective within a learning community and in relation to both research and our daily practices—for individual and collective interactively constructed understanding. The eco-reflexive approach includes an understanding of interconnectedness and holistic thinking. Through responsive decision-making processes, individuals need to be able to navigate in our multifaceted world. Therefore, both teacher and student teachers need to establish a professionally critical reflection capability—since a reconceptualisation of HESD with the eco-societal systems is essential. Herranen et al. (2021) consider meta-skills such as critical thinking, creativity, resiliency, and self-efficacy important when challenging situations occur. In planning for a sustainable society, Herranen et al. (2021) emphasise resiliency will be required to overcome the unavoidable challenges ahead. One way to obtain a holistic view of HESD, instead of fragmented pieces of knowledge (Nordén & Avery, 2021), is to adopt a system-oriented approach—such as WSA—to understand the challenges of achieving sustainable development. It permits us to recognise the system components of WSA also within HESD, the integration of subject knowledge within and between different disciplines.

20.5.1 Lessons Learned and Suggestions for Strengthening WSA from an HESD Perspective

This case study provides a tool to address the didactical WSA questions *of why, how,* and *what.* Powerful HESD knowledge in a WSA is placed in the centre, supported by critical knowledge capability (Nordén, 2016). From an eco-reflexive Bildung perspective, the WSA-influenced HESD is framed by a socio-eco-critical awareness. One of the key lessons from this case study for course and curriculum design is that implications are concerned with the quality of learning associated with the variation in the nature of knowledge that teacher education institutions intend their students to learn. The categories of description of the object of study vary, including a focus on topics rather than the discipline as a whole (i.e. WIA), on knowledge as being unproblematic rather than disputed or questionable, and on specific practical skills rather than analytical, enquiry-based skills (Nordén, 2016). Consequently, the didactic modelling (Fig. 20.2) could be further developed and tested in various WIA contexts, especially empirically with student teachers specialising in teaching students from different age groups and different backgrounds.

When students share their knowledge and experiences of Anthropocene sustainability challenges, they develop critical sustainability literacy. Tentative results in this study indicate that student teachers developed critical eco-reflexive thinking focusing on learning and holistic understanding—seeing a bigger picture while taking responsibility for their own student-led learning. This case study could be considered as (and compared to) a whole school approach since the investigated course could also be seen as involving prospective schools where the students will act in their roles as teachers in their future professional life, thereby influencing and implementing criteria similar to the WSA criteria (Bianchi et al., 2022).

The key findings from the study generate recommendations for WSA policy and practice in higher education, HESD, and teacher education institutions. For instance, educative and critical eco-reflexive pedagogy for inclusive teacher education could be useful. Hopefully, the methodological strengths and limitations of this study could inspire future research. The TfS course aims to enable higher education to deal with the complex socio-ecological and economic questions and uncertainties of our time (Wals, 2021).

Understanding internationalisation as an empowering process that fosters collaboration between universities and society (cf. Mindt et al., 2017) may be a valuable starting point to develop WSA-influenced international sustainability-oriented curricula.

The WSA is not a tool or a prescription for employing a specific agenda like ESD but a means to encourage schools—including higher education institutions—to think about educational innovation generally. A WSA allows multiple themes to be simultaneously addressed within the overarching umbrella of "sustainability" or "sustainable development", not by reducing them to "learning tasks" but by using them as entry points to a different way of working and living. As such, a WSA represents a transition perspective (Van Mierlo et al., 2020), in that it does not intend to optimise mainstream education. Rather, it seeks to re-orient it by anchoring it in different principles and values that contribute to an education that is more relevant, responsive, resourced, reformative, and re-imaginative in the light of urgent global challenges.

Acknowledgement Special thanks to *Studies in Science, Environmental and Mathematics Education* (SISEME) for research funding (autumn 2022) from the Department of Natural Science-Mathematics-Society at the Faculty of Education and Society, Malmö University.

Informed Consent Statement: Informed consent was obtained from all subjects involved in the study.

Data Availability Statement: Data available on request due to ethical restrictions. The data presented in this study can be available on request from the corresponding author.

Conflicts of Interest: The author declares no conflict of interest.

References

Alexander, H. A. (2018). What is critical about critical pedagogy? Conflicting conceptions of criticism in the curriculum. *Educational Philosophy and Theory, 50*, 903–916. https://doi.org/10.1080/00131857.2016.1228519

Avery, H., & Nordén, B. (2021). We can only do it together: Addressing global sustainability challenges through a collaborative paradigm. In W. L. Filho et al. (Eds.), *Universities, sustainability and society: Supporting the implementation of the sustainable development goals*. World Sustainability Series. Springer. https://doi.org/10.1007/978-3-030-63399-8_16. Book chapter, pp. 239–252.

Bianchi, G., Pisiotis, U., & Cabrera Giraldez, M. (2022). In Y. Punie & M. Bacigalupo (Eds.), *Green Comp The European sustainability competence framework*. EUR 30955 EN, Publications Office of the European Union. https://doi.org/10.2760/13286. ISBN 978-92-76-46485-3, JRC128040.

Bladh, G. (2020). GeoCapabilities, Didaktical analysis and curriculum thinking—Furthering the dialogue between Didaktik and curriculum. *International Research in Geographical and Environmental Education, 29*(3), 206–220.

Bowden, J. A. (2004). Capabilities-driven curriculum design. *Effective Teaching and Learning in Engineering, 2*(1), 36-47.

Brooks, J., McCluskey, S., Turley, E., & King, N. (2015). The utility of template analysis in qualitative psychology research. *Qualitative Research in Psychology, 12*(2), 202–222.

Carlgren, I. (2020). Powerful knowns and powerful knowings. *Journal of Curriculum Studies, 52*(3), 323–336. https://doi.org/10.1080/00220272.2020.1717634

Christersson, C., Melin, M., Widén, P., Ekelund, N., Christensen, J., Lundegren, N., & Staaf, P. (2022). International journal of innovative teaching and learning in higher. *Education, 3*(1), 1–14.

Goldberg, J. M., Sommers-Spijkerman, M. P. J., Clarke, A. M., Schreurs, K. M. G., & Bohlmeijer, E. T. (2022). Positive education in daily teaching, the promotion of wellbeing, and engagement in a whole school approach: A clustered quasi-experimental trial. *School Effectiveness and School Improvement, 33*(1), 148–167. https://doi.org/10.1080/09243453.2021.1988989

Henderson, K., & Tilbury, D. (2006). *Whole-school approaches to sustainability: An international review of sustainable school programs*. Australian Research Institute in Education for Sustainability: Australian Government.

Herranen, J., Yavuzkaya, M., & Sjöström, J. (2021). Embedding chemistry education into environmental and sustainability education: Development of a Didaktik model based on an eco-reflexive approach. *Sustainability, 13*(4), 1746. https://doi.org/10.3390/su13041746

Kemmis, S., et al. (2014). *Changing practices, changing education*. Springer.

Leijon, M., Gudmundsson, P., Staaf, P., & Christersson, C. (2021). Challenge based learning in higher education—A systematic literature review. *Innovations in Education and Teaching International, 59*(5), 609–618. https://doi.org/10.1080/14703297.2021.1892503

Li, K. W. E. (2021). *Researching inclusive pedagogy with teachers in Hong Kong*. Doctoral dissertation, University of Cambridge.

Malmö University. (2022). *Teaching for sustainability*, 15 credits [Syllabus]. Retrieved May 14, 2023, from https://utbildningsinfo.mau.se/kurs/kursplan/NM164E/20222

Marton, F., & Booth, S. (1997). *Learning and awareness*. Lawrence Erlbaum.

Mathie, R. G., & Wals, A. E. J. (2022). *Whole school approaches to sustainability: Exemplary practices from around the world*. Education and Learning Sciences/Wageningen University. https://doi.org/10.18174/566782

Mindt, L., Disterheft, A., & Rieckmann, M. (2017). Development of an International Master's Programme on "Sustainability-driven Entrepreneurship": Using a European multistakeholder approach. In H. Casper-Hehne & T. Reiffenrath (Eds.), *Internationalisierung der Curricula an Hochschulen* (pp. 250–262). Konzepte, Initiativen, Maßnahmen.

Nordén, B. (2016). *Learning and teaching sustainable development in global-local contexts*. Doctoral Diss., Malmö university/Lund University.

Nordén, B. (2018). Transdisciplinary teaching for sustainable development in a whole school project. *Environmental Education Research, 24*(5), 663–677. https://doi.org/10.1080/13504622.2016.1266302

Nordén, B., & Avery, H. (2021). Global learning for sustainable development: A historical review. *Sustainability, 13*(6), 3451. https://doi.org/10.3390/su13063451

Nussbaum, M. (2010). *Not for profit: Why democracy needs the humanities*. Princeton University Press.

Olsson, D., & Gericke, N. (2016). Eco-schools in Sweden and the effects on students' sustainability consciousness. In *The ECER (European Conference on Educational Research) conference held by EERA (European Educational Research Association) in Dublin, Ireland*, 22–26 August, 2016.

Pauw, J. B.-D., & Van Petegem, P. (2018). Eco-school evaluation beyond labels: The impact of environmental policy, didactics and nature at school on student outcomes. *Environmental Education Research, 24*(9), 1250–1267.

Pauw, J. B.-D., Gericke, N., Olsson, D., & Berglund, T. (2015). The effectiveness of education for sustainable development. *Sustainability, 7*(11), 15693–15717. https://doi.org/10.3390/su71115693

Peirce, C. S. (1934). Collected papers V of Charles Sanders Peirce. In C. Hartshorne & P. Weiss (Eds.), *Pragmatism and Pragmaticism* (Vol. 5). Harvard University Press.

Scott, W., & Vare, P. (2018). *The world we'll leave behind: Grasping the sustainability challenge*. Routledge. https://doi.org/10.4324/9781351242936

Sjöström, J. (2022). Didaktik modelling illustrated by sustainability teaching arrangements in preschool. *Educare, 2022*, 5. https://doi.org/10.24834/educare.2022.5.8

Sjöström, J., & Eilks, I. (2020). The Bildung theory—From von Humboldt to Klafki and beyond. In T. J. Kennedy & B. Akpan (Eds.), *Science education in theory and practice* (pp. 55–67). Springer.

Sjöström, J., & Talanquer, V. (2018). Eco-reflexive chemical thinking and action. *Current Opinion in Green and Sustainable Chemistry, 13*, 16–20. https://doi.org/10.1016/j.cogsc.2018.02.012

Sjöström, J., Frerichs, N., Zuin, V. G., & Eilks, I. (2017). Use of the concept of Bildung in the international science education literature, its potential, and implications for teaching and learning. *Studies in Science Education, 53*(2), 165–192.

Sörlin, S. (2021). *Till bildningens försvar. Den svåra konsten att veta tillsammans*. Natur and Kultur.

Steiner, G., & Posch, A. (2006). Higher education for sustainability by means of transdisciplinary case studies: an innovative approach for solving complex, real-world problems. *Journal of Cleaner Production, 14*(9–11), 877–890.

Sund, P., Gericke, N., & Bladh, G. (2020). Educational content in cross-curricular ESE teaching and a model to discern teacher's teaching traditions. *Journal of Education for Sustainable Development, 14*(1), 78–97.

Tilbury, D., & Galvin, C. (2022a). *European Commission Input Paper: A whole school approach to learning for environmental sustainability* (p. 11). Expert briefing paper in support of the first meeting of the EU Working Group Schools: Learning for Sustainability. European Commission. Retrieved July 5, 2022, from https://education.ec.europa.eu/document/input-paper-a-whole-school-approach-tolearning-for-environmental-sustainability

Tilbury, D., & Galvin, C. (2022b). *European Commission Input Paper: A whole school approach to learning for environmental sustainability* (p. 23). Expert briefing paper in support of the first meeting of the EU Working Group Schools: Learning for sustainability. European Commission. Retrieved July 5, 2022, from https://education.ec.europa.eu/document/input-paper-a-whole-school-approach-tolearning-for-environmental-sustainability

Trigwell, K., & Prosser, M. (2014). Qualitative variation in constructive alignment in curriculum design. *Higher Education, 67*(2), 141–154. https://doi.org/10.1007/s10734-013-9701-1

UNESCO. (2020). *Education for sustainable development: A roadmap*. UNESCO publishing.

Van Mierlo, B., Halbe, J., Beers, P., Scholz, G., & Vinke-de Kruijf, J. (2020). Learning about learning in sustainability transitions. *Environmental Innovation and Societal Transitions, 34*. https://doi.org/10.1016/j.eist.2019.11.001

Wals, A. (2021). The power of transgressive learning. In *GTI Forum: The pedagogy of transition*. Great Transition Initiative.

Yueh, M. C. M., & Barker, M. (2011). Framework thinking, subject thinking and "Taiwan-ness" in environmental education. *Australian Journal of Environmental Education, 27*(1), 134–148.

21 Whole-School Project-Based Approach to Promote Education for Sustainability: A Mauritian Experience

Anwar Rumjaun and Shakeel M. C. Atchia

Key Message
This chapter documents a contextualised case study foregrounded in selected primary schools where teachers had to conceptualise and implement a whole-school project on relevant sustainability issues. It uncovers the challenges faced therein and the 'disconnectedness' among the key components which constitute the whole-school approach. The chapter laid the foundation for further studies to consolidate or to review the model of whole school as an approach in mainstreaming sustainability in teaching and learning.

21.1 Introduction

Schools have a crucial role in empowering our youth with the necessary sustainability competences and culture needed as the future workforce and to adeptly face current and future challenges (Atchia, 2022; Murdoch, 2004). The whole-school approach (WSA) to sustainability, which lies at the heart of effective learning for sustainability, is an effective way to develop sustainability competences and culture among the stakeholders of the school community and consequently among the society at large.

Since its emergence, the WSA has been in response to global calls to reorient the management and practice of formal education to contribute in addressing inequalities and building a better world (UNCED, 1992; UNESCO, 2002). The WSA for sustainability, which is grounded within the scholarship of school effectiveness and school improvement, encourages schools to innovate and showcase changes in practice not only within the different formal spheres of schools (student achievement, teaching and learning, ethos and climate, school infrastructure and school leadership) but also within the wider community. The WSA provides the necessary structure to holistically integrate the calls for sustainable development and sustainability, through its integration in the formal curriculum as well as through sustainable school operations such as integrated governance, stakeholder and community engagement, long-term planning and sustainability monitoring and evaluation (Kohl et al., 2022). The WSA includes the active participation and partnerships of all stakeholders within schools (teachers, pupils, management and administrative staff), and the community (organisations, business, industry and governments).

Moreover, the WSA to sustainability views school beyond its role as a training ground to foster appreciation of the environment to showcase it as a site of good practice for the community.

A. Rumjaun (✉) · S. M. C. Atchia
Science Education Department, Mauritius Institute of Education, Moka, Mauritius
e-mail: a.rumjaun@mie.ac.mu;
shakeel.atchia@mie.ac.mu

A. E. J. Wals et al. (eds.), *Whole School Approaches to Sustainability*, Sustainable Development Goals Series, https://doi.org/10.1007/978-3-031-56172-6_21

Aligned to international conferences, documents, recommendations and commitments, it questions the dominant approach of educating 'about' the environment only and instead encourages the need for educating 'for' sustainability. The WSA to sustainability does not only engage students and other stakeholders in the community to critically reflect on their current lifestyles and actions, but also provide opportunities to act on informed decisions to bring changes toward a more sustainable world. The focus has shifted from the 'what to teach students?' and 'how are they behaving?' to seeing schools as a focal point where children, adults and the community interact to learn and develop competences together (Tilbury & Wortman, 2004). The pedagogy underpinning the WSA focuses on the development of skills, competences and dispositions for partnerships, participation and action within learners. This shift has had implications for how to conceptualise and approach issues such as school governance, pedagogical approaches, curriculum, extracurricular activities, resource management, the use of school grounds and community partnerships (Henderson & Tilbury, 2004).

Thus, as a potential means to empower the Mauritian youth with the necessary sustainability competences and culture, a case study methodology underpinned by the UNESCO model of WSA, as depicted in Fig. 21.1, was used.

21.2 Policies on Whole-School Approaches and Education for Sustainability in Mauritius

The formal education sector has been the focus for change toward sustainability since the 1970s. This focus has been driven by authoritative international documents and commitments, such as the Rio Earth Summit (1992), Agenda 21, World Summit on Sustainable Development in Johannesburg (2002), Tbilisi Declaration, Dakar Framework for Action and the WESSA conferences, which advocated for educational reform or reorientation to reflect the sustainability agenda. Evidence of educational institutions reflecting the principles of sustainability started to appear mainly in the United Kingdom, North America and Europe to eventually reach other parts of the world, including Mauritius. Mauritius, a small island developing state, is a member of several regional and international organisations. Therefore, as a member state of such organisations, Mauritius has signed the relevant conventions and agreements related to sustainability. In this respect, our national policy documents, which are analysed in this chapter, are all inspired by drawing from regional and international conventions, agreements, programmes and documents. Table 21.1 depicts some of the key national conferences, documents, recommendations and commitments that support the use of WSA and 'Education for Sustainability' in Mauritius.

However, it is important to note that though several initiatives of using whole-school approaches for EfS have emerged in Mauritius during the past decades as depicted in Table 21.1, many have not been sustained. For instance, the Green School Project was a highly acclaimed initiative at both primary and secondary school levels for the first few years but unfortunately went to oblivion. In fact, we believe that though the development of policies is crucial in providing the necessary environment, the sustainability culture can only be developed if associated with the right approach and the proper mindset, which are often the missing link to reach the targets of EfS. Responding to this call, the case study described in this chapter showcases the use of the WSA model where a structured pathway is used to implement whole-school sustainability projects (WSSP).

21.3 A Mauritian Case Study

This section summarises the context, methodology and sampling before presenting the findings and discussions of the Mauritian case study which unpacked the experiences of using the WSA by 30 primary school teachers in their respective schools.

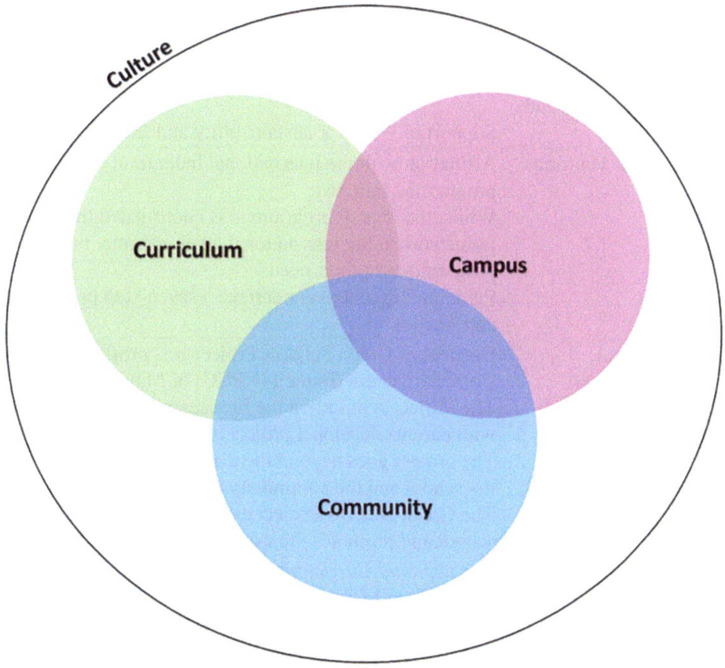

Fig. 21.1 UNESCO WSA framework. (UNESCO, 2014)

21.3.1 Context of the Case Study

The primary education in Mauritius, which operates under the aegis of the Ministry of Education, lasts for 6 years (grade 1 to grade 6) following 2 years of early childhood education. All primary school students sit for the 'Primary School Assessment Certificate', which is the national examination marking the end of the primary schooling. The result of the examination is used to allocate seats to the students in regional secondary schools.

Teachers working in primary schools, though recruited by the Ministry of Education, have to successfully complete a pre-service 'Teacher's Diploma' course at the 'Mauritius Institute of Education' (the only tertiary institution specialised in teacher education in Mauritius) to become full-time primary teachers. As part of the course, the prospective teachers have to complete a 45-h module entitled 'Education for Sustainable Development', where they are engaged in the implementation of a sustainability project in the schools at which they are posted. The sustainability project aimed at mainstreaming sustainability into all aspects of school life by rethinking the curriculum, defining campus operations, embracing organisational culture, enabling student participation, promoting leadership, and establishing community relationships (UNESCO, 2014).

21.3.2 Methodology and Sampling

A case study methodology was used to unpack the experiences of using the WSA by 30 primary school teachers in their respective schools. Using a convenience sampling, 30 primary school teachers enrolled in a professional learning course on 'Education for Sustainable Development' at the Mauritius Institute of Education (MIE) constituted the sample of this case study. As part of the course requirement, they had to implement a small-scale project with pupils of grade 5, aged 9–10 years old. Fig. 21.2 summarises the steps trailed by each participating school of this case study.

As depicted in Fig. 21.2, once the grade 5 section for implementation of the sustainability project was finalised, the participating teacher organised a brainstorming session with his/her respective pupils to identify a potential theme for

Table 21.1 Selected policy documents that support the use of WSA and Education for Sustainability (EfS) in Mauritius

National conferences, documents, recommendations and commitments	Support to WSA for sustainability and EfS
Eco-School Mauritius	Affiliation with the international federation of environmental education eco-school programme initiative Whilst the overall programme is coordinated through a common framework at the international level, member nations have flexibility to tailor the programmes to their needs Participating schools undertake a seven-step process to work towards Green Flag certification
Green School Mauritius (2009)	Concept of Green Schools Project was proposed by the National Productivity and Competitiveness Council (NPCC) of Mauritius The Project is based on the fundamentals of Green Productivity where schools together with parents develop a project based on the premise of 'practice what you preach' The project goes beyond the teaching-learning transaction to permeate as a way of life in the school and the community as well The Green Schools Project envisages self-assessment of the environmental practices of the school and promotes the use of environmentally friendly technologies in the school
National Curriculum Framework (2016/2017)	The *National Curriculum Frameworks* for pre-primary, primary and lower secondary education in Mauritius provide opportunities for using holistic and integrated approaches to develop sustainability skills, competences and dispositions in learners The *National Curriculum Framework*, through the 'Teaching and Learning syllabi', contributes to the integration of sustainability learning in the formal curriculum
Green Paper (2008)	The paper conceptualised the vision of making Mauritius a Sustainable Island ('Maurice Ile Durable') The 'Maurice Ile Durable' concept includes a participatory approach towards elaborating a national strategy for sustainable development aiming at taking on board the whole society in the implementation of this ambitious project. Special focus is laid on the role of the Mauritian schools in reaching the 'Maurice Ile Durable' targets
Maurice Ile Durable Policy (Green Paper, 2011)	Developed from the Green paper, this included a 10-year Strategy Plan and an Action Plan for the 'Maurice Ile Durable' Project The policy laid the foundation for WSA through (i) the educational system promoting the development of 'sustainable development' skills, knowledge and values through lifelong learning to ensure the holistic development of the citizen, (ii) our formal and informal education systems fostering responsible, green and civic values in all age groups to achieve a caring society and (iii) the potential of our human resource fully tapped to foster social equity along with enhancing economic, political, environmental and cultural well-being
Inspiring Every Child (2016)	This policy document forms the basis of the current 'Nine Years Continuous Basic Education' reform The policy document is in line with the international commitment to the UN sustainable development and international commitments for sustainability The document sets the foundation to engage students as ambassadors of sustainability
Sustainability in early childhood development (SECD), 2021	Align with the SDG target 4.2 calls for 'quality pre-primary education' for all children to achieve universal, inclusive and equitable education, the education system prepares children for school and to promote lifelong learning This policy document recommends the introduction of the sustainability concept through means such as storytelling, songs, teacher-guided play and art activities, whilst simultaneously enhancing the kids' feelings and moods
Environmental charter document, 2022	The master charter provides a wide array of examples of actions to foster sustainability in the community Schools are requested to customise this document to produce their own charter to adhere to the pledges Together with students, schools prepare the 'Student Environmental Charter' which represents pledges towards fostering sustainability in the community

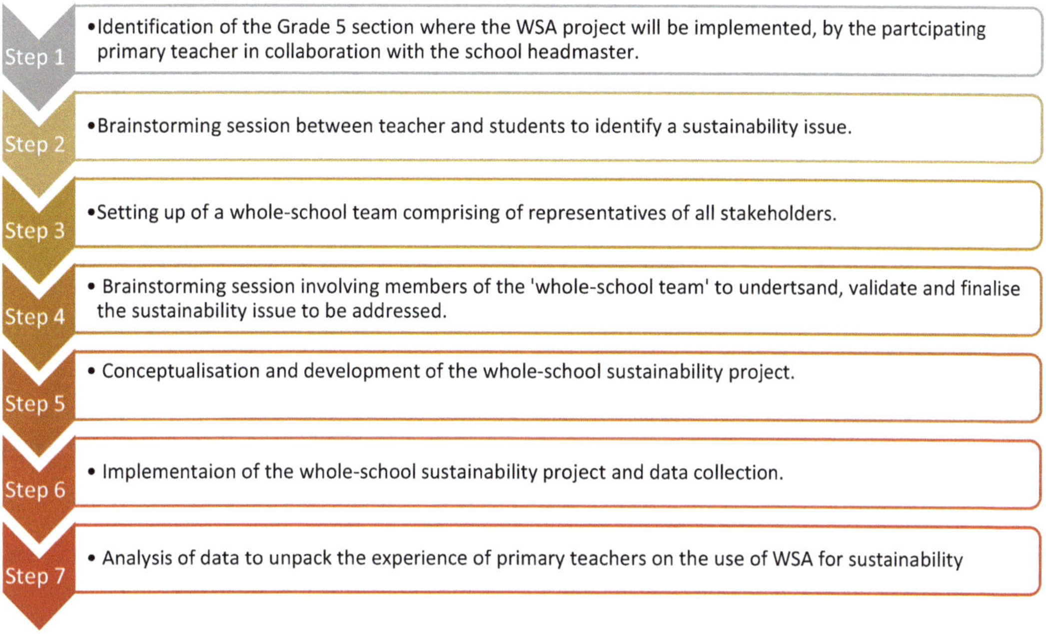

Fig. 21.2 Summary of methodology used to unpack the experience of primary teachers on the use of WSA for sustainability

the project, that is aligned to the content prescribed in the primary school curriculum. Following the brainstorming session, a 'whole-school team' consisting of the following representatives was set in consultation with the headmaster of each of the 30 sampled schools:

- Headmaster or Deputy Headmaster (Head of the school and chairperson of the project school team)
- Holistic Education teacher (teacher responsible to teach non-core subjects related to the holistic development of pupils, as per the current National Curriculum Framework)
- Information and Communication Technology teacher
- School administrative staffs including the clerical officers and the school attendants
- President or Secretary of the 'Parent-Teacher Association'
- Representative of a local non-governmental organisation
- Representative of the local authority (village or municipal council)
- Representative of government organisations (Central Water Authority, Ministry of Environment, and other relevant organisations related to the topic under study)
- Representative of a private company in the area (bank, supermarket, filling station, …)
- Five pupils representing the class who are selected by the participating teachers under the guidance of the chairperson of the project

The 'whole-school team' had as guiding aim compliance with sustainability ideas and principles such as equity, good governance and community engagement. Following the setting up of the 'whole-school team', the participating teachers together with the pupils were engaged to (i) brainstorm on the topics they would like to focus on in their whole-school sustainability project (WSSP), (ii) elicit the idea in relation to sustainability engagement and (iii) conceptualise and develop the whole-school sustainability project. These steps were significant to 'hook' the pupils on the topic and create a sense of ownership towards the project. The feasibility of the project was then discussed with the headmaster (also the chairperson of the 'whole-school team'), before presenting the same to the 'Parent-Teacher Association' meeting. The whole-school sustainability project was then presented in the 'whole-school team' meeting and amendments were brought wherever required.

It is important to note that though the sustainability issue was identified during the brainstorming session involving only the class teacher and the pupils, validation and conceptualisation of the projects involved all members of the 'whole-school team'. Beyond the provision of feedback and validation, the whole-school team had the responsibility to seek financial support needed to shape and implement the whole-school sustainability project, either from the 'Parent-Teacher Association' or from sponsors from the local community. In fact, all members of the 'whole-school team' were given specific roles and responsibilities to ensure that team spirit and collective engagement are maintained throughout the 'whole-school sustainability project'.

The sustainability project became the school project, where all members of the 'whole-school team' met at least once a week to ensure the proper monitoring and evaluation of the implementation phase. This was done through structured follow-up to (i) identify issues that emerged during the implementation phase, (ii) analyse the issue, (iii) amend the implementation plan, if needs be and (iv) mobilise support and resources.

The data collected in the form of teacher diaries were then collected and qualitatively analysed to unpack the experiences of primary teachers who used the WSA for sustainability.

21.4 Findings and Discussions

Table 21.2 depicts the projects, categorised under topics/concepts which are included in the school curriculum, that emerged from the 30 participating schools.

Referring to Table 21.2, we concluded that all the selected themes were actually issues being faced in the Mauritian context, namely waste production, water pollution, loss of biodiversity, health and energy issues.

Analysis of the teacher diaries revealed that though the identification of issues and planning of the implementation phase were done within the boundaries of the classrooms, the implementation strategies used by the participating schools varied from one context to another. It was actually noted that most of the 30 participating schools used interactive and participative out-of-classrooms approaches to address sustainability issues. The following quote from a participant supports the importance of out-of-classrooms activities:

> The walk instilled a strong sense of awareness about water recycling and conservation in my students. As the learning experiences did not only take place within the four walls of a classroom, pupils were happy and excited. Student-centred learning took place. And the walk would be memorable for them.

The participating teachers expounded that pupils' engagement in the WSA project empowered them with skills, know-how and dispositions that will help them to cope with national issues and realities. The interactions and engagement of the main participants (pupils and teachers) with other stakeholders add value to the school project and provided opportunities to bridge the gap between formal and informal learning in a range of contexts and concepts. Moreover, the activities planned by the teachers to engage their pupils in the sustainability project were rich in terms of learning experiences and learning by doing. Pupils were made to ask questions, develop and apply critical-thinking skills and analyse their own thinking in the project. One participant stated the following:

> These tasks enabled my pupils to develop skills that will accompany them throughout their life as responsible citizens. For instance, they were able to collaborate in a civic activity in groups, thus developing in them interpersonal, critical thinking and problem-solving skills. In fact, the National Curriculum Framework 2015 aims at developing knowledge, skills and values beyond the boundaries of the 5 learning areas in order to prepare learners to meet the challenges of daily life and deal with unpredictable situations like the one recently faced with the COVID-19 lockdown.

The participating teachers also highlighted that by engaging and immersing themselves in the local community, the pupils developed sustainability skills and values. The WSA project

Table 21.2 Whole-school sustainability projects

Water	Pollution	Health	Energy	Biodiversity
Rainwater harvesting as part of water shortage management practice Water scarcity (3)[a] Water recycling and reuse (2) Sustainable approach towards water management	Managing paper waste at school School composting (3) Implementation of the 3Rs at school Waste management at school Waste management in school, including waste reduction and valorisation (3) Paper waste at school	School health project Promoting sexuality education Community health care at school Health care at school post-Covid-19 Health promotion and well-being	Installation of solar photovoltaic cells on school's rooftop School Greenhouse project Home gardens	Vertical Gardening project Herbal and medicinal plants School garden project (2) Becoming self-sufficient in food crops (vegetables and herbs) by creating their own greenhouse home garden

[a]The number given in bracket stands for the number of schools with the same whole-school project

helped them to become creative, good observers, investigators, communicators and critical thinkers. The whole-school projects also provided opportunities for the pupils to conduct surveys, collect and analyse data, examine the current context and propose solutions to the selected problems. Through engagement in the project, pupils developed higher order thinking needed to sort out problems in real life situations. In fact, their contributions in the project had dual benefits for the pupils. They were provided with various learning opportunities to extend their knowledge on the topic in addition to developing skills and values needed to sustain the environment.

In addition to the above opportunities, the whole-school sustainability projects provided several benefits related to the four components of the holistic 4C model depicted in Fig. 21.1, namely enacted curriculum, campus, community and culture, as detailed below.

Enacted Curriculum Through the whole-school sustainability projects, participating teachers and pupils were involved in interdisciplinary learning. The WSA enabled students to explore the relationships, connections and integration of different subjects at primary education level, such as Science, History, Geography, Languages, Maths and Arts. In fact, Mathie and Wals (2022) highlighted that the interdisciplinary approach is associated with a reduction in the curriculum coverage time as it effectively brings together knowledge of different disciplines, reducing the load of individual subjects. The following quote from a participant supports the above statement:

> My classroom was transformed into a mini laboratory to encourage pupils to participate in learning where knowledge and experiences gained from different subject areas are brought together to spread the need for the protection and conservation of the 'blue gold', that is water…

Through the whole-school sustainability projects, the teachers were able to achieve several curriculum objectives in Maths, Science, Arts and History. In fact, each participating teacher incorporated both their curricular aims and externally mandated achievement aims into the project.

Campus School ethos and school vibrance are powerful dimensions in providing a conducive learning environment. The morning assembly, the meetings (either weekly or fortnightly) and the interactions with relevant resource persons at the school helped in contributing to the school ethos and climate. The visits by stakeholders in the schools contributed to a congruence which is not usually seen and experienced on the school campus.

Community The WSA project also helped to bring parents and other stakeholders into the school community, which has been a difficult task for long in the Mauritian education system. Through the projects, parents were not only given the opportunity to take cognisance of the learning processes of their children at school levels but were also given the opportunity to become part of the decision-making process. In fact, by involving parents who are the child's first teachers, schools created a dynamic home-school partnership which enabled close monitoring of the pupils' learning and progress. In fact, through the projects, schools were able to build strong relationships between all key stakeholders.

Culture Implementation of the WSA projects has showcased the development of a sustainability culture in the participating schools, which encompasses the achievement of objectives related to enacted curriculum, campus and community, as depicted in Fig. 21.1. In fact, engagement in the WSA projects developed a sustainability culture, impregnated with citizenship and sustainability values and attitudes towards the environment. All members of 'whole-school team', including pupils, teachers, school administrators and members of the community, highlighted that engagement in the projects led to the development of key values and attitudes, such as trustworthiness, respect, altruism, integrity, generosity, courage, gratitude, adaptability, self-respect and flexibility. The project contributed to community building in the school and provided a purpose for all actors in the participating schools. However, it should be noted that this was not the case for all the 30 schools involved in the case study. Variations in the understanding of sustainability led to alternatives in the operationalisation of the WSA to sustainability.

Challenges Encountered During the Implementation Phase

Some of the challenges encountered during the adoption of a whole-school approach were limited communication between the stakeholders, limited resources to monitor tasks implementation and accountability. It was also not clear to whom the actors were accountable as the management practices are shifted from one single person to a collective decision-making panel. Thus, the issue of accountability jeopardised the whole-school approach in some of the participating schools. Moreover, not all the members of the whole-school approach had the skills to work and operate in a team. Another important challenge was the time constraints which impacted on the teaching and learning process and the compliance with the school curriculum.

Though the 'whole-school approach' was well received by all the key actors within the 'whole-school team', of all the participating schools in this case study, the major challenge faced by the schools was the alignment of the teaching and learning practices with the assessment practices. In Mauritius, education is very competitive due to national examinations at the end of primary schooling, which enable students to secure their seats in 'star' secondary schools. Therefore, for the parents of these students, engaging in activities out of school is considered as a waste of time.

21.5 Alignment of the Mauritian Experiences of Using the Whole-School Approach to Education for Sustainability to Other Contexts Documented in the Literature

The whole-school approach provides a framework for a holistic, systemic, collective and reflexive efforts towards education for sustainability. It is an attempt to embed sustainability issues from multiple perspectives in an integrated and relational way (Mathie & Wals, 2022). In the Mauritian case study, the use of the whole-school approach through the implementation of a WSA project, as implemented in 30 primary schools, showcased good practice in working holistically to promote students' learning and well-being. It was also acknowledged that school community, created through the 'whole-school team', impacted positively on pupils' socio-psycho-pedagogical development by providing useful and meaningful learning experiences. By adopting this approach, the participating schools increased their engagement within the community, securing sustainable improvements. The approach allowed different members of the school community, namely students, staff, parents and other community members coming from diverse sectors (both public and private), to work collaboratively. This is reflected in the following statement of an educator:

> ……the project allowed us to strive to be a caring learning community where staff, pupil, parents work together with mutual respect to fulfil our God-given potential with integrity, compassion and a sense of honour in the generous pursuit of a better world for all to share.

In fact, the whole-school approach defined the entire school community as the unit of change and involved coordinated action between three interrelated components, namely (i) curriculum, teaching and learning; (ii) school ethos and environment and (iii) family and community partnerships (Goldberg et al., 2019). The whole-school approach successfully integrated conceptual knowledge, skills and values development into the daily interactions and practices using collaborative efforts that include all staff, teachers, families and children (Jones & Bouffard, 2012; Meyers et al., 2015). The case study has also showcased coordination among the three interrelated components of the WSA highlighted by Goldberg et al. (2019).

However, it was also noted that the nature and effectiveness of such coordination depends on the commitment of the relevant stakeholders and school leadership and what makes it be a good governance. In some cases, despite extensive investment in whole-school interventions, the nature of the interventions and their effectiveness remained unclear. Durlak et al. (2011) and Langford et al. (2015) also stated that whole-school interventions adopting a whole-school approach are failing to show impacts. However, this study showed that though interventions from key stakeholders of some of the whole-school teams were not aligned to the requirements of the national curriculum framework, the other schools showcased laudable alignments. Moreover, the Mauritian case study initiated some key transformations in teaching, role of parents and 'Parent-Teacher' associations, and even broadened the concept of school leadership and governance.

The key elements and pathways of the whole-school approach were supported by factors such as informal teaching, reinforcement and adequate time for relationships development embedded within the whole-school orientation. The results of this study are used to formulate a theoretical model of how the whole-school approach builds school connectedness (Rowe & Stewart, 2011). The whole-school approach provides a framework for a comprehensive education for sustainability and the qualities recognised to consolidate education within it are coherence, appropriate policies, transparency, practice and continuing professional development (Morgen et al., 2019).

Curriculum approaches that promoted inclusive student participation through real-life activities and student-centred activities in which students had a say in their learning were important in promoting school connectedness in this study. These approaches demonstrated direct links to school connectedness, which is strongly verified by the literature (Patton et al., 2006; Simovska, 2007). However, what was more important is that it engaged students in a greater number of interactions with other students in their own class, in other classes, with school staff, and with members of the broader school community. In this respect, it is important to note the benefits of linking real-life, student-centred curriculum activities to a whole-school community activity, for the promotion of connectedness (Rowe & Stewart, 2011), which act as a cementing material to tighten the WSA.

Whole-school approach also contributed in positive reinforcement, especially when school staffs interact with students. School staff (teaching and non-teaching) established working and interpersonal relationships with students in a variety of contexts, either during classroom or out-of-classroom activities. Moreover, the informal interactions among school community members promoted mutual reciprocity. This development of informal interactions was key at merging the formal and non-formal learning processes. The whole-school approach therefore created a potential framework for accommodating non-formal learning within the formal education system.

21.6 Recommendations

It has been theoretically suggested that the school organisation is one of the most crucial facilitators of 'Education for Sustainable Development' implementation in practice (Nikel & Lowe, 2010). Global sustainability agenda challenges traditional pedagogies and calls for a school education that fosters awareness of the complexities and uncertainty of the surrounding world. It calls for a holistic approach to address global and complex problems thereby developing the necessary competences and sustainability culture among our youth. The way that schools are organised and operate in Mauritius or elsewhere does not do justice to the importance of sustainability concepts and principles and does not create the environment required for sustainability to be embedded in the teaching and learning cultures. Expanding and extending the school governance by 'roping' in key actors in the decision-making process and its translation in the school operations is a must. A school organisation that promotes such changes can be regarded as reflexive in relation to social learning and new social movements (Lotz-Sisitka et al., 2015). This allows us to explore how different ways of organising a school relate to different 'Education for Sustainable Development' implementation strategies and offer some recommendations for future 'Education for Sustainable Development' implementation efforts based on a framework rooted in school improvement research.

21.7 Conclusion

Academic literature and policy documents consistently advocate the adoption of a 'whole-school' approach as one of several necessary components in the successful pursuit of education for sustainability. However, the implementation of whole-school approaches remains partial and disconnected, with the focus being placed on distinct topics or projects, rather than attempting to identify the underlying commonalities that make up the broader tenets of this approach. Thus, it would be relevant to rethink the teacher education models by transiting from a campus-based to a more school-based teacher education programme implementation. Our study has also indicated that commitment of key actors at school to bring a transformation in the teaching and learning practices can contribute to establishing a whole-school approach to sustainability.

References

Atchia, S. M. C. (2022). A shift in discourse towards 'Public Understanding of Sustainability Science'(PUSS). *Journal of Education and Research, 12*(1), 57–86.

Durlak, J. A., Weissberg, R. P., Dymnicki, A. B., Taylor, R. D., & Schellinger, K. B. (2011). The impact of enhancing students' social and emotional learning: a meta-analysis of school-based universal interventions. *Child Development, 82*(1), 405–432.

Goldberg, J. M., Sklad, M., Elfrink, T. R., et al. (2019). Effectiveness of interventions adopting a whole school approach to enhancing social and emotional development: a meta-analysis. *European Journal of Psychology of Education, 34*, 755–782. https://doi.org/10.1007/s10212-018-0406-9

Green Paper. (2011). *Towards a national policy for sustainable development*. Ministry of Environment and Sustainable Development, Mauritius.

Green School Mauritius. (2009). *Guidance manual by national productivity and competitiveness council*. Green School Mauritius.

Henderson, K., & Tilbury, D. (2004). *Whole-school approaches to sustainability: An international review of sustainable school programs*. Australian Research Institute in Education for Sustainability: Australian Government.

Inspiring Every Child. (2016). *Ministry of education and human resources*. Tertiary Education and Scientific Research.

Jones, S. M., & Bouffard, S. M. (2012). Social and emotional learning in schools: From programs to strategies: social policy report. *Society for Research in Child Development, 26*(4), 3–22.

Kohl, K., Hopkins, C., Barth, M., Michelsen, G., Dlouhá, J., Razak, D. A., …, & Toman, I. (2022). A whole-institution approach towards sustainability: A crucial aspect of higher education's individual and collective engagement with the SDGs and beyond. *International*

Journal of Sustainability in Higher Education, 23(2), 218–236.

Langford, R., Bonell, C., Jones, H., Pouliou, T., Murphy, S., Waters, E., Komro, K., Gibbs, L., Magnus, D., & Campbell, R. (2015). The World Health Organization's Health Promoting Schools framework: A Cochrane systematic review and meta-analysis. *BMC Public Health, 15*(1), 130.

Lotz-Sisitka, H., Wals, A. E., Kronlid, D., & McGarry, D. (2015). Transformative, transgressive social learning: Rethinking higher education pedagogy in times of systemic global dysfunction. *Current Opinion in Environmental Sustainability, 16*, 73–80.

Mathie, R. G., & Wals, A. E. J. (2022). *Whole school approaches to sustainability: Exemplary practices from around the world*. Wageningen University, Education and Learning Sciences.

Meyers, D. C., Gil, L., Cross, R., Keister, S., Domitrovich, C. E., & Weissberg, R. P. (2015). *CASEL guide for schoolwide social and emotional learning*. CASEL.

Morgen, A., Gericke, N., & Scherp, H. Å. (2019). Whole school approaches to education for sustainable development: A model that links to school improvements. *Environmental Education Research, 25*(4), 508–531.

Murdoch, K. (2004). Teaching and learning to live with the environment. In *International perspectives on natural disasters: Occurrence, mitigation, and consequences* (pp. 341–358). Springer.

National Curriculum Framework (2016/2017). *Nine-year continuous basic education*. National Curriculum Framework.

Nikel, J., & Lowe, J. (2010). Talking of fabric: A multi-dimensional model of quality in education. *Compare, 40*(5), 589–605.

Patton, G. C., Bond, L., Carlin, J. B., Thomas, L., Butler, H., Glover, S., et al. (2006). Promoting social inclusion in schools: a group-randomized trial of effects on student health risk behavior and well-being. *American Journal of Public Health, 96*(9), 1582–1587.

Rowe, F. N., & Stewart, D. E. (2011). Promoting connectedness through whole-school approaches: a qualitative study. *Health Education, 109*, 396–413.

Simovska, V. (2007). The changing meanings of participation in school-based health education and health promotion: the participants' voices. *Health Education Research, 22*(6), 864–878.

Tilbury, D., & Wortman, D. (2004). *Engaging people in sustainability*. IUCN.

UNCED-United Nations Conference on Environment and Development. (1992). *Agenda 21*. Rio de Janeiro.

UNESCO. (2002). *Education for sustainability – From Rio to Johannesburg: Lessons learnt from a decade of commitment*. UNESCO Education Sector.

UNESCO. (2014). *Shaping the future we want – UN decade of education for sustainable development (2005–2014) – report*. UNESCO.

22 Exploring a Whole-School Approach at Nonformal Folk High Schools in Norway Through Common Goals and Action Learning

Brita Phuthi

Key Message
Folk high schools (FHSs) are living examples of how schools can practice a Whole-School Approach (WSA) to sustainability through learning by doing, dialogue and democratic participation, in a cross-disciplinary and cross-sectorial learning environment. Fostering sustainable development is a process, for staff and students, and requires regular reflections and evaluations. Having common whole-school sustainability goals within the greater FHS community, with the liberty for each school to find their way of working toward these goals, is important in this process.

22.1 Introduction

The Folk High Schools (FHSs) originated in Denmark in 1844, and the FHS idea was spread across the Nordic countries. In Norway, the FHSs celebrate their 160th anniversary in 2024 (NOU, 2001: 16). The FHSs in Norway offer a wide range of courses, promoting lifelong learning. They are nonformal educational institutions, with no fixed curriculum, exams, nor grades. The FHSs receive government subsidies, and students get two study credits for attending a full year at a FHS (Folkehøgskolene, undated-a). Currently, over 7000 young adults with different backgrounds come to live and learn at the 84 FHSs across Norway for 9 months every year, where most students are between 18 and 23 years old (Folkehøgskolene, undated-b). As stated in the Norwegian Folk High School Act from 2002, the schools' objectives are "allmenndanning" *(bildung)* and "folkeopplysning" (*public enlightenment*) (my translation). These concepts are dynamic and not easily translated (see e.g. Lövgren, 2015; Lysgaard et al., 2023). For the purpose of this chapter, bildung is understood as democratic, and personal development through engaging with a community, and public enlightenment as the general knowledge individuals need to engage in society – locally and globally. Each FHS is autonomous, owned by either Christian or Liberal organizations/foundations, and has its own value basis (Folkehøgskolene, undated-c). The value basis may vary from one school to another, as each school decides its own, but it is founded on the FHS pedagogy (see next section).

In 2015, when the global community agreed on the 17 Sustainable Development Goals (SDGs), forty FHSs in Norway joined a project called *Action Research and Sustainability—the folk high schools for the future*, with a goal of implementing sustainability in the school's operation and teaching through action learning and research. Four years later, the five national FHS

B. Phuthi (✉)
Faculty of Education and International Studies, Oslo Metropolitan University, Oslo, Norway
e-mail: Brita.Phuthi@oslomet.no

A. E. J. Wals et al. (eds.), *Whole School Approaches to Sustainability*, Sustainable Development Goals Series, https://doi.org/10.1007/978-3-031-56172-6_22

organizations (two labor unions, two information and marketing organizations, and an umbrella organization, representing the four organizations and the FHSs) (Folkehøgskolerådet, undated-a), agreed on a common decision with four sustainability targets for all the 84 FHSs to contribute to reaching the SDGs (see next section). As discussed in this chapter, while a Whole-School Approach (WSA) to sustainability is not a common term among the FHSs in Norway, the approaches and pedagogies used throughout the FHS initiatives resonate and align with the characteristics of a WSA.

With the Norwegian FHSs implementation of sustainable development as a case, this chapter draws upon examples from collective initiatives as well as examples from individual FHSs, to critically analyze and reflect on processes the schools have engaged in over the last 9 years. A WSA is applied as an analytical lens to explore the challenges and opportunities that a WSA brings in the context of the FHSs.

22.2 Characteristics of Norwegian FHSs and Their Involvement in Sustainability Issues

The FHSs have a long history of being involved in sustainability issues in Norway and across the world (Phuthi, 2013). Not having to follow a fixed curriculum, the FHSs have the liberty to dive into relevant issues with their students and address emerging topics, such as global politics, human rights, and sustainable development. Learning by doing and learning outside the classroom, are characteristics of everyday life at the FHSs. The FHS's pedagogical approach is based on the philosophy of Nicolai F. S. Grundtvig, the founder of the FHS idea in Denmark. "The central principle in Grundtvig's pedagogical philosophy is captured in the concept of 'living interaction'" (Lövgren, 2022, p. 3). A central component of this interaction is dialogue-based learning, where the teacher guides the students through conversations and questions.

The FHSs in Norway are boarding schools, where most students share a room in dormitories. The *social pedagogics* focus on students learning to live in a community and are a central part of the FHS pedagogy. Most FHSs have weekly dormitory meetings, which are facilitated by a teacher, and focus on both practical reality and philosophical importance of socially living in a community. Grundtvig argued that "true enlightenment can never only aim at developing students' understandings of themselves but must always have an outward focus as well" (Lövgren, 2022, p. 3). The pedagogical aim is also to develop the self, in relation to others, including relevant political issues, concerning local, national as well as global matters. FHSs were originally founded as a school for the people (*folk*), primarily for peasants and factory workers, "to gain the personal development and knowledge needed to become active citizens" in the growing Nordic democracies (Lövgren, 2022, p. 3). In today's globalized world, Lövgren (2022) argues that contemporary FHSs have reoriented from their initial focus on nation building, to emphasize global citizenship education based on human rights.

The Norwegian FHSs cooperate and meet regularly through conferences, seminars, and courses arranged by one or more of the five national FHS organizations. The action learning and sustainability project, mentioned in the introduction, was one example of the FHS organizations, together with an NGO, inviting the schools to participate. For participation in the project, each school joined with a whole-school team comprised of two students, one operational staff, one teacher, and either the principal or deputy. The goal was for sustainability to be implemented in the schools' operation and teaching through action research and learning. The schools were also encouraged to connect with their local community through activities. The action research project was running from 2014 to 2017.

The other example, the common sustainability decision from 2019, was a result of initiatives from some schools and the five national FHS organizations. The decision, made by the boards of the five national FHS organizations, included the following four targets supporting the SDGs, emphasizing on SDG 4.7 and SDG 13:

- Reduce the total of the FHS's carbon footprint by 40% within 2030, based on measures from 2019, with emphasis on reducing airplane travel.
- Include sustainable development as the main topic in one of the school's annual self-evaluation reports within the year 2023.
- By 2021: Apply the Global guidelines ("Globalvettreglene," created by the FHSs' International Committee in 2017) when planning study tours, and in connection with the school's global involvement.
- All schools draw up an annual local plan and strategy to achieve the SDGs.
(Folk high school council, undated).

These four targets were not legally binding for the schools, as the five national FHS organizations that decided on these targets do not hold legal authority over the schools. However, through the following analysis in the next section, it is evident that the schools sympathized with the decision and worked to support these goals in *their own ways*, a central characteristic of the WSA to sustainability.

22.3 The WSA Flower Model as an Analytical Lens

By adopting the WSA flower model by Wals and Mathie (2022) as an analytical lens, I critically reflect, as a practitioner, on the abovementioned initiatives directed at the Norwegian FHS, from a macro and a micro perspective. The analysis gives insight into how the FHS pedagogies, and shared FHS sustainability initiatives can contribute toward research and literature on WSAs focused on sustainability. The WSA flower model includes six key components (Fig. 22.1). In the following section, each component of the WSA flower is considered in relation to the FHS initiatives and sustainability targets. The analysis is based on two project reports from the action research project, a survey from 2022 on the 2019-decision involving 79 FHSs, a FHS's self-evaluation report for the academic year 2021–22, and conversations with FHS staff regarding their process of following up the 2019-decision at their schools.

22.3.1 Vision, Ethos, Leadership, and Coordination

As previously mentioned, the national FHS organizations do not hold legal authority over the FHSs in Norway, which means that the 2019-decision is not binding for the 84 FHSs. It is therefore interesting to see that a recent internal survey shows that the schools support this decision (Phuthi, 2022). This is not to say that all schools equally implemented the four targets, but through the school's responses in the survey, it is evident that they support the *vision* of the decision and think that it is in alignment with the FHS's objectives and *ethos*. The supportive framework that each FHS finds themselves within, in this case the 2019-decision, is thus relevant for each school's values and *leadership*.

Findings from the two reports from the action research project show that it is essential that the schools' leadership prioritizes sustainability, also when the school has dedicated staff and students, for the implementation to be a reality (Tiller et al., 2017). To further strengthen the school's efforts and vision of becoming more sustainable, some schools involved their respective school boards, which was perceived as a successful process by the schools' management in implementing necessary changes together with the staff (Tiller et al., 2017). This measure has also been part of the process of following up the 2019-decision by several schools (Phuthi, 2022).

Wals and Mathie (2022) argue that a key to a WSA to sustainability education "is that all stakeholders in a school are involved in developing a vision of what a sustainable school entails and invites" (p. 4). The FHSs are based on a culture of involving all staff in processes regarding the school's development, as the FHS's value basis is democratic participation (Tiller et al., 2017). Both the action research project and the 2019-decision support this structure. However, as a recent study of three FHSs shows, even if

Fig. 22.1 The WSA flower model by Wals and Mathie (2022)

the schools have prioritized sustainability, they have not necessarily taken the time to go in-depth to achieve a common understanding of what Education for Sustainable Development (ESD) entails for their school (Isungset, 2021). As much as the FHSs have the liberty to form their own learning programs founded on the school's values, some schools experience that it is not always easy to find the time where all the staff can meet and discuss relevant issues in a busy everyday life (Isungset, 2021). Hence, the *structures* to allow time for these meetings must be in place.

One school that managed to develop this structure, was Lofoten FHS, where the deputy principal lead a process at her school to follow up on the 2019-decision (Frilynt folkehøgskole, 2022). She followed a five-step process within the time frame of one school year: (1) mapping, (2) structure, (3) vision, (4) long-term plan, and (5) communication and marketing. In the first step of mapping, all staff were interviewed, and information was shared about what the school was already doing regarding sustainability initiatives. This information was presented as a map that outlined the structure (2) of the school's operation, learning program, and the school's communication/marketing. In step 3, students were invited to participate in a visioning-workshop, in which they imagined the school in 2027 (5 years later). The ideas from this workshop then formed the basis of the development of a long-term plan (step 4), where all the staff were involved in formulating concrete targets for the school's sustainability plan. Step 5 was then based on communicating the long-term plan (Frilynt folkehøgskole,

2022). Lofoten FHS is an example of how a FHS can act in their *own way* upon the *shared* sustainability decision from 2019. As evidenced by Lofoten FHS, this was achieved through setting aside time and involving all staff, as well as students, in a whole-school process.

22.3.2 Capacity Building for All Staff

According to Wals and Mathie (2022), the WSA model suggests that staff consider "more integrated, existential, and relational forms of teaching and learning" (p. 6). The FHS model is similarly based on this educational approach, however, through capacity building, the FHSs may further strengthen their integrated teaching and learning.

Findings from the action research project show that ownership is a success criterion for students' and staffs' involvement, as well as the importance of the cooperation between operational staff and teaching staff (Tiller et al., 2017). One school wrote that,

> last year the project was 'forced' on the kitchen staff, with limited success. This year the kitchen staff found a solution for themselves. From this we concluded that the people who are nearest the measures, often have the best solutions, and this affirms the importance of ownership. In projects, which staff are told to participate, they tend to be less motivated than projects in which the staff have been consulted for their ideas on developing the project. (Tiller et al., 2017, p. 45, my translation).

Ownership may start by a sense of connection to the issue and relevance to the work. This is a process, and as Rønningen FHS writes in their self-evaluation report from 2021–22, the first step for a teacher to become engaged in the school's vision for sustainable development is having ownership of the vision. Perhaps by being honest about that it is sometimes challenging—and by finding a way forward together (Rønningen FHS, 2022, p. 29). Hence, capacity building can be an internal process at the school, where staff supervise each other through integrated learning approaches, reflecting on how to foster the desired learning experiences for the students.

22.3.3 Locally Designed Learning Programs (No Fixed Curriculum)

As mentioned, the FHSs do not have a fixed curriculum, and each teacher designs the learning program for their class, which is then approved by the principal. Each class has a chosen subject, like art, outdoor life, sports, music, etc., and is centered around the FHS objective of fostering *bildung* and *public enlightenment*.

The content, as well as methodology in these programs is highly relevant for the implementation of sustainability at each school. Wals and Mathie (2022) argue that an open curriculum, which is dialogical and cocreated, rooted in the lifeworld of the learners, is more conducive to a WSA. FHS teachers design learning programs where the students are encouraged to share the topics and questions they want to investigate. This methodology also challenges the banking model (Freire, 1972), which many students have experienced through their formal schooling. Students must therefore challenge themselves in finding questions they want to ask and consider why they want to direct their learning in particular directions. Unlearning previous methods, and how things have been done before, becomes part of this process for the students. In finding more sustainable ways, we cannot solve our problems by applying the same thinking we used when we created them, as Einstein famously said. Hence, unlearning and rethinking are important skills to foster sustainable development.

The teaching at FHSs depends on the competencies of each teacher. Hence, knowledge of and interest for sustainability in an educational context varies from one teacher to another. Lack of competencies of ESD, as Isungset (2021) found, may be a challenge in meeting the targets of the 2019-decision. One way to address the lack of ESD competencies within a school is to actively recruit staff who have this knowledge, which we have seen examples of recently in FHS's hiring practices. However, even more important, is to find structures and time for the schools' staff to engage in the topic through the FHS model of democratic participation, which is central to the WSA.

22.3.4 Folk High School Pedagogy and Lifelong Learning

> In a WSA the pedagogical environment a teacher and the school create tends to be one of trust, curiosity, collaboration, participation, and democracy. Much of the learning does not take place inside the classroom but also in other spaces in the school building, as well as on the school grounds and in the local environment (...) (Wals & Mathie, 2022, p. 5).

This quote could be a description of FHS pedagogy, where the teacher is more of a facilitator and the classroom is the school's community, which is situated within and connected to (through study tours) various regions in Norway as well as other regions in the world.

Öhman and Östman (2019) advocate for a pluralistic approach to education, which goes beyond the knowledge-based and the normative approach to education, and views sustainable development as political. Teaching should enable students to become democratic citizens, who engage critically with how (un)sustainability problems may be solved. This approach resonates with the FHS ideology and pedagogy, but as discussed earlier in this chapter, WSA to sustainability is a continuous process, and there is always room for improvement. Education *for* sustainable development may lead FHS teachers to promote certain normative attitudes through predetermined activities. Isungset (2021) found that some FHS teachers had not reflected properly on how to understand ESD together with their fellow colleagues, but at the same time stated that they would like to change the students' attitudes, hence the risk for the teaching to become more instrumental and normative.

The FHS pedagogy aims at fostering lifelong learning by incorporating the mind, the heart, and the hands—the whole person. Such a holistic approach may strengthen the WSA to sustainability. However, one challenge in sustainability education is that there are issues that need to be dealt with urgently, such as the climate crisis as well as growing inequalities—and that calls for action-oriented education. Many young people are demanding actions, which we have seen through school strikes for the climate across the world (Sinnes, 2022). FHSs have the liberty to take initiatives together with their students for the climate and equality. At the same time, it is central to the FHS pedagogy to not rush into new ideas, but to take the time to gain knowledge, reflect and discuss, to make sure that the initiatives are contributing positively and do not have unintended effects—as advocated by the FHSs Global guidelines, one of the targets of the 2019-decision (Folkehøgskolerådet, undated-b).

22.3.5 Institutional Practices: Cross-Disciplinary Cooperation and Learning

"Walking the talk" is how Wals and Mathie (2022) describe institutional practices in the WSA flower model. The school must practice what they teach and follow up on the values that the school is based upon. Living in dormitories and learning together through various activities at the FHS invites different approaches in implementing sustainable lifestyles, as documented in the action research-project report (Tiller et al., 2017). Both the school's operation and the community the school is located within require integrating "thinking" with "acting" (Wals & Mathie, 2022). Action research is one way of making this connection, as shown through the example below:

One student-initiated action research project focused on changing the students' and staffs' attitudes toward vegetarian food at their FHS (Tiller & Phuthi, 2017). Their first action was to conduct a survey among the students, followed by the kitchen making a vegetarian surprise dinner. Afterward, the students followed up with a new survey and in-depth interviews with a selection of students. It turned out that the students did not appreciate the surprise dinner and being "forced" to eat vegetarian food. Reflecting on their findings, the students decided to try a new approach. They invited an NGO to speak about food and sustainable development, next they cooperated with the kitchen staff and organized a voluntary vegetarian cooking course, making "awesome veggie burgers." This turned out to be a hit, and many students attended the course. In a final survey, 9 out of 10 students said that they would like to eat more vegetarian food after they had attended the course. The students concluded with the following: "Through voluntary courses, interesting lectures and good information, we can change the students' and staffs' attitudes to vegetarian food." (Tiller & Phuthi, 2017, p. 14, my translation). This example shows how the school can change toward more sustainable practices, and when students take initiatives, it is crucial that the staff support these initiatives. By involving the wider school community, kitchen staff, the janitor, or the financial manager, cross-disciplinary learning, involving multistakeholders, is thus central to make *thinking* meet *action*. Cross-disciplinary cooperation, however, will be difficult if structures and common visions are not in place for *all* staff to see the relevance of these actions.

In their self-evaluation report from 2021–22, Rønningen FHS reflects on this process of integrating thinking and action. Although two of Rønningen FHS's values are "green" and "global," these values will not be followed up through actions if all staff are not onboard. Hence, the school chose "sustainability" as the topic for their self-evaluation. Most of the staff and several students contributed by writing their reflections on institutional practices regarding sustainability. As Wals and Mathie (2022) point out, the WSA is an ongoing process, and by "interrogating, rethinking, and redesigning institutional practices, the 'hidden curriculum of unsustainability' that is often present, can also be exposed and addressed." (p. 5). This kind of process is one of the strengths of the self-evaluation report—*if* the school ensures that the report is not a one-staff job, but a process for the whole personnel, and a whole-year process.

Being part of a greater FHS community is also central to the changes in FHS's institutional practices. When a FHS sees that neighboring schools make changes for sustainable development, it is easier to initiate similar measures at their own school (see Stoknes, 2015), hence the importance of common goals for the FHSs in Norway.

22.3.6 Community Connections

There are numerous examples of how FHSs involve their students through community-oriented sustainability activities, such as harvesting apples from neighbors that do not pick these themselves or picking up the remaining potatoes in the field after the farmer has completed the harvest. Students go to the shores to collect garbage; they organize local demonstrations for the climate; or they invite local politicians to the school for discussions. Being involved in the local community, as well as cooperation with partners across the world, is an important part of the learning programs at the FHSs.

Community connections through WSA may also be understood to include the greater FHS community. Over the past few years, the effects of the common sustainability project and the 2019-decision can be seen through increased sustainability initiatives across the FHSs (Phuthi, 2022). Being part of a joint project, seeing that neighboring schools make changes, shows that the community and common goals of the FHSs matter and become a motivator. The national FHS organizations support the schools through supervision, and by developing tools, such as a FHS customized carbon footprint calculator. So far, three schools have been awarded the "Folk High School Sustainability Prize," to inspire and motivate all the FHSs.

Lysgaard (2020) emphasizes the importance of the FHS community in relation to sustainability initiatives and argues that an FHS-year is primarily shaped by participation in the "binding community" at the school. Life at a FHS is not about only attending classes and then withdrawing to one's room, but about each student participating in the community at the school—sometimes in smaller groups and other times all together (Lysgaard, 2020). It is not up to each person to transform the world; rather sustainable development is about finding new ways *together*. And in this space, the focus is not necessarily on quick fixes, but asking the big existential questions and searching for answers together, as a community.

22.4 A WSA to Sustainability through Democratic Confidence?

As discussed in this chapter, the FHS pedagogical model engages with all six elements of the WSA flower model (Wals & Mathie, 2022), and can be argued to be an example of schools that practice the *process* of a WSA to sustainability. The emphasis on process is important, as the FHSs are constantly developing; through their annual self-evaluation and dialogues with students, and in cooperation with the larger FHS community.

Education *as* sustainable development emphasizes how education may be carried out in a sustainable way, through actions and activities that promote sustainability (Sinnes, 2015, p. 51). In many ways, one can argue that the WSA is education *as* sustainable development, through its holistic approach by learning through theory and practice, and how the learning emphasizes cross-sector cooperation and cross-disciplinary learning.

Being a school that is founded on democratic values, such as participation and dialogue, I argue that there is a need to go one step further in the WSA to sustainability. Van Poeck and Vandenabeele (2012) argue for an alternative approach to ESD: Learning *from* sustainable development, which "shifts the focus from the competencies that citizens must acquire to the democratic nature of educational spaces and practices" (p. 13). They further argue that,

> This alternative approach to ESD focuses on how people may learn, again and again, in response to the ambiguities and differences they encounter when facing contemporary sustainability issues. This is not a process of schooling but an educational practice, acknowledging the plurality of voices and the controversy surrounding many sustainability issues without resorting to an 'anything goes' relativism. (Van Poeck & Vandenabeele, 2012, p. 584).

The FHS's main objective is to foster *bildung* and *public enlightenment* through dialogue and democratic participation. From this perspective, the FHS model may be argued to be in a place between education *as*, and learning *from* sustainable development, where the emphasis is on *democratic bildung*. Hansen (2019) argues that we are currently experiencing a democratic crisis. One reason, he argues, is that many have lost faith in their ability to make a difference. Young people have great knowledge about democracy and democratic processes, but low *democratic confidence* (Hansen, 2019). Current structural problems need common solutions (Haltli, 2021). If young people have been taught to think that problems should be solved at the individual level, there's a risk for resignation. Fostering democratic confidence is about seeing and experiencing that there are alternatives to individual solutions (Haltli, 2021). A stronger emphasis on learning *from* sustainable development through fostering *democratic confidence*, among students as well as staff, might be the way forward in further developing a WSA to sustainability.

In conclusion, I will end with a FHS reflection on how to evoke the students' engagement and ownership to the sustainability agenda, through staff being open to, and able to relate to the students' realities: "We must dare to walk in front, to walk alongside and to walk behind the students" (Rønningen FHS, 2022, p. 46, my translation):

> *Walk in front,* by showing possibilities, and why it is important to get involved.
>
> *Walk alongside*, to avoid that sustainability is forced on the students. Rather, show how it can be meaningful and inspirational.
>
> *Walk behind,* to support the students' own initiatives, and be willing to be challenged and led by the students.

References

Folkehøgskolene. (undated-a). *Dette er folkehøgskole.* Retrieved January 13, 2023, from https://www.folkehogskole.no/om-folkehogskole

Folkehøgskolene. (undated-b). *Facts about the Norwegian Folk High Schools.* Retrieved August, 28 2022, from https://www.folkehogskole.no/en/about

Folkehøgskolene. (undated-c). *Christian and liberal schools.* Retrieved January 13, 2023, from https://www.folkehogskole.no/en/about/christian-and-liberal-schools

Folkehøgskolerådet. (undated-a). *About the Folk High School organizations*. Retrieved May 23, 2023, from https://www.folkehogskoleradet.no/kontakt

Folkehøgskolerådet. (undated-b). *The sustainability decision*. Retrieved July 28, 2022, from https://www.folkehogskoleradet.no/berekraftvedtak

Freire, P. (1972). *Pedagogy of the oppressed*. Penguin Books.

Frilynt folkehøgskole. (2022, June 24). *Lofoten folkehøgskole leverer på bærekraft*. Retrieved August 2, 2022, from https://www.frilyntfolkehogskole.no/blogg/2022/06/lofoten-folkehogskole-leverer-pa-baeekraft/

Haltli, T. (2021). Folkelige bevegelser, demokrati og motstand. In *Folkehøgskolen* 5/21. Folkehøgskoleforbundet. (pp. 16–19).

Hansen, B. (2019). Tag ansvar. Number 3 in the series 'Højskolens ti bud', Folkehøgskoleforeningen i Danmark (FFD) Publisher, Copenhagen.

Isungset, F. (2021). *Education for sustainable development in Norwegian folk high schools*. Master's thesis, Western Norway University of Applied Sciences.

Lövgren, J. (2015). Norwegian folk high schools redefine their role as value-based institutions—Analysis of value documents from two folk high schools. *Nordisk Kulturpolitisk Tidsskrift, 18*(2), 199–217. https://doi.org/10.18261/ISSN2000-8325-2015-02-05

Lövgren, J. (2022). From nation building to global citizenship: Human rights education in the Nordic folk high schools. *Human Rights Education Review, 5*(2), 77–97. https://doi.org/10.7577/hrer.4470

Lysgaard, J. A. (2020). Bæredyktig udvikling og folkehøjskoler. In J. A. Lysgaard & H. J. Jørgensen (Eds.), *Bæredygtighedens pædagogik – forskningsperspektiver og eksempler fra praksis* (pp. 226–241). Frydenlund Academic.

Lysgaard, J. A., Maribo, P. W., & Albers, S. (2023). Sustainable bldung – Danish folk high school teachers in times of change. In J. Lövgren, J. A. Lysgaard, R. K. R. Rahbek, & A. Hallqvist (Eds.), *The Nordic folk high school teacher: identity, work and education*. LIT Verlag.

NOU. (2001: 16). *Frihet til mangfold – Om folkehøgskolens rammevilkår*. Kirke-, utdannings- og forskningsdepartementet. https://www.regjeringen.no/no/dokumenter/nou-2001-16/id378021/

Öhman, J., & Östman, L. (2019). Different teaching traditions in environmental and sustainability education. In K. Van Poeck, L. Östman, & J. Öhman (Red.), *Sustainable development teaching – Ethical and political challenges* (pp. 70–82). Routledge. https://doi.org/10.4324/9781351124348-6

Phuthi, B. (2013). *Folkehøgskolenes internasjonale engasjement: Internasjonalt utvalg for folkehøgskolen 30 år*. Folkehøgskolerådet.

Phuthi, B. (2022). Folkehøgskolene på vei til bærekraft. In *Folkehøgskolen*, 2/22. Folkehøgskoleforbundet, (pp. 38–39).

Rønningen Folk High School. (2022). *Selvevaluering 2021–22: Bærekraft*. https://ronningen.fhs.no/om-skolen/kvalitetsutvikling-og-selvevaluering/

Sinnes, A. (2015). *Utdanning for bærekraftig utvikling. Hva, hvorfor og hvordan?* Universitetsforlaget.

Sinnes, A. T. (2022). *Action, takk! Hva kan skolen lære av unge menneskers handlinger for bærekraftig utvikling?* Gyldendal.

Stoknes, P. E. (2015). *What we think about when we try not to think about global warning*. Chelsea Green Publishing.

The Folk High School Act. (2002). Retrieved August 02, 2022, from https://lovdata.no/dokument/NLE/lov/2002-12-06-72

Tiller, T., & Phuthi, B. (2017). *Røtter og vinger*. Kortrapport fra prosjektet «Aksjonsforskning og bærekraft – folkehøgskolen for framtiden». Folkehøgskolene og Framtiden i våre hender.

Tiller, T., Bjerke, A., Phuthi, B., Lindberg, K., Wiland, M., & Vinje, S. F. (2017). *Aksjonsforskning og bærekraft – folkehøgskolen for framtiden*. Sluttrapport. Folkehøgskolene og Framtiden i våre hender.

Van Poeck, K., & Vandenabeele, J. (2012). Learning from sustainable development: Education in the light of public issues. *Environmental Education Research, 18*(4), 541–552. https://doi.org/10.1080/13504622.2011.633162

Wals, A. E. J., & Mathie, R. G. (2022). Whole School responses to climate urgency and related sustainability challenges. In M. A. Peters & R. Heraud (Eds.), *Encyclopedia of educational innovation*. Springer. https://doi.org/10.1007/978-981-13-2262-4_263-1

23 A Whole-School Approach to Building Communities of Practice of ESD: A Case Study of Two Japanese High Schools

Hiroko Shibakawa

Key Message
The transformative learning required for ESD emerges in the process of forming communities of practice for building a sustainable society; communities of practice that promote ESD transcend the boundaries between school and community, and are formed through the multidimensional and multi-layered engagement not only of students but also of teachers and communities. WSA needs a 'structure and mechanism' to create ESD communities of practice in the future.

23.1 Introduction

This study identifies the structures and requirements of Whole-School Approach (WSA) practices that promote the formation of Communities of Practice (CoPs) in ESD. The core of ESD learning theory is 'learning to transform oneself and society', rooted in lifelong learning (UNESCO, 2012). The question then becomes: how do we create this learning environment while transforming the existing education system? Lave and Wenger (1991) suggest that every person is a 'legitimate peripheral participant' in some CoPs. In light of this, the essential and transformative learning that ESD requires would emerge in the process of 'legitimate peripheral participation' in a CoP working towards sustainability (ESD-CoP). Moreover, ESD is a broad practice that goes far beyond the framework of formal education. It is a collaborative effort between a wide and diverse range of stakeholders, blurring the boundaries between formal, non-formal and informal education (Wals, 2011). ESD seeks a 'convivial' (Illich & Lang, 1973), a sustainable society where ESD-CoPs are formulated and interact. The primary requirement of WSA is, therefore, not only to introduce systemic change in schools but also to promote the formation of ESD-CoPs in the wider community. Hence, this study focuses on the transformation of teachers as ESD practitioners and identifies the structures of practice that facilitated their organisation and engagement in ESD-CoPs.

23.1.1 Promotion of ESD and WSA in Japan—From a Teacher Education Perspective

In Japan, the principle of ESD has been clearly stated in the Curriculum Guidelines; consequently, ESD efforts in public education are thriving. Educators in Japan are aware of the importance of WSA (Asia-Pacific Cultural Centre for UNESCO [ACCU], 2016). As Tejima (2017),

H. Shibakawa (✉)
Graduate School of Education, Okayama University, Okayama, Japan
e-mail: hirokoshibakawa@okayama-u.ac.jp

A. E. J. Wals et al. (eds.), *Whole School Approaches to Sustainability*, Sustainable Development Goals Series, https://doi.org/10.1007/978-3-031-56172-6_23

ACCU (2017) and others have shown, a large number of good practices of ESD have also accumulated in Japan. Hence, as Yoshida and Natori (2020) noted the absence of studies that examined in detail the organisational structure and transformation of schools that promote ESD using WSA and the factors involved in schools' interactions with their surrounding communities. In addition, ESD has been promoted in a small number of exceptionally privileged schools and the most important factor has been the lack of understanding of ESD by teachers (ESD Special Subcommittee, Education Committee, Japanese National Commission for UNESCO, 2015). ESD is a 'lens' through which knowledge is not imparted but acquired in practice. If teachers need to deepen their understanding of ESD, it is necessary to create a learning environment in which they can grow and develop their identity as practitioners of ESD. Furthermore, training teachers responsible for promoting ESD to develop their capacity to engage with the local community is a challenge in many countries where school-based ESD practice has been promoted (Okayama University ESD Promotion Centre, 2020). Wals and Mathie (2022) also suggest a transformation of the school environment through WSA for ESD, with teachers acting as agents of change. A new stage of WSA needs to include mechanisms and procedures to formulate ESD-CoPs in which teachers are the legitimate peripheral participant. As Schallcross (2005) and many others have shown, there are many challenges involved in implementing ESD with WSA. Hargreaves (2008), Gericke (2022) and other WSA studies suggest a particular challenge is to take an already rigid schooling system from within. It is necessary to open it up to the outside, to continue and develop pluralistic interactions and to allow diverse stakeholders to realise transformative learning. In a way, the process of realising WSA itself is ESD. Not only from external demands, but also from internal ones: the questions 'What is education for?' or 'What can education do for the planet?' are unanswered, and this question drives teachers to advance ESD. On the other hand, a foundational model for teachers to continue exploring, growing and transforming that question with diverse others is still under development. This study focuses on the Flower model of WSA by Wals and Mathie. The model is not intended to provide teachers/schools with a model or idealised image of a crested practice, but rather a reference framework for reviewing all systems, including the hard-to-find status quo, and to be used as a thinking tool to initiate and guide an ongoing dialogue by diverse actors. From this perspective, it is also a response to the aforementioned challenges of in-service teacher education: how the practice of ESD in WSAs can help teachers to form ESD-CoPs. Accordingly, the reference will be made to three elements of the Flower model; INSTITUTIONAL PRACTICES, COMMUNITY CONNECTIONS AND VISION and LEADERSHIP and COORDINATION that are particularly relevant to this task.

23.1.2 Challenges of WSA Through ESD—Periods for Integrated Studies as a Key

MEXT (2021) suggests in *A Guide to Promoting ESD* to 'create a unit plan while connecting different subjects and areas, with the Period for Integrated Studies (PIS) as the hub' particularly to 'ensure the repeated and developmental process of inquiry.' PIS is central to Japan's education reform, based on the lifelong learning philosophy of the Delore's Report (1996), and consists of problem-solving and experiential learning that enables creativity in each school (Yamanaka & Suzuki, 2020). In addition, PIS is positioned as the core of curriculum management and is recommended in the Curriculum Guidelines to be set in the local community and be constructed in such a way that collaborative experiential learning with local residents is linked to other subject studies. For these reasons, the promotion of ESD using PIS has been advocated in Japan (Miyagawa et al., 2018). In Japanese high schools, PIS is part of the curriculum within the 74-credit graduation requirement, which is set to account for between 3 and 6 credits. Students usually attend approximately 35 h of classes per week, 7 h per day, with 1–2 h allotted for PIS.

PIS requires independence and autonomy of each school, cross-curricular and comprehensive nature and self-help operation of peripheral resources, and requires curriculum management that goes beyond the descriptive methodological level (Tabuchi & Harada, 2018). In particular, 'a group of teachers must work together to create a new practice based on a new philosophy from diverse perspectives' (Kurebayashi, 2000). However, neither a social support system nor an educational system from this perspective has yet been established, and it is pointed out that many teachers are becoming increasingly burdened and busy with the introduction of this integrated learning (Tabuchi and Harada, ibid; Bjork, 2007). Many teachers have pointed out that the introduction of integrated learning has increased their workload and busyness (Nakatome & Soga, 2015) and that there is a danger that learning will remain superficial or become a mere formality. In promoting ESD under these circumstances, it is an extremely important issue how to empower teachers to proactively build up their practice together with the local community. This is a perspective that should be further deepened in the WSA-ESA.

23.1.3 Research Method

Both high schools examined in this chapter have begun whole-school ESD initiatives, using the introduction and reform of PIS as an opportunity to make ESD a distinctive educational feature of their schools. As a result, their unique PIS programmes, developed in the local community, are also gaining national attention and recognition. One of the reasons for this is that in Japan, in addition to the difficulty of implementing the essential practices of PIS itself, especially in high schools where subject specialisation tends to be emphasised, it is often difficult to promote mainstream ESD in public schools where students aim to enter universities (Ichinose, 2019). On the other hand, while the practice of PIS and the associated mainstreaming of ESD are currently in the hands of teachers, there are no long-term empirical studies on the relationship between their transformation as ESD practitioners and WSA-ESD from observed practices.

This study refers to the ethnographical research method of Fujita and Kitamura (2013) and the long-term procedure is shown in Fig. 23.1. The study derived structural characteristics of WSA that promoted the formation of ESD-CoP from comparing two schools' and practitioners' narratives based on the Grounded Theory Approach (Glaser & Strauss, 1967). The author has links to both schools as a former colleague and as an adviser to a wider network of learning exchanges to which both schools belong. The study was conducted in accordance with the research ethics code of the research institution of the author.

23.2 Outcomes of WSA-ESD Initiatives in Two High Schools

23.2.1 High School A: Reformation of PIS and ESD with WSA in Response to School and Community Crises

Faced with the depopulation of the region, declining student numbers and related difficulties in educational activities, High School A introduced small-group problem-solving learning based on students' career aspirations within the framework of PIS in 2003 to improve their career prospects. In 2011, the learning theme was changed to 'Achieving Sustainable Local Communities', and the school placed it at the centre of its educational activities. All groups were required to work closely with various stakeholders in the community. The groups were also divided into ten major themes, for example, under the theme 'To think about the structure of Japanese society', with additional keywords such as 'justice', 'legislation', 'administration', 'economic recovery', 'contemporary social issues', 'family' and 'social advancement of women'. The overall structure of their practice is shown in Fig. 23.2.

High School A made a special effort to include all students from Years 1–3 in one group and

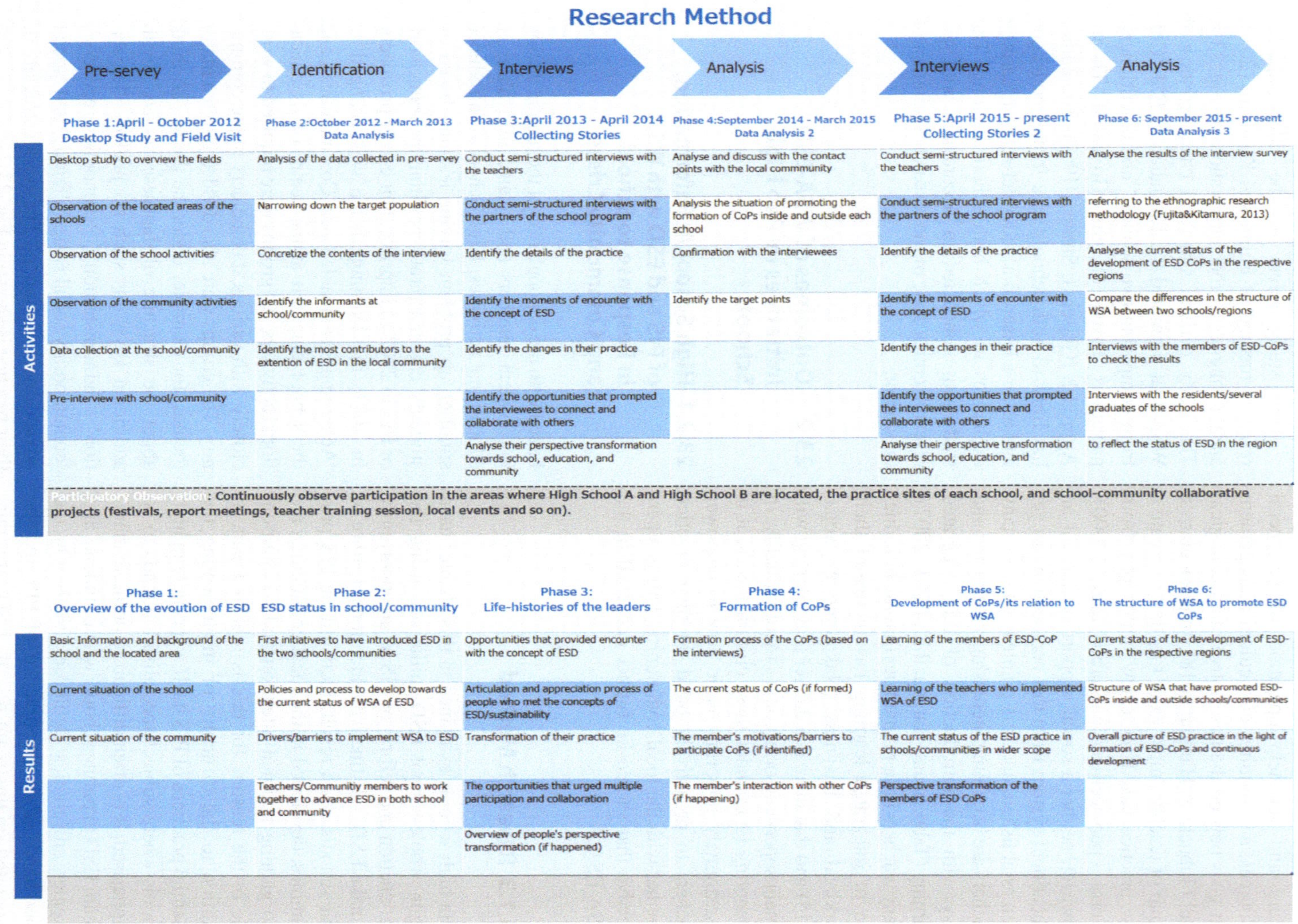

Fig. 23.1 Research procedure

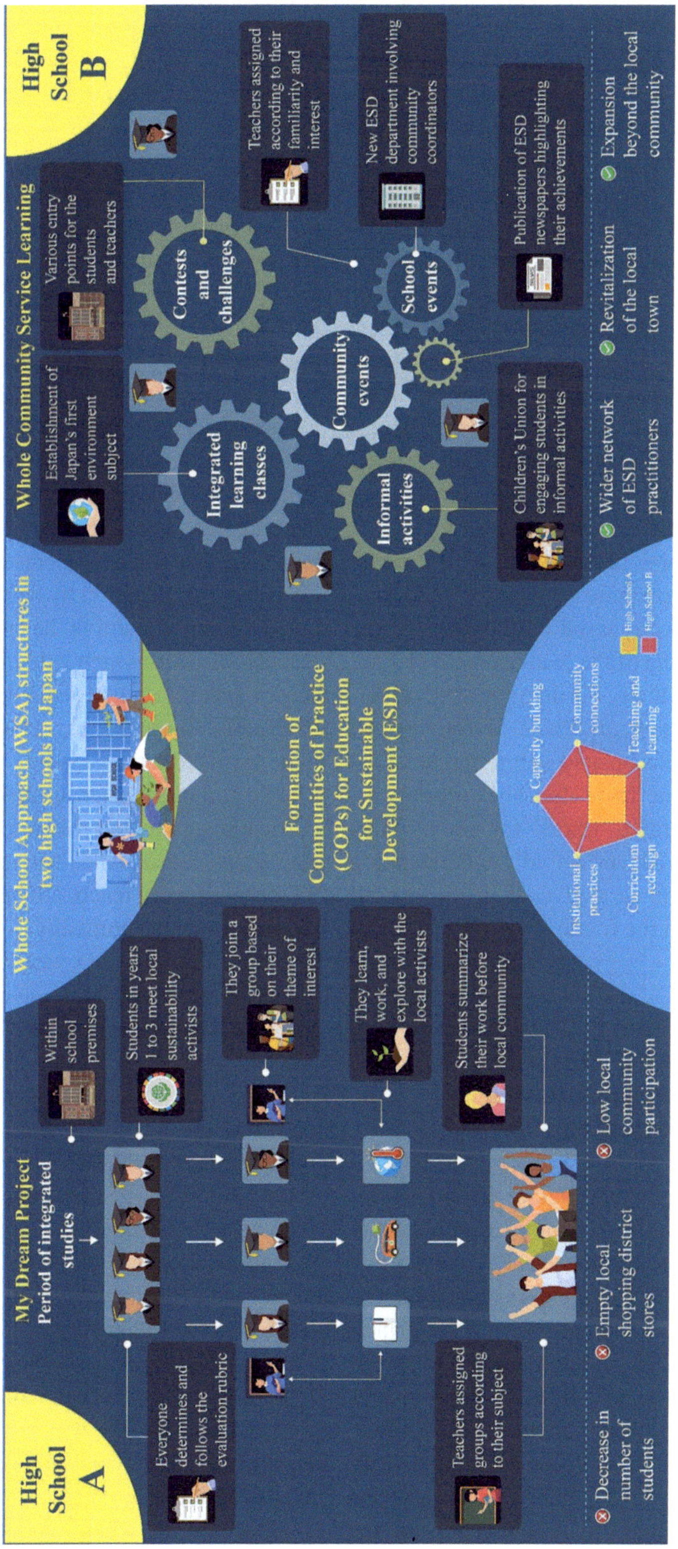

Fig. 23.2 Structures of WSA-ESD in high school A and high school B

introduce activities at the beginning of the school year to become acquainted with the community, including inviting local experts on subjects of interest to each group to the school for a 'hands-on' experience. In addition, rubric evaluation is used as an evaluation method and students themselves identify and record the skills they want to develop through ESD. What is innovative is that in order to make the collaborative project a continuous one, different age groups of students participate simultaneously, thus anticipating the growth of the students themselves. Another special feature is that the students themselves are made aware of the skills to be developed in ESD (communication skills, multifaceted thinking, systems thinking, etc.) and involved as part of the self-evaluation criteria.

During the course of the research, it became clear that the programme's objectives and promotion methods were based on the ideas of the principal, with some of the lead teachers making efforts to improve them. Furthermore, the principal created a system of 'Place of Encounter', 'Local Festival' and opened debriefing sessions to the public in the hope that high school students would become a 'nexus' for practitioners in the community. The principal's understanding of WSA-ESD is integrated well in the overall programme within the framework of PIS.

23.2.2 High School B: Conjunction of PIS and ESD Starting from Teacher-Initiated Experiential Learning

In High School B, extracurricular activities initiated by Teacher 1 to learn through essential experiences were reflected in their PIS curriculum and subsequently led to developing a unique new subject titled 'Environment', with ESD principles at its core. ESD in High School B spread throughout the town and beyond, relationships with local partners were stabilised and a town-wide mechanism to support the associated WSA was established. From the findings, it is clear that the structure is rooted in an awareness of the limitations of formal and public education among the leading teachers and that ESD is not just about improving the school and its surrounding community but has a broader perspective and benefits all involved, including teachers themselves.

Teacher 1 described briefly: "*At the same time, I felt 'What have we been doing in school education?' I have never experienced this in my usual classe~~s~~ high school students expressing something with such an emotional attitude. Then I thought, 'I am sure such an experience will touch the heart of the students', which was the first reason I began environmental education.*" (Okayama University ESD Promotion Centre, 2020, p. 6)

From this transformative experience, Teacher 1 began to emphasise the importance of students' learning outside of school while at the same time believing that teachers themselves could make a difference in education by being there. In exploring this kind of learning, he encountered the philosophy of ESD, which led him to develop a new school-specific course centred on it with the help of experts. Teacher 1 expected the entire school to be involved, understanding that 'environment' is not a part of the science curriculum or an extra-curricular activity but the entire environment that surrounds people, and that teachers, no matter what subject they teach, are involved in the environment in which humanity lives. Exposed to this understanding of ESD, several of the school's teachers, including the founder and the chief of the ESD department: Teacher 2, who would later succeed Teacher 1, became committed to creating a system to share this philosophy widely, even though they were in completely different subjects and positions. The structure of High School B's WSA-ESD is shown also in Fig. 23.2.

Teacher 1's successors were aware of the following challenges to ESD in Japan's public schools and consciously created systems to overcome them as follows:

1. Limitation of ESD practice only within the framework of PIS, which is left to the competence of individual teachers.
2. Disruption of collaborative community projects in public schools due to teacher transfers every school year.

3. ESD philosophy and principles are likely to disappear if teachers do not create an environment where they can continue to learn repeatedly.
4. Teachers' level of interest in the local community is not uniform, nor is it defined by subject specialisation.
5. The teachers' main concern, student growth and achievement through ESD, should be widely shared.

This means that there is a wide range of entry and exit points for both students and teachers at High School B, and the field of practice extends not only to schools (formal education), but also to libraries and other social education institutions (non-formal education) and local children's voluntary associations (informal education).

23.3 The Process of Forming ESD-CoPs in Two High Schools

23.3.1 Transformations Induced by the PIS of High School A

It was evident that these innovations provided an opportunity to start the formation of an ESD-CoP. The study identified four main groups of local partners that most contributed to the programme development of High School A. Within 2 years, these four groups became interested and involved in each other's efforts through the practices of High School A, and two or three of them began to work together on new projects. For example, Group T and.

Group M organised homestays for overseas youth, giving them opportunities to experience the life of agriculture and the issues in remote areas; Group M and Group H jointly organised an English camp for children in the mountains combined with nature experiences with high school students; Group S co-hosted an exchange event with Group M for high school students and local residents, and a number of other collaborative projects were initiated in the community.

It is clear that the students acted as a knot connecting these intensive teams and generating their ideas for sustainability. As Engeström (1987) suggests in his theory of learning by expanding and knot-working, all students involved began to see their own activities as a practice related to the future of the community and education, rather than as mere volunteering, hobbies or livelihoods, through contact with their aspirations and consciousness. The teachers, on the other hand, did not continue to be involved in these new projects. Finally, the school welcomed the members of Group H into the school and put them in charge of the new Regional Studies course. This is the only relationship with the local community that continues to this day.

23.3.2 Challenges in Sustaining the Quality of the Programme

In addition, it was observed that these joint activities were disbanded after a few years. The following factors emerged from the interviews;

1. The fact that the original philosophy of the development of the programme was not maintained and shared as teachers moved on and community partners changed due to life events, resulting in both parties feeling that they were going in 'different directions'.
2. Despite the requirement to include Years 1–3 in the group to enhance the continuity of the project, both students and teachers may change their aims and preferences at the start of the project. They can look back on last year's project and, if they do not find it appropriate, can suspend or change it for the coming year. On the other hand, community practice partners do not always accept these changes in a positive way.
3. The activities are part of compulsory education, i.e. PIS, and school security issues make it difficult for High School A to carry out the activities in an informal, out-of-school setting.

4. The damage to the community caused by the words and actions of less proactive students is greater than the school expects.

23.3.3 Transformations Induced by the Programme of High School B

On the other hand, in High School B, almost 20 years after the start of the practice, the following results of the practice were observed (Fig. 23.3). First, High School B began to emphasise education in collaboration with the local community, and teachers were encouraged to hire a dedicated regional coordinator with funds specifically provided by the city. This is rare in Japan and is only possible with the understanding of the local community. Second, the town itself has become more vibrant. Evidence of this is that despite the community suffering the same population decline as High School A, conservation work in the historic area of the town is progressing, with major events continued to be held actively participated by students from High School B. Teachers now actively take their students out and are involved in these events, and there are many occasions when they meet with local residents to discuss education.

The students' activities have spread outside the school and continued to involve not only high school but also elementary and junior high school students in their community development activities, which have won awards in the field of ESD and environmental education. The children's union, co-founded by Teacher 1 and a resident, is well connected to High School B through the teachers and is encouraged to engage in activities outside the school. The co-founding resident has become the community coordinator of High School B, whose role was well succeeded by another member now. In addition, a trans-regional network of adults has been formed to support these child-centred activities, and study sessions are being held to create a system to enhance informal learning opportunities. One of the most notable achievements has been the establishment of an ESD department within the school, the first in Japan. Leaders from different departments within the school, including regional coordina-

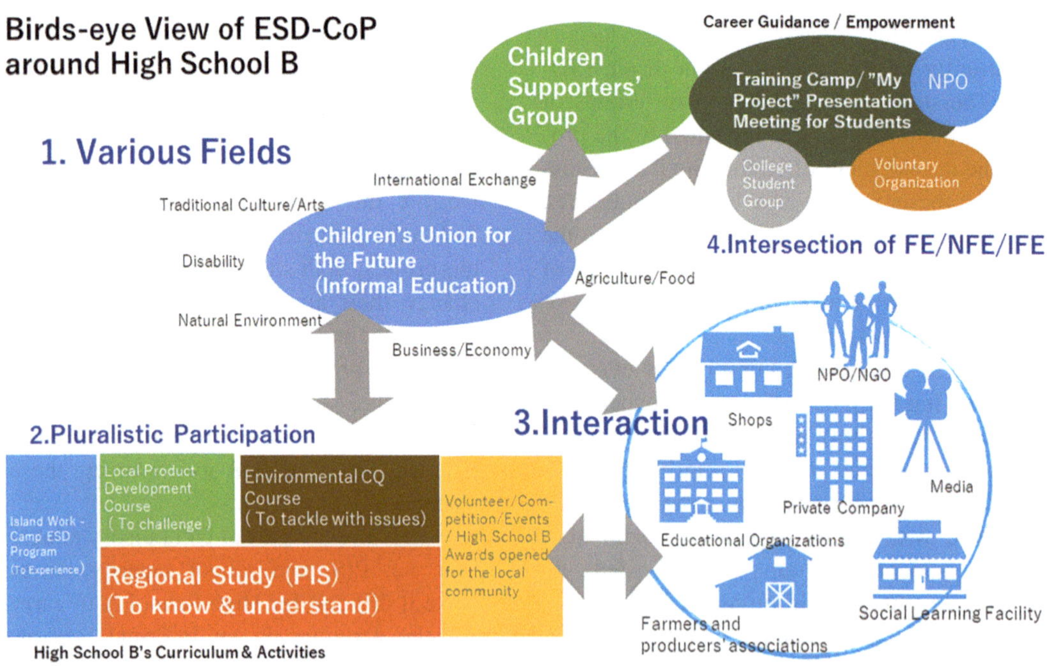

Fig. 23.3 The structure of ESD-WSA of high school B and the formation of ESD-CoPs

tors, meet to share information about ESD and discuss how to develop its implementation. Finally, the town was ranked as one of the most liveable and happy cities in the Chugoku region of Japan in 2022.

23.3.4 Growth as a Practitioner of Teachers Through Promoting ESD-WSA

Teacher 2, who initiated these reforms one by one with other colleagues, observed a significant change in himself. In the interview, he admitted that he was initially resistant to ESD, and reflected that this was because teachers were usually reluctant to teach students themselves about ESD that they had not experienced themselves. However, he learned ESD with his students outside of school. He added that he grew as a teacher and with recognition outside of school. He made sense of his experience as such learning provided him with the opportunity to promote WSA for ESD at his own school and to actively collaborate with diverse actors in the local community (OUESDPC, ibid., p.21).

Teachers who contributed as leaders in promoting WSA-ESD at High School A gradually left their positions or transferred to other schools. In the interviews with them, the following words were heard: Teacher 3 says, 'ESD is a novelty and a hot topic, and my efforts were covered by various media, but I couldn't make others understand why I was involved in ESD. I gradually found it too hard to commit and could no longer continue'. Teacher 4 says 'We (unlike the B high school) have not been able to sow the seeds in our community'. What these words suggest is that they have worked hard to survive and make 'High School A' a better place through WSA-ESD, but they have failed to truly form a community of practice. This indicates that ESD is not grounded in their own internal experiences, but is merely a proposition given to them from the outside.

23.4 Elements of WSA Promoting the Formation of ESD-CoPs

From the case studies of both schools, the structure of the WSA that facilitated the formation of ESD-CoP is analysed with reference to the WSA model of Wals and Mathie (2022). First, the difference in the area of CAPCITY-BUILDING seems to have a significant impact on the formation of ESD-CoPs: in High School A, the focus is solely on students' competency, and teachers are automatically assigned to PIS groups according to their subject specialties. In contrast, teachers of High School B are encouraged to go outside the school at every opportunity, while a mechanism for allocating staff is created according to their own level of interest in the community and to deepen their participation as Legitimate Peripheral Participants in the ESD-CoPs. This also creates a difference community connection between schools. High School A only focuses on COMMUNITY CONNECTIONS within the PIS framework, while School B actively collaborates with various organisations that provide informal learning in the classroom, extracurricular activities and events. This has resulted in a multilayered and multidimensional connection between students, teachers and the community. Teachers are able to engage not only formally in the classroom, but also in a variety of contexts, such as community events, volunteer work and councils for community development in product development. In particular, we note that teachers, who have experienced both the limitations of schooling and the possibilities of non-formal/informal education, value their students' willingness to go out into the community and 'fail'. In community partnerships, teachers become committed to and coordinate 'not letting students fail'. A similar point is made in ESD and environmental education, where curricularised programmes such as service-learning, for example, cannot be truly radical, transformative or emancipatory practices (Butin, 2006). To achieve the 'transformation' sought by ESD, the practice must be 'driven by ideology' (Op. cit., p. 478)). The transformation of teachers themselves does not occur within the training or the classroom.

Rather, the process of mutual transformation occurs 'outside the framework' (Jackson, 2011), and it is the discomfort created by exposure to different perspectives that elicits reflection and meaningful learning (Jickling & Wals, 2008). In other words, the community connections themselves may need to be (re)imagined as a richer 'reciprocal relationship' rather than fixed as a strong partnership. As Henry and Breyfogle (2006) suggest, schools and society as 'partners' in an identity will need to be transformed into a broader definition of 'community', where all participants in collaboration to solve common problems are understood as stakeholders.

In High School B, all subjects and school activities, including school climate building, are positioned as ESD and not limited to PIS. PIS is positioned as the 'gateway' to ESD from the formal education side, and classes and activities are prepared not only to learn about the community through competitions, etc., but also to address local concerns and issues together with the community. In addition to this foundation, there are classes and activities that address local concerns and issues together with the local community. In addition to this foundation, teachers are not completely separated from or left alone with informal education outside of school, but are connected in a perfect balance and play a role in supporting students' activities in the community. The systematic PIS at High School A was merely a mechanical matching of students with many local actors based on the expectation and interest that they would be the link to the community. The systematic PIS at High School A was merely a mechanical matching of students with many community activists based on their expectations and interests as community ties. However, from a LEADERSHIP and COORDINATION perspective, the community coordinator at High School B, in an interview, respected these teachers' strengths as 'a perspective that teachers can only have' and 'a job that only teachers can do', and had a better understanding of their work (OUESDPC, p. 16). In order to realise an education that is centred on students, not on the needs of the school, it is an important task for teachers to engage steadily with the community while 'crossing the border' between school and society, and this may suggest a new type of leadership.

23.5 Conclusion

In Lave and Wenger's (1991) 'legitimate peripheral participation' and Wenger's (1998) 'Communities of Practice' and Engeström's (1987) 'learning by expanding', the process of forming voluntary groups and learning organisations, learning to participate and share experiences in them and full communication and interaction between diverse groups are essential for the development of identity and empowerment. It is understood that the power of individuals and groups to create a sustainable society is demonstrated in the mutually transformative learning experiences of communities that are open to others and diverse. Therefore, it is essential to understand that ESD is not completed in a specific unit or class but is developed and expanded from its starting point. Based on the interests of students and teachers, it is necessary to allow multiple entry points in different disciplines and at different levels and to enable multifaceted participation in them. This is because they can be change-makers for a sustainable society.

Future research will continue to investigate the effectiveness of the structure that facilitated the transformation of teachers at High School B in bringing about systemic change in high school education itself, and how it will change and develop in the future should be closely monitored. One limitation of this study is that the students themselves were not focused on, and their development as practitioners should continue to be investigated. They are themselves legitimate peripheral participants in the ESD-CoP.

References

Asia-Pacific Cultural Centre for UNESCO (ACCU). (2016). *Let's think about the future of UNESCO Associated Schools – Whole School Approach – As the focal point for promoting ESD, that has expanded, deepened and become more connected:*

UNESCO Associated Schools in Japan as Bases for Promoting ESD – Current status and way forward. ACCU. ISBN978-4-946438-97-4.

Asia-Pacific Cultural Centre for UNESCO (ACCU). (2017). Kirari Hasshin! Sustainable School - Whole School Approach de egaku mirai no gakko -. ACCU. ISBN978-4-946438-99-8. (in Japanese).

Bjork, C. (2007). Imagining Japan's 'relaxed education' curriculum: Continuity or change? In D. B. Willis & J. Rappleye (Eds.), *Reimagining Japanese education: Borders, transfers, circulations, and the comparative* (Oxford studies in comparative education). Symposium Books.

Butin, D.W. (2006). The Limits of Service-Learning in Higher Education. *The Review of Higher Education, 29*, 473–498.

Engeström, Y. (1987). *Learning by expanding: An activity-theoretical approach to developmental research.* Orienta-Konsultit Ltd.

ESD Special Subcommittee, Education Committee, Japanese National Commission for UNESCO, Ministry of Education, Culture, Sports, Science and Technology: MEXT. (2015). *For further promotion of Education for Sustainable Development (ESD).* https://www.mext.go.jp/en/news/topics/detail/__icsFiles/afieldfile/2016/09/16/1377397_001_1.pdf. Last accessed 23 Jan 2023.

ESD Special Subcommittee, Education Committee, Japanese National Commission for UNESCO, Ministry of Education, Culture, Sports, Science and Technology: MEXT. (2021). *A guide to promoting Education for Sustainable Development (ESD) (Provisional Translation).* https://www.mext.go.jp/component/a_menu/other/micro_detail/__icsFiles/afieldfile/2018/04/11/1369326_02.pdf. Last accessed 13 Feb 2023.

Fujita, Y., & Kitamura, A. (2013). *Gendai ethnography: Atarashii field work no Riron to Jissen (Word Map), Shinyosha.* (in Japanese).

Henry, S. E., & Breyfogle, M. L. (2006). Toward a New Framework of "Server" and "Served": De(and Re)constructing Reciprocity in Service-Learning Pedagogy. *The International Journal of Teaching and Learning in Higher Education, 18*, 27–35.

Gericke, N. (2022). *Implementation of education for sustainable development through a whole school approach* (pp. 153–166). Education for Sustainable Development in Primary and Secondary Schools. https://doi.org/10.1007/978-3-031-09112-4_11

Glaser, B., & Strauss, A. (1967). *The discovery of grounded theory: Strategies for qualitative research.* Sociology Press.

Hargreaves, L. (2008). The Whole-school Approach to Education for Sustainable Development: From Pilot Projects to Systemic Change. *Policy & Practice – A Development Education Review, 6*, 69–74.

Ichinose, T. (2019). An analysis of the teachers' recognition of the ESD school practices by the Questionnaire Survey. *Journal of ESD Research, 2*, 3–12. (in Japanse).

Illich, I., & Lang, A. (1973). *Tools for conviviality.* Harper and Row.

Jackson, M. G. (2011). The real challenge of ESD. *Journal of Education for Sustainable Development, 5*(1), 27–37.

Jickling, B., & Wals, A. E. J. (2008). Globalization and environmental education: looking beyond sustainable development. *Journal of Curriculum Studies, 40*(1), 1–21.

Kurebayashi, N. (2000). Kawaru Curriculum to Kyouiku Jissen – Kyoushi no kanousei toshiteno 『Period for Integrated Studies』, In S. Nagai & M. Koga (Eds.), *Kyoushi toiu shigoto=work*, Gakubunsha. (in Japanese). 19–45.

Lave, J., & Wenger, E. (1991). *Situated learning: Legitimate peripheral participation.* Cambridge University Press.

Miyagawa, H., Isobe, M., & Nakashima, Y. (2018). Perspective and expansion of ESD (Education for sustainable development) on the period for integrated study. *Bulletin of Contemporary Education, Chubu University, 10*, 39–46. (in Japanese).

Nakatome. T., & Soga, E. (2015). *Curriculum management no aratana chousen – Integrated Studies ni okeru renkansei to kyoudousei ni shouten wo atete -*, Kyouikukaihatsu Kenkyusho. (in Japanese).

Okayama University ESD Promotion Centre. (2020). *Supplementary reader 'ESD video learning material "Teachers engaged with the local community".* https://edu.okayama-u.ac.jp/wp-content/uploads/2024/05/Supplementary-reader.pdf. Last accessed 14 May 2024.

Shallcross, T. (2005). *Whole school approaches to education for sustainable development through school focused professional development.* The SEEPS Project.

Tabuchi, K, & Harada, T. (2018). Current Situation and Issues of "the Period for Integrated Studies" from the Perspective of Curriculum Management, *Kwassui Bulletin Faculty of Wellness Studies, 61*, 37–52. (in Japanese).

Tejima, T. (2017). *Gakkouhatsu ESD No Manabi.* Kyoiku-Shuppan. (in Japanese).

United Nations Educational, Scientific and Cultural Organization. (2012). *Shaping the education of tomorrow: 2012 report on the UN decade of education for sustainable development, abridged.* UNESCO.

Wals, A. E. J. (2011). Learning our way to sustainability. *Journal of Education for Sustainable Development, 5*(2), 177–186.

Wals, A. E. J., & Mathie, R. G. (2022). Whole school responses to climate urgency and related sustainability challenges: A perspective from northern Europe. In M. Peters & R. Heraud (Eds.), *Encyclopedia of educational innovation.* Springer.

Wenger, E. (1998). *Communities of Practice : Learning Meaning and Identity*. Cambridge University Press.

Yamanaka, S., & Suzuki, K. H. (2020). Japanese education reform towards twenty-first century education. In F. Reimers (Ed.), *Audacious education purposes* (pp. 81–103). Springer.

Yoshida, T., & Natori, Y. (2020). Development of school management utilizing whole school approach design sheet: A case study of ESD practice at KARAKUWA elementary school in KESENNUMA. *Journal of ESD Research, 3*, 18–28. (in Japanese).

It Takes a Whole School: A Synthesis

24

Birgitte Bjønness, Ingrid Eikeland, Astrid Sinnes, and Arjen E. J. Wals

The whole-school approach offers a structure and language to exchange and celebrate diverse educational perspectives, priorities, and practices. Our goal in assembling this collection is not to offer a blueprint, a narrowing solution to problems of schooling in contemporary times, but to share promising and inspiring cases, to provoke dialogue and open conversations, but also to question, resist, struggle, and break with restricting traditions and orders. Of course, all schools are unique, contextually rich, complex, dynamic, and ever-evolving. They are educational communities, education homes assembled through scholarly expertise, professional practices, institutional structures, and lots of dedication, love, and care. Professional growth, Maxine Greene writes, requires a "quest for a better state of things for those we teach and the world we share" (Greene, 1995, p. 1). This quest brings together the innovative practices outlined in this collection. Here, we showcase diverse professional communities reflexively engaged in working together, teaching and learning together, in particular contexts and times. We offer this book as an invitation to think differently about schools and schooling. Above all else, this invitation is deeply educational. In this closing chapter, we highlight tensions, common threads, and noticeable absences that might inform the future development of the WSA in our pursuit of a world that is more sustainable than the one currently in prospect.

B. Bjønness · I. Eikeland · A. Sinnes
Norwegian University of Life Sciences, Ås, Norway
e-mail: birgitte.bjonness@nmbu.no; ingrid.eikeland@nmbu.no; astrid.sinnes@nmbu.no

A. E. J. Wals (✉)
Wageningen University & Research, Wageningen, The Netherlands
e-mail: arjen.wals@wur.nl

24.1 Transitioning Versus Optimizing

This book has brought together a wide range of perspectives on what a whole-school approach to sustainability can entail. Whereas there is the common thread of a more systemic approach to addressing sustainability, rather than an "add-on" approach, there are different ideas about the meaning of both *systemic* and *sustainability*, but also differences in where to emphasize bringing about *deep change*. Roughly speaking, there are those who use one of the petals of the WSA flower as an entry point with some starting with professional development of staff, some with the school's architecture, some with the agency and empowerment of the members of the school community, some with a concrete sustainability topic like the school's energy supply or addressing climate change, some more on school–community relationships. There are some who seem more comfortable in working within the existing

A. E. J. Wals et al. (eds.), *Whole School Approaches to Sustainability*, Sustainable Development Goals Series, https://doi.org/10.1007/978-3-031-56172-6_24

structures and frameworks, tweaking them a bit, optimizing them, incrementally improving matters, while others seem convinced that a whole-system redesign based on different values, principles, and ways of working is the only way a school can systemically work toward sustainability.

The idea of sustainability as something to embody, to enact, and to live, connects with this redesign perspective. One could say there are at least two distinct pathways: the optimization pathway and the transition pathway. The optimization pathway takes and accepts the current system as a given, while the transition pathway considers the current system as highly problematic. Incremental change simply won't suffice from this perspective and, in fact, makes matters worse as it solidifies these existing systems and structures risks and might even distract from what is really needed.

The optimization pathway might be more successful in getting school-wide support as it stays closer to people's comfort zones and might be the only way to change often conservative schools and the communities of which they are part. The more radical transition pathway risks not having the support base and the commitment from all stakeholders involved and running into resistance. The transition pathway only seems to have a chance in a somewhat special environment, with a special kind of leadership, a high level of joint commitment among all stakeholders, and at least some policy support.

In line with the optimalization approach, we see some efforts to present the WSA as an implementation problem in that there can be steps, indicators, benchmarks, and assessment schemes that can guide this implementation. More in line with the transition approach, we see schools and their stakeholders considering the WSA as something that needs to unfold guided by continuous monitoring and reflection where people in the school community together try to figure out what sustainability means and how schools can contribute to it coming into being. This also implies that attention is guided towards giving voice and developing agency among the key actors in the system, including the young people, as well as considering matters of diversity, also inviting dissonant perspectives, and inclusion, bringing in those who tend to be side-lined, and thereby too often silenced. For people who like to plan, manage, and avoid surprises, as many school leaders and teachers do, the transition approach is a rather unsettling and uncomfortable path to embark on.

Culture and related policy environments also play a role here. At the meso-level, school culture influences the possibilities for enacting a WSA to sustainability but so does a country's culture in terms of space for autonomy, voice, and freedom to co-design. In more egalitarian societies with tendencies toward deliberative democracy, a WSA might be realized easier than in more authoritative societies with hierarchical decision-making structures. Attempts to develop a WSA in the latter context might receive more resistance. This is not to say that developing a WSA in the former context is easy, as we see resistance there too and how fragile innovations can be (see, for example, Chap. 15, Leite). When a culture of testing and measurement and the ambition to score high on the PISA rankings drive educational policy at the national level, the environment for developing a WSA tends to be rather unfriendly. In countries where such a culture of accountability is under pressure, as student and teacher motivation are low, and the relevance of education for today's world is being questioned, spaces for a WSA do open up.

24.2 Boundary Crossing

No matter what pathway is chosen, a WSA calls for boundary crossing between different parts of the educational system and society through promoting collaborative cultures and participation. A WSA in its very nature calls for an "all on board" mentality to change schools in fundamental ways. These aspects are frequently brought up and grappled with throughout the contributions in this book, both through a theoretical and practical lens. Thus, making boundary crossing a bold spot in the WSA discourse.

The contributions include examples of multi-stakeholder collaborations, spanning students, teachers, school leaders, other staff, parents, researchers, politicians, NGOs, and other community members. To create connections and synergies between different stakeholders in the educational ecosystem is an ideal vision to strive for in a WSA, coming together with different perspectives to shape and promote sustainability education (see, for example, Chap. 12, Eikeland and Sinnes; Chap. 14, Manni et al.; and Chap. 21, Rumjaun and Atchia). This entails new structures promoting horizontal power relations and distributed leadership, collective decision-making, and creating shared visions, to name a few. To practice boundary crossing thus calls for cultures where it is unavoidable, and necessary, to become uncomfortable together and deal with uncertainty (see for example Chap. 18, Hargis, and Chap. 23, Shibakawa). Furthermore, to sustain these collaborations demand resources, time, a high level of coordination, competency in practicing collective decision-making and creating environments that nurture a culture of sharing and learning from and with each other (see for example Chap. 11, Borg et al.; Chap. 4, Gericke et al.; Chap. 21, Rumjaun and Atchia; Chap. 10, Vanderveen et al.; and Chap. 5, Verhelst et al.).

The support given to the schools needs to be context-sensitive to the individual school's ethos and culture (Koh & Askell-Williams, 2021). Multiple contributions in this book point to the fact that schools need to lead their own journey toward sustainability and educational innovation, rather than this being a fixed path provided from the outside. Furthermore, that schools should be provided with the support to build their own reflexivity competency, being able to continuously rethink and reorient their educational practice, to adapt to a constantly changing nature of sustainability issues (see for example Chap. 7, Field et al.; Chap. 2, Mathie; Chap. 19, Phillips and Howard; and Chap. 13, Shintesetseg et al.).

To practice context sensitivity and collaborative cultures spanning a diverse group of actors, calls for a movement toward co-research (Reason & Bradbury, 2008) where participants to a large degree have an active role in shaping and influencing the knowledge and experience built around a WSA (see for example Chap. 22, Phuti).

24.3 Student Voice

A critical factor in a WSA to sustainability is the voice of the students (Torsdottir et al., 2023). The book includes cases where students are included in a direct way through observations, interviews, and written feedback (see for example Chap. 17, Ballegeer et al.; Chap. 16, Mathisen and Johansen; Chap. 6, Morin et al.; Chap. 15, Leite; and Chap. 20, Nordén), or in an indirect way through teachers' voice (see Chap. 9, Hugo and Iversen). There are also many chapters that highlight the importance of student engagement through theoretical and policy-based lenses of student-centered education through ownership, empowerment, inclusion, participation, and democratic citizenship (see for example Chap. 8, Rončević and Rieckmann, and Chap. 3, Zachariou et al.).

Looking back at the chapters in this book, one might conclude that student voice in the sense of a student's capacity to access and influence decision-making processes, especially beyond the classroom, does not feature prominently. To further promote a focus on student voice is also strengthened by the backdrop painted by the contributions in this book reporting that students have a very limited sense of empowerment when it comes to ESD issues, and where students, teachers, and others struggle with getting used to new roles and power dynamic when practicing a student-centered education. This might be seen as a blind spot or a weak spot as student voice, agency, and participation are not only returning themes in environmental and sustainability education, but they are also recognized in article 12 of the UN Convention on the Rights of the Child (UNCRC).

In his dissertation on student voice in the classroom, Jeroen Bron (2018) points out that student voice means more than simply learning to speak and/or use freedom of speech. Rather it refers to a student's capacity to access and influence decision-making processes. Based on a literature review Bron identifies five arguments for giving students a voice in education and curriculum development:

1. "Normative: Young people are entitled to the right to have a voice in matters that affect them.
2. Developmental: Many children and young people assume responsibility and exercise autonomy outside school. However, they are seldom offered this opportunity within. We regard students as citizens with developmental readiness to participate and assume responsibility within school.
3. Political: Inviting students to participate in curriculum design changes the power paradigm, providing opportunity for voices that are often marginalized to speak and for those in positions of power to listen and hear.
4. Educational: Participation in negotiating and decision-making processes has educational benefits, contributing to the development of citizenship and twenty-first century skills.
5. Relevance: Involving students in curriculum development adds significant stakeholders in the curriculum discussion, improving the relevance of curricula." (Bron, 2018, p242).

In a WSA, student voice can be or should come in when considering all elements of the flower, not only in curriculum and education but also in school ethos, leadership, identifying professional development needs of staff, shaping school-community relations, and in finding ways to make the school itself more sustainable. The fact that student voice in decision-making processes remains somewhat hidden in the book, seems to indicate that schools either do not see its importance or struggle to find ways to open up spaces for the systemic embedding of student voice. In a study on development toward a WSA in Mongolian schools (see Chap. 13, Shinetsetseg et al.), student voice was particularly found to be a struggle as it is a completely new culture. Interestingly, several of the schools in this case did manage step by step to create a culture for collaborative decision-making especially enjoying the student voice as part of the process.

24.4 Decolonization

As we have entered a time of sensitivity toward phenomena of exclusion, abuse of power, revisional histories, and marginalization of indigeneity, along with associated polarization and radicalization, schools will also need to grapple with a heavy and loaded topic of decolonizing of education, especially but not exclusively, in the Global North. This is a complex and ongoing process that involves recognizing and addressing historical and ongoing colonial legacies, biases, and inequalities within educational systems. While the importance of inclusivity and diversity is acknowledged in some of the chapters, only few contributions are entering this terrain by emphasizing the importance of including perspectives from historically marginalized groups, including indigenous knowledge, literature, history, and contributions from various cultures and backgrounds (see Chap. 7, Field et al.; Chap. 18, Hargis; Chap. 19, Philips and Howard; and Chap. 8, Rončević and Rieckmann). The decolonizing perspective also points at the revitalization of indigenous and local languages within educational systems. In terms of capacity building for staff, cultural competencies might need attention as well so that teachers and school leaders can better understand and respect the cultures, traditions, and histories of marginalized communities.

In terms of pedagogy, a more critical pedagogy, one that encourages students to critically analyze and challenge colonial narratives, stereotypes, and power structures, might be needed. Such a critical approach aims to expose tensions, inequalities, and exploitation, as well as the underlying mechanism and processes (Andreotti, 2007, 2011). Sund and Pashby (2020) argue that education needs to explicitly interrogate coloniality as a central condition of today's global issues, and to acknowledge it as a key element of, what they refer to as, ethical global issues pedagogy that centers on "delinking as a decolonial praxis." This delinking has several activities, including: the exploration of multiple perspectives that reflect different worldviews and narratives and explore and engage with the complexities and contractions between them, and "denaturalizing" dominant one-sided narratives (on progress, development, consumption, etc.) and recognize how these concepts are socially and politically constituted.

24.5 Posthumanism and Rewilding

Another emerging trend in environmental and sustainability education that still needs to find its way in the WSA discourse is the rise of posthumanism (e.g., Duobliene & Vaitekaitis, 2021), rewilding (e.g., Carver et al., 2021) and wild pedagogies (e.g., Blenkinsop et al., 2022). Posthumanism encourages humans to extend moral consideration to nonhuman entities, fostering a greater sense of responsibility toward the natural world. This shift arguably can lead to more ethical and sustainable ways of living. Nature-inclusive education and wild pedagogies place nature at the center of the learning process and emphasize experiential learning in natural environments that fosters a deep connection to the natural world. These parallel streams that have quite different roots but likely will intersect in the years to come, advocate a decentering of the human and recognizing and strengthening our entanglement with nature. Posthumanism and wild pedagogies offer complementary insights into how we can foster a more sustainable relationship with the natural world (Paulsen et al., 2022). Posthumanism encourages a shift away from anthropocentrism, while wild pedagogies and rewilding provide a practical framework for nurturing a deeper connection to nature and promoting sustainable practices rooted in the life world (Jickling et al., 2018).

Whereas the WSA to sustainability tends to emphasize problem-solving and working on issues around health, well-being, climate, inclusivity, biodiversity, and so on, not much is being said about our relating to and our relationship with other species, our surroundings, and the affordances of immersing ourselves with other beings without being inhibited by misplaced superiority. The idea of more eco-centric and relational ways of being and creating schools that invite such a way of being is still in its infancy.

24.6 Levers and Ways Forward to Make a WSA Possible

As educators we are obligated to ask what schooling ought to mean in an era of social and ecological uncertainties. We cannot lose schools entirely to discourses of management, accountability, and efficiency. A WSA seeks to recognize that schools matter way beyond PISA scores and individual performance. To energize projects and prospects of schooling is to awaken and listen attentively to those involved, to embrace dialogue and institutional questions, and reflect on institutional histories and responses.

This book provides a collection of diverse educational perspectives and practices from educational communities worldwide. There are promising practices and perspectives that hopefully inspire ways forward. Throughout the book, we find examples of the importance of educational policy building, social and material innovation, and practicing holistic and connected rather than siloed approaches to sustainability. The continuous reflection and dialogue between different practitioners in community of practices is highlighted as vital to shape and promote a WSA.

We also find blind spots revealing a need to pay closer attention to those traditionally cast to the shadows, marginalized, and thereby too often silenced. A WSA calls for a more diverse and inclusive perspective which "highlights the connections between viable interdependent ecosystems and viable interdependent communities—and that our future depends on maintaining the widest possible diversity in cultural approaches to sustainable living" (Bowers, 2005, p. 148).

In the end, we are talking about good education here. In an ideal world, we should not need to call for the importance of paying attention to sustainability in education or the need to advocate a whole-school approach *to sustainability*, as it would be what schools are doing by default. As we are approaching 2030, the year the SDGs and the Paris Agreement should be realized, we can only hope that a new normal in education will be emerging, one where a whole child can enjoy a whole school in a whole community nested in a whole Earth. 2030 might be a bit too optimistic, but at least we are seeing niches unfold that show promise in realizing a much-needed transition in education and society at large.

References

Andreotti, V. (2007). An ethical engagement with the other: Spivak's ideas on education. *Critical Literacy, Theories and Practices, 1*(1), 69–79.

Andreotti, V. (2011). (Towards) decoloniality and diversality in global citizenship education. *Globalisation, Societies and Education, 9*(3–4), 381–397.

Blenkinsop, S., Morse, M., & Jickling, B. (2022). Wild pedagogies: Opportunities and challenges for practice. In M. Paulsen, J. Jagodzinski, & S. M. Hawke (Eds.), *Pedagogy in the Anthropocene: Re-wilding education for a new earth* (pp. 31–52). Palgrave Macmillan. https://doi.org/10.1007/978-3-030-90980-2

Bowers, C. A. (2005). How the ideas of Paulo Freire contribute to the cultural roots of the ecological crisis. In C. A. Bowers & F. Apffel-Marglin (Eds.), *Rethinking Freire: Globalization and the environmental crisis* (pp. 133–150). Lawrence Erlbaum Associates, Inc. Publishers.

Bron, J. G. (2018). *Student voice in curriculum development: Explorations of curriculum negotiation in secondary education classrooms* [Doctoral dissertation]. Universiteit voor de Humanistiek.

Carver, S., Convery, I., Hawkins, S., Beyers, R., Eagle, A., Kun, Z., …, Soulé, M. (2021). Guiding principles for rewilding. *Conservation Biology, 35*(6), 1882–1893.

Duobliene, L., & Vaitekaitis, J. (2021). A Posthumanist approach to human/child-centred education. *Journal of Futures Studies, 26*(2), 37–50. https://doi.org/10.6531/JFS.202112_26(2).0003

Greene, M. (1995). Art and imagination: Reclaiming the sense of possibility. *The Phi Delta Kappan, 76*(5), 378–382.

Jickling, B., Blenkinsop, S., Timmerman, N., & De Danann Sitka-Sage, M. (Eds.). (2018). *Wild pedagogies: Touchstones for re-negotiating education and the environment in the Anthropocene* (pp. 1–22). Springer Verlag.

Koh, G. A., & Askell-Williams, H. (2021). Sustainable school-improvement in complex adaptive systems: A scoping review. *Review of Education, 9*(1), 281–314. https://doi.org/10.1002/rev3.3246

Paulsen, M., Jagodzinski, J., & Hawke, S. M. (Eds.). (2022). *Pedagogy in the Anthropocene: Re-wilding education for a new earth.* Palgrave Macmillan. https://doi.org/10.1007/978-3-030-90980-2

Reason, P., & Bradbury, H. (Eds.). (2008). *The SAGE handbook of action research: Participative inquiry and practice.* SAGE Publications Ltd.. https://doi.org/10.4135/9781848607934

Sund, L., & Pashby, K. (2020). Delinking global issues in northern Europe classrooms. *The Journal of Environmental Education, 51*(2), 156–170. https://doi.org/10.1080/00958964.2020.1726264

Torsdottir, A. E., Sinnes, A. T., Olsson, D., & Wals, A. E. J. (2023). Do students have anything to say? Student participation in a whole school approach to sustainability. *Environmental Education Research.* https://doi.org/10.1080/13504622.2023.2213427

Index

A. Wals et al. (eds.), *Whole School Approaches to Sustainability*, Sustainable Development Goals Series, https://doi.org/10.1007/978-3-031-56172-6

MIX
Papier aus verantwortungsvollen Quellen
Paper from responsible sources
FSC® C105338

If you have any concerns about our products,
you can contact us on
ProductSafety@springernature.com

In case Publisher is established outside the EU,
the EU authorized representative is:
Springer Nature Customer Service Center GmbH
Europaplatz 3, 69115 Heidelberg, Germany

Printed by Libri Plureos GmbH
in Hamburg, Germany